Carbohydrate Chemistry

Carbohydrate Chemistry: Proven Synthetic Methods
Series Editor: Pavol Kováč
National Institutes of Health, Bethesda, Maryland, USA

Carbohydrate Chemistry: Proven Synthetic Methods, Volume 1
by Pavol Kováč

Carbohydrate Chemistry: Proven Synthetic Methods, Volume 2
by Gijsbert van der Marel and Jeroen Codee

Carbohydrate Chemistry: Proven Synthetic Methods, Volume 3
by René Roy and Sébastien Vidal

Carbohydrate Chemistry: Proven Synthetic Methods, Volume 4
by Christian Vogel and Paul Murphy

Carbohydrate Chemistry: Proven Synthetic Methods, Volume 5
by Paul Kosma, Tanja Wrodnigg, and Arnold Stütz

Carbohydrate Chemistry

Proven Synthetic Methods, Volume 5

Edited by

Paul Kosma, Tanja Wrodnigg, and Arnold Stütz

CRC Press
Taylor & Francis Group
Boca Raton   London   New York

CRC Press is an imprint of the
Taylor & Francis Group, an **informa** business

First edition published 2021
by CRC Press
6000 Broken Sound Parkway NW, Suite 300, Boca Raton, FL 33487-2742

and by CRC Press
2 Park Square, Milton Park, Abingdon, Oxon, OX14 4RN

CRC Press is an imprint of Taylor & Francis Group, LLC

Library of Congress Cataloging-in-Publication Data

Names: Kosma, Paul, editor. | Wrodnigg, Tanja M., editor. | Stütz, Arnold E., editor.
Title: Carbohydrate chemistry / edited by Paul Kosma, Tanja M. Wrodnigg and Arnold Stütz.
Description: Boca Raton : CRC Press, 2020. | Series: Proven synthetic methods ; volume 5 | Includes bibliographical references and index.
Identifiers: LCCN 2020020204 (print) | LCCN 2020020205 (ebook) | ISBN 9780815367888 (hardback) | ISBN 9780367561734 (paperback) | ISBN 9781351256087 (ebook)
Subjects: LCSH: Carbohydrates.
Classification: LCC QD321 .C2734 2020 (print) | LCC QD321 (ebook) | DDC 547/.78--dc23
LC record available at https://lccn.loc.gov/2020020204
LC ebook record available at https://lccn.loc.gov/2020020205

ISBN: 978-0-8153-6788-8 (hbk)
ISBN: 978-0-367-56173-4 (pbk)
ISBN: 978-1-351-25608-7 (ebk)

Typeset in Times
by KnowledgeWorks Global Ltd.

This series is dedicated to Sir John W. Cornforth, the 1975 Nobel Prize winner in chemistry, who was the first to publicly criticize the unfortunate trend in chemical synthesis, which he described as "pouring a large volume of unpurified sewage into the chemical literature."*

---

* Cornforth, J. W., Aust. J. Chem. 1993, 46, 157–170.

Contents

PART I SYNTHETIC METHODS

Contents   xi

Foreword

THE AIM AND SCOPE OF THIS SERIES, AND INFORMATION FOR PROSPECTIVE AUTHORS

AIM AND SCOPE

The objective of *Carbohydrate Chemistry: Proven Synthetic Methods* (later, *Proven Synthetic Methods, Proven Methods,* or this series) is to compile reliable protocols for the preparation of intermediates for carbohydrate synthesis or other substances that can be expected to be useful in the glycosciences. To ensure reproducibility, an independent, reasonably skilled artisan (checker) will verify the experimental part involved by repeating the protocol. When the article is published, the checker's name will appear within the list of authors. Because authors must find a qualified checker capable and willing to do the work, they should keep in mind that it may be difficult for them to find qualified chemists willing to act as checkers when the sequence of transformations is too long.

MANUSCRIPT REQUIREMENTS, I.E., WHAT WE PUBLISH

Authors may submit any number of contributions. The experimental protocols described should be preferentially one-step transformations, or the number of steps should be minimal. Articles submitted to this series may describe new methods or improvements to an existing, previously described one. The working protocols may also be a more detailed version of a protocol published previously but not duly recognized or one that was published by others in the distant past or a not readily available source. Particularly useful could be republication, after passing the checking process, of protocols published in the now defunct journals (such as *Journal of Carbohydrates, Nucleosides, Nucleotides* or *Carbohydrate Letters*), or low-circulation national chemistry journals, especially those published in languages other than English. The original or other authors may submit also protocols of interest to carbohydrate chemists that were published in the past but deemed "buried" within other topics. Methodology published within about five years may also be submitted, provided that these methods are applied to the preparation of compounds other than those described in the original publication. Our requirements for purification and characterization of new or published compounds still hold (more on this topic later). Thus, when the method was originally applied to the preparation of a compound that has not been obtained in the analytically pure state and properly characterized, that preparation may be republished (preferentially with more experimental details), provided proper characterization is now included. The final decision regarding the suitability of a subject for inclusion in *Proven Methods* will be at the discretion of editors. There is no formal deadline for submissions because if a manuscript does not make it for publication in the volume in preparation, it will be shifted to the next volume. There are many reasons for a manuscript not making it for publication in the volume in preparation; for example, the volume is full, the checker is tardy, the manuscript could not be checked for nomenclature in time, or some other technicalities.

MANUSCRIPT STRUCTURE

Please check the most recent volume in this series for the format of individual chapters. Essentially, the body of the text of every chapter starts with the lead scheme, followed by the introductory part (without the title "Introduction"), which should be concise (remember, chapters in this series are not reviews on the method described) and contain only those references that are relevant to the preparation described. Often, one or two references to a relevant review are sufficient. The Experimental part consists of General Methods and individual working protocol(s). The latter must describe in detail all nonstandard operations and observations (*e.g.*, change in color, the formation of precipitates, and generation of heat) so that there are no surprises when one wants to reproduce the protocol. General Methods should list commercial sources of specialty chemicals and starting materials. The preparation of carbohydrate intermediates should be referenced here *and* in the experimental protocols where they are used. For further information, please refer to the Foreword in Volume 4, which is available from the editors upon request. The Experimental part is followed by Acknowledgements, if any, copies of 1D 1H and 13C spectra and References. For formatting references, see previous volumes in this series.

IDENTIFICATION AND CHARACTERIZATION OF NEW AND PREVIOUSLY PUBLISHED SUBSTANCES

Proper proof of identity, purity, and characterization of compounds whose preparation is described must be provided for all intermediates and final products.

New compounds. In this series, we consider a compound to be "new" when it has not been properly identified and characterized before. To identify a new compound means to provide data from which the structure can be deduced, such as assigned NMR resonances (at least the structurally significant ones), MS (not necessarily HRMS), and perhaps other spectral data. The structure deduced in this way must be supported by elemental composition. In exceptional situations, when to provide the proof of purity by correct combustion analysis figures is not required (see below), the elemental composition determined by HRMS is acceptable. For characterization, we require a reliably determined melting range ("mp" is the conventional term but "range" is what we determine and report) of crystalline compounds and $[\alpha]_D$ for chiral compounds. Other characteristics may also be provided, *e.g.*, refractive index and IR spectral data. Characteristic constants must be determined with material for which proof of purity has been provided because these characteristic values should be possible to use for identification purposes of newly prepared, known compounds. Thus, to report the boiling point of a volatile substance is meaningless if the distillation setup does not allow measuring the temperature and the pressure at the proper location within the system. Similarly, reliable and reproducible mp can be established only when the compound was crystallized from a suitable solvent and recrystallized to constancy (mp of the recrystallized material should not change by more than 1–2°C) and the temperature of the sample during mp determination raises at the conventionally agreed-upon rate of ~4°C/min. Needless to say, for a chapter to be acceptable for publication in *Proven Methods*, characterization or identification does not have to include all possible criteria, but accurate and convincing identification and characterization of new compounds must be included. The absence of combustion analysis data must be justified in an acceptable way; for example, the compound is unstable/very reactive or demonstrably hygroscopic, such as very polar amorphous materials, compounds containing multiple, *e.g.*, hydroxyl, thiol, carboxyl, amino groups, and often carbonyl groups (read on about how to deal with some of these situations).

COMPOUNDS PREVIOUSLY SYNTHESIZED AND PROPERLY CHARACTERIZED

Identification and providing proof of purity for newly synthesized compounds belonging to this category is relatively simple: matching physical constants found for the newly synthesized compounds with the data in the literature. The latter values, however, had to be measured with material that passed the test of purity by combustion analysis. Providing additional support for identity is welcomed, *e.g.*, mode of synthesis, MS, HRMS, chromatographic behavior, or assigned NMR data, especially when the NMR spectra of the newly synthesized compound were measured at a higher field than when the substance was described for the first time. Thus, authors who want to use the physical characteristics for their material as proof of purity and identity must verify that the literature data they use as reference were measured with material that produced correct combustion analysis figures. If previous authors did not provide correct combustion analysis figures, the new authors have to do that, thereby showing that the physical constants they found for the material are reliable. A special situation arises when, for example, you prepare a known compound that was reported amorphous, and you manage to obtain the material in the crystalline state. When you then measure the specific optical rotation and your value differs considerably (more than 5% relative) from that previously reported, you have to prove that your data are correct. You can do that only by providing correct analytical figures for your material. If your material is stable and belongs to the category of substances where we do not require analytical figures, then you must measure $[\alpha]_D$ immediately after exhaustive drying. The amount of moisture held tenaciously by your (hygroscopic) material will not be such that it would affect $[\alpha]_D$ considerably, although it may be sufficient to affect the combustion analysis figures. If correct analytical figures for compounds containing hydroxyl, thiol, carboxyl, or amino groups cannot be obtained, it is desirable that these compounds be converted to readily made derivatives, and characterized as, *e.g.*, the corresponding esters. Such compounds are generally more stable, and they can be often crystallized and/or dried at a higher temperature and produce correct analytical figures.

NMR Spectra

Regardless of whether the compound is new or one that has been previously reported and characterized, copies of ^1H and ^1H-decoupled ^{13}C NMR spectra must be part of the manuscript that describes the preparation of an organic compound. Do not include 2 D spectra, unless the compound is new *and* you want to prove by such spectra something that cannot be proved otherwise. Copies of the spectra must include the spectral window starting from 0 ppm to the last low-field resonance. It is desirable to show also expansions of the most relevant part of the spectrum. Spectra should be recorded with sufficient signal-to-noise ratio ($S/N \geq 10$). Relevant signals must be of reasonable intensity (strong singlets may go off scale). The assignment of resonances must be listed in the Experimental for at least structurally significant signals. Integration values must be included only when they are used for identification purposes. All spectra must show the structure and compound numbering as in the body of the manuscript. For data presentation, consult previous volumes in the series or editors.

Submitting the Manuscript and the Checking Process

Authors should submit their manuscripts to the volume editor from whom they have received the invitation. Before submitting the actual manuscript, verify with the editor that your topic is suitable for publication in this series. Send the editor the reaction scheme and information about the novelty or other justification for publication, such as improvement or experimental simplicity of the new protocol, more complete characterization or spectral assignments, crystallization of compounds previously known only as amorphous, and preparation of a compound in a not-an-easily-available journal or a language other than English. When submitting, inform the editor who will be the checker of the laboratory protocol. Authors are responsible for finding a qualified person willing to be the checker, *i.e.*, repeat at the bench all protocols described and verify that all protocols described are safe and reproducible in their entirety (yields, physical constants, and spectral properties). Nevertheless, the editors reserve the right to entrust the checking function to someone else. Checking comes with the benefit of becoming a full-fledged author when the checked chapter is published. Any chemist competent at the bench can be the checker of any number of preparative protocols. Many students or technicians at any level, when properly supervised, can become reliable checkers. The supervisory role of the more senior scientist is always indicated in the published work, to add credibility to the checking process. Checkers are not required to send their products for combustion analysis when the manuscript already includes such data. On the other hand, when authors do not include these figures, it is the checker who has to provide any missing information. The checker is supposed to be able to reproduce data found by authors. Thus, all the characterization data, including mp, solvent for crystallization, optical rotation, and confident assignment of structurally significant NMR resonances, must be in the manuscript before it goes to the checker.

The checker must not be associated in any way with any of the authors of that particular chapter and cannot be from the same institution as any of the authors. Authors are supposed to provide the checker with intermediates that are not readily available and/or expensive reagents, and the two parties may discuss the needs in this regard. When a transfer of chemicals is necessary, these should be provided in the amounts sufficient to run experiments in duplicate. Checkers should run/verify the synthesis on the same scale as originally described (at least 1 mmol of the carbohydrate starting material. Authors should be reasonable in this regard; it is desirable to show that the protocol is scalable by describing it on a larger than 1-mmol scale, when the compound described can be expected to be an intermediate for an early step in a multistep synthesis and will likely be needed in large amount). When the checker's findings do not agree with any aspect of the protocol described, or when the checker finds that something could be improved, he/she discusses the matter with the authors and they together may modify the original protocol. In any case, the two parties must agree with the final version of the manuscript.

There is only one checker per chapter. As stated before, the number of steps within a chapter should be kept small and, therefore, there is no need for a team of chemists to check a one-step protocol or a very short sequence. Consequently, only one checker's name may be added to the list of authors. After completion of the work, the checker sends a report to the Volume Editor involved. It can be a simple statement, when everything works the way described, or more elaborate when the checker's findings warrant explanations or suggestions for improvement.

Checkers are supposed to do the work in a timely fashion. As with any scientific material, it should be in everyone's interest that publication not be delayed on account of a tardy checker.

REVISION OF THE MANUSCRIPT DURING PUBLICATION PROCESS

Checkers deserve to receive manuscripts where everything is clearly stated so that they experience no surprises when they exactly follow the working protocol. Similarly, readers deserve clear, concise text free of irrelevant material. Experience with manuscripts submitted to previous volumes has shown that most manuscripts submitted left a lot to be desired in that regard. Consequently, several revisions had been necessary, before and after the checking process, before the manuscripts could be accepted for publication. While we have no formal deadline for submission of manuscripts, the need for multiple revisions slows down the publication process. Needless to say, the "cleaner" that the manuscript submitted is, the faster the new volume appears in print. To avoid unnecessary delay in seeing your contribution cited by others, please adhere to the suggestions that follow:

- Chemical conversion yield and the actual yield of target compounds are two different things. Practically, a 100% yield can be claimed only when equimolar amounts of synthons are used to make a nonvolatile product and the conversion is complete; if one synthon is used in excess that reagent must be volatile. Then, there is no aqueous workup or chromatography, and the only operation involved in obtaining the product is the concentration of the reaction mixture. When you are a practicing chemist, you must agree that such a situation is extremely rare. If other operations are involved in the workup, which is a more likely scenario in the real world, yields over 94% are unrealistic[1] and, therefore, unacceptable in this series.

- It is normal and perfectly within the unwritten but accepted rules for reporting results of organic synthesis to prove the identity of alternatively synthesized known compounds by comparing physical constants found with those reported in the reliable literature. With chiral compounds, a comparison of specific optical rotation with the value reported is a must. Comparisons of other physical constants may be added. However, for such comparison to be acceptable, the literature value had to be determined with material that passed the test of purity by combustion analysis. With carbohydrates, physical constants most suitable for identification purposes are the melting range and specific optical rotation. When the values found differ considerably [mp more than ± 3 °C; $[\alpha]_D$ >5% (relative) from the reported value measured in the same solvent at the same/close concentration] from those reported in the literature for material that passed the proof of purity by combustion analysis, the authors must present the proof of the purity of the newly synthesized material. The reliability of physical constants for compounds that have not produced correct combustion analysis figures is anybody's guess. To report optical rotation data on such compounds is acceptable only when that particular compound belongs to the kind for which we do not require proof of purity by combustion analysis (see elsewhere in this writing). Then, it is understood that the authors did the best they could.
- In situations when the proof of purity for new compounds by combustion analysis is required, it is unacceptable to argue that such figures are not provided because a sufficient amount of material was not available. Between authors and the checker, there always has to be enough material left for combustion analysis, especially when the conversion is run on the required or larger than 1-mmol scale. When authors do not provide correct analytical figures for whatever reason, it is the checker's responsibility to do so and include the data in the manuscript.
- Do not evaporate solutions, concentrate them. Authors often confuse these terms. These terms mean different things and are far from synonymous. Solvents are evaporated, solutions are concentrated. Solutions consist of solvents and solutes. Solvents are evaporated from the solutions to obtain solutes. The same can be accomplished by the concentration of solutions. Solutions are normally not evaporated, only when the solute also evaporates, *e.g.,* during distillation or at a very high temperature, like at ground zero of a nuclear explosion.
- It is often useful to include the following statement within General Methods: Solutions in organic solvents were dried over (some drying agent) and concentrated at (some temperature and pressure). When you follow this advice, do not repeat essentially the same information in the body of individual working protocols.

---

[1] Wernerova, M.; Hudlicky, T. *Synlett,* 2010, 2701–2707.

- While we do not agree with many policies of some mainstream journals, we agree with objecting in titles to claims of priority, originality, convenience, effectiveness, or value. Thus, we strongly advise against trying to attract attention to one's work, which is never as important as one thinks, by using in titles adjectives that express emotions, such as, *e.g.* exceptional, (highly) efficient, useful, unprecedented, and (very) convenient. Let readers decide about these attributes.

- Report preparative protocols on a realistic, useful, real-life scale. When you are reporting a protocol for making a synthetic intermediate, the product of your synthesis is likely to be used for further transformation(s). Therefore, rather than describing the conversion on a few milligrams scale, be realistic and describe the protocol on a scale you would perform the reaction if you wanted to go on with the product and use it to finish your project successfully. Few exceptions aside where a smaller scale can be justified, preparation of every compound in this series should be described on at least 1-mmol scale. This means that when the sequence comprises, for example, three steps, it is advisable to start the first step with a sufficient amount of the starting material that you have at least 1-mmol material available to start the last step. To grant exceptions to the requirement that the last step must start with at least 1 mmol of the starting material is at the discretion of editors. It is worth mentioning that it was a serious oversight when this requirement was not specified in the past, which led to the description of many preparations on a much smaller scale.

- Do not confuse terms glycosylation and glycosidation. These terms are not fully synonymous. To glycosylate means to introduce a glycosyl group (similarly, *e.g.*, to acetylate means to introduce an acetyl group). To glycosidate means to convert to a glycoside. When you make a glycoside/oligosaccharide from a glycosyl donor and a glycosyl acceptor, you glycosylate the acceptor and glycosidate the donor.

- The list of things to avoid would not be complete without mentioning a few violations of good laboratory practices found in print, also in journals that pride themselves with a very high Golden Calf Factor.[2] A selection of such *gems* of laboratory practice resulting from poor training by ignorant advisors at some contemporary, sometimes also famous graduate schools are listed below.

The crystallinity of organic substances is a serious topic, and everything about it must be dealt with responsibly. Crystallization is a way of separating material into two portions: the crystalline material and a solution (mother liquor) of some amount of the same material plus soluble impurities, which are normally always present in the original, amorphous sample. When the compound is collected from a chromatography column as a solution and the solution is concentrated, the residue is sometimes obtained as a crystalline solid. However, in this case, the operation(s) that led to the

---

[2] Kováč, P. In *Carbohydrate Chemistry: Proven Synthetic Methods*; Kováč, P. Ed.; CRC Press/Taylor & Francis, Boca Raton, FL, 2015, Vol. 3.

crystalline residue is not "crystallization" because there is no mother liquor by way of which the impurities, which still may be present (if not other than the residue after evaporation of the solvent), would be separated from the crystalline material. Thus, the crystalline residue must be crystallized from a suitable solvent. With a seed crystal at hand, it should be an easy task, and the only explanation or reason for not doing so in the past was the utmost ignorance or laziness of the practitioner. When melting points of amorphous materials or crystalline residues after evaporation of solutions appear in print, the blame for such violation of common sense and good laboratory practices belongs to poor/unqualified referees and journal editors. When the residue after evaporation is a solid that does not look like crystals, it is wrong to assume that the material is amorphous or that it cannot be crystallized. There are many examples in the literature to the contrary. Attempts to crystallize the material from a suitable solvent should always be made, especially when the solid is a reasonably pure material. A small amount of such material should be saved and used as seed crystals even when the material looks to the naked eye as an amorphous powder. It often consists of or contains microcrystals and can be useful as an aid for initial crystallization. If several attempts to induce crystallization fail, a note should be added to the manuscript: "Attempt to crystallize the material from common organic solvents failed."

To obtain "recrystallized" material means to purify by crystallization at least twice. The crystalline but not *crystallized* material, such as that obtained by concentration of the solution mentioned before, is not suitable for the melting point measurement because the value recorded would not be reliable. Such value is virtually always lower or wider, or both, than the one recorded for the same material after (re)crystallization from a suitable solvent followed by thorough drying and is not acceptable. Equally unacceptable are values without mentioning the solvent for crystallization. It does not happen often, but it has been observed that compounds sometimes crystallize from different solvents in different crystalline modifications and show different, sometimes vastly different, melting characteristics. Such materials produce identical NMR spectra, and when these are chiral compounds, they show the same $[\alpha]_D$ values. It is equally inappropriate and unacceptable to report that a compound was obtained as amorphous solid and then report mp, as is to report that the compound was obtained as a crystalline solid and not to report mp of the material that was properly crystallized to constancy. Melting range should be reported after the observed values have been rounded to the next half of 1°, *e.g.*, report 105.5–107°, instead of 105.3–106.8°, unless the melting range was determined by the thermogravimetric analysis, and if so, this must be specified in General Methods. In this context, we emphasize that it is unacceptable to adjust molecular formulas of amorphous compounds for solvation to fit combustion analysis data. Such adjustment is acceptable only with crystalline compounds. When the solvent is organic, the amount of solvent claimed as part of the crystalline lattice should be possible to prove by integration of the respective signals in the NMR spectrum.

When an organic compound forms a solid upon concentration of its solution, as, for example, when collected after chromatography, it is good laboratory practice to attempt crystallization from a common solvent or solvent mixtures. Regardless of what you find in your favorite high Golden Calf Factor journal, it is inappropriate to report products isolated after chromatography simply as (white) solids. When you

do, it implies to every well-trained conscientious chemist that you tried crystallization but failed. Nevertheless, you may want to add a note: Crystallization from common organic solvents failed. In any case, to attempt crystallization is useful. Nobody
can feel proud if later someone else publishes the preparation of the same compound
and say that they obtained the compound crystalline for the first time and cite the
original authors as those who obtained the same substance amorphous.

   Measuring optical rotation is simple and nondestructive to the sample. The value
of the specific optical rotation is an important criterion of purity of chiral compounds, and it is often the only practicable way to assess the purity. In other words,
there are situations when regardless of the field your NMR spectrometer operates
at or to how many decimals you measure your HRMS data, you cannot prove the
purity and identity of your compound without providing correct combustion analysis figures and $[\alpha]_D$ value for the substance. Thus, when you submit a manuscript
and you do not include the $[\alpha]_D$ values for new carbohydrates whose identity or
purity you want to prove, you would better provide the data as required, because if
you do not, it will delay the publication of your manuscript (if the required data are
not provided in time, your contribution will be shifted to next volume). When the
compound is a known chiral, amorphous substance and you want to avoid having
to prove its purity by combustion analysis, the $[\alpha]_D$ value matching the one in the
literature is the only physical property you can use as a criterion of purity, when
the literature value was obtained with the analytically pure substance. Therefore,
reporting $[\alpha]_D$ and the comparison with the published value observed with material
that passed the test of purity by combustion analysis is a must with chiral, amorphous, known compounds. For the data to be suitable for comparison, the optical
rotation must be measured at the same wavelength (usually the sodium line D at
589 nm), and at the concentration very close to that at which the reported value
was determined. Before you calculate $[\alpha]_D$, make sure that the measured (observed)
absolute numerical value of rotation (α in the formula to calculate $[\alpha]_D$) is less than
180, unless you use a modern, computerized instrument, where the software takes
care of the important correction. With less than contemporary instruments, values
larger than 180 could be ambiguous because the less sophisticated polarimeter cannot distinguish between rotations, for example, +270° and −90°. With old-fashioned,
although possibly very precise polarimeters that do not include the proper correction
to observed values, the rotation must be measured at a lower concentration, or with
a shorter path cell. Because concentration is one of the variables in the calculation
of $[\alpha]_D$, in these situations, the final absolute value of the calculated specific optical rotation may be and often is larger than 180. For the reported compound to be
considered of acceptable purity, the observed $[\alpha]_D$ must fit the reported value within
5% (relative, measured in the same solvent and at the concentration close to that
originally reported with a substance that passed the test of purity by combustion
analysis). When the compound is new, the authors are free to choose any solvent
but, as with choosing a solvent system for TLC, some solvents are more appropriate
than others. We strongly advise against using CH_2Cl_2 or any solvent of such volatil-
ity as a solvent for measuring $[\alpha]_D$ values, because the intended concentration may
not stay constant during the operations involved between sample preparation and
the actual measurement. When any physical constant differs considerably from that

published, a comment is necessary. The best solution is to provide correct combustion analysis figures and proper proof of identity. In this way, authors can prove that their material has the claimed structure and is pure. The published values could be a misprint, and the newly published values correct the erroneous records in the literature. Report values of $[\alpha]_D$ to one decimal place only and include not only the minus sign but also the plus sign, if that is what is wished to be conveyed. In this context, you may find it useful to consult the chapter "In lieu of Introduction", Volume 2 in this series.

R_f values such as, *e.g.*, 0.48 are hardly ever reproducible. They vary, hopefully only within one-tenth, not only when going from a lab to another lab but also within the same lab depending on many factors. Therefore, when reporting TLC mobility, report data to one decimal place only, and round the found R_f values to the nearest tenth. Human error during weighing and measuring volumes is usually responsible for the inconsistency and, thus, it makes a lot of sense to report the $[\alpha]_D$ values and the c associated with it in the same way. For format for reporting other data, please refer to previous volumes in this series.

Pavol Kováč, Series Editor

Introduction

Carbohydrate Chemistry: Proven Synthetic Methods founded by Pavol Kováč has evolved into a highly appreciated series for chemists active in glycosciences.

The main objective of the series is to provide a reliable source of laboratory protocols which have been scrutinized by an external checker and thus sets a high standard with respect to the state-of-the art procedures and ensures reproducibility of results, including yields, purity, as well as full physical and spectroscopic characterization and correct nomenclature of synthesized materials. Unfortunately, these data are frequently not adequately covered or even missing in many current chemistry journals, leading to an overall decay of the scientific quality of manuscripts.

We were thus very pleased to have been invited by the Editor-In-Chief, Pavol Kováč, as guest editors for Volume 5 for *Carbohydrate Chemistry: Proven Synthetic Methods* in continuation of the preceding Volume 4, edited by Christian Vogel and Paul V. Murphy.

Volume 5 contains 33 chapters written by experts in the field, some of which are contributions from authors who are not regularly involved in carbohydrate synthesis. The collection of protocols covers common and scalable synthetic methods as well as the preparation of building blocks with appropriate protecting group patterns, suitable for further transformations needed in oligosaccharide synthesis. The scope of modifications at the anomeric center ranges from *O*-, *S*-, and *N*-glycosides to the formation of *C*-glycosyl compounds, 1-iodo glycals, and anomeric phosphates. Several examples illustrate the introduction of fluorine, azide, and nitro groups among other modifications. Protocols provide sufficient details for starting materials, set-up, and work-up of reactions, safety precautions as well as physical characterization that should also allow novices in the field to successfully perform the synthetic transformations. Whenever checkers could not reach the yields given by the authors, the lower yields have been listed in the experimental section to ensure reproducibility. Purity of products is demonstrated by combustion analysis data as well as copies of ^{1}H and ^{13}C NMR spectra.

We would like to express our sincere thanks to the checkers and when appropriate to their respective supervisors. In that context, the series also fulfills an important educational role for the younger colleagues and Ph.D. students, emphasizing the importance of scientific rigor for their future professional career. Special thanks go to Francesco Nicotra who took care of in-depth surveillance of correct nomenclature used in every chapter.

Last, but not least, we are grateful to our Editor-In-Chief, Pavol Kováč, for valuable critical comments and detailed reviews of submitted manuscripts that were instrumental to ensure the high scientific quality of contributions to Volume 5 of this series.

Paul Kosma
Arnold Stütz
Tanja Wrodnigg

About the Series Editor

Pavol Kováč, Ph.D., Dr. h.c., with more than 40 years of experience in carbohydrate chemistry and more than 300 papers published in refereed scientific journals, is a strong promoter of good laboratory practices and a vocal critic of publication of experimental chemistry lacking data that allows reproducibility. He obtained his M.Sc. in Chemistry at Slovak Technical University in Bratislava (Slovakia) and Ph.D. in Organic Chemistry at the Institute of Chemistry, Slovak Academy of Sciences, Bratislava. After postdoctoral training at the Department of Biochemistry, Purdue University, Lafayette, Indiana (R. L. Whistler, advisor), he returned to the Institute of Chemistry and formed a group of synthetic carbohydrate chemists, which had been active mainly in oligosaccharide chemistry. After relocating to the United States in 1981, he first worked at Bachem, Inc., Torrance, CA, where he established a laboratory for the production of oligonucleotides for the automated synthesis of DNA. He joined NIH in 1983, where he later became one of the Principal Investigators and Chief of the Section on Carbohydrates (NIDDK, Laboratory of Bioorganic Chemistry). The section was originally established by the greatest American carbohydrate chemist, Claude S. Hudson, in 1914 and was under Dr. Kováč's leadership arguably the world's oldest research group continuously working on chemistry, biochemistry, and immunology of carbohydrates. Dr. Kováč is currently Scientist Emeritus at NIDDK (LBC), and his main interest is in the development of conjugate vaccines for infectious diseases from synthetic and bacterial carbohydrate antigens.

About the Editors

Paul Kosma is the professor of Organic Chemistry at the Department of Chemistry at the University of Natural Resources and Life Sciences in Vienna, Austria. He earned his Ph.D. in organic chemistry in 1980 from the University of Technology, Vienna in the field of heterocyclic chemistry. Starting with postdoctoral studies at the former SANDOZ-Research Institute in Vienna under the guidance of Frank Michael Unger and a research stay at the N.D. Zelinsky Institute of Organic Chemistry in Moscow with Leonid Backinowsky, he entered the field of carbohydrate chemistry. In 1988, he earned readership (Habilitation) at the University of Natural Resources and Life Sciences in Vienna and was promoted to full professor of Organic Chemistry in 1991. From 1998 to 2006, he served as the director of a Christian Doppler Laboratory on Pulp Reactivity and as a member of the European Polysaccharide Network of Excellence with a focus on cellulose analytics and modification. In 2009, he was chair of the 15th European Carbohydrate Symposium in Vienna and served as representative of Austria in the European and International Carbohydrate Organizations (ECO, ICO) until 2019. He was presented with a prestigious award (Ehrenkreuz für Wissenschaft und Kunst 1. Klasse) in 2018. His research interests—resulting in more than 280 publications—are in synthesis of complex oligosaccharides and neoglycoconjugates of biomedical importance from bacteria, viruses, and parasites, as well as biosynthesis and structural analysis of bacterial glycans by NMR spectroscopy.

Tanja M. Wrodnigg holds a professorship at the Institute of Chemistry and Technology of Biobased Systems at Graz University of Technology, Austria. She earned her Ph.D. in organic chemistry (1999) at Graz University of Technology, Austria under the supervision of Arnold Stütz, working on the synthesis of iminosugar-based ligands for glycoside-processing enzymes. She joined in 2001 the group of Stephen Withers at the University of British Columbia, Vancouver as a postdoctoral fellow, to obtain insights into the biochemical field of the foregoing enzyme class. Back at TU Graz (2003), she was awarded a Hertha Firnberg Fellowship from the Austrian Science foundation (FWF), to build up her own research group in collaboration with Glycogroup focus on the design and synthesis of glycomimetics as ligands and therapeutics for (bio)-medicinal applications in the context of glycoside-processing enzymes. In 2007, she went for a sabbatical to Danish Technical University (DTU) Lyngby, Denmark to work with Prof. Inge Lundt. In 2008, she was appointed Associate Professor at TU Graz, and since 2020 she is a full professor and established the new Institute for Chemistry and Technology of Biobased Systems at TU Graz with the focus on the design and synthesis of bioactive glycoconjugates and glycomimetics as tools for glycoprocessing enzymes and modification of carbohydrate-based biomolecules for technological applications.

Arnold Stütz was born in Graz, Austria, in 1957 and received his Ph.D. at Graz Technical University in 1983. After postdoctoral work with Profs. Yoshito Kishi (1984/85, Palytoxin) and Robert J. (Robin) Ferrier (1986/87, Carbasugars), he returned to Graz and initiated the start of Glycogroup at the Graz University of Technology. He was appointed the professor for Organic Chemistry in 1993. His stays abroad, including his guest professorships at Danish Technical University (DTU), have contributed to a range of fruitful long-term international collaborations. His main research interests have covered aminocyclitol antibiotics, numerous types of glycosidase inhibitors, synthetic applications and mechanism of xylose isomerase, and other means of isomerization of free sugars as well as added-value products from cheap renewable sources, just to mention a few. Current work is focused on cyclopentanoid sugar analogs, nonnatural glycolipids as well as amino acid—sugar adducts.

Contributors

Sanaz Ahmadipour
Lennard-Jones Laboratory, School of Chemical and
Physical Sciences, Keele University,
Keele, United Kingdom

Rosalino Balo
Centro Singular de Investigación en Química Biolóxica e
Materiais Moleculares (CiQUS),
Departamento de Química Orgánica,
Universidade de Santiago de Compostela
Santiago de Compostela, Spain

Marek Baráth
Institute of Chemistry, Slovak Academy of Sciences,
Bratislava, Slovakia

Kedar N. Baryal
Department of Chemistry, Michigan State University,
East Lansing, Michigan, United States

Caecilie M. M. Benckendorff
Lennard-Jones Laboratory,
Keele University,
Keele, United Kingdom

Jonathan Berry
Otto Diels Institute of Organic Chemistry,
Christiana Albertina University of Kiel,
Kiel, Germany

Benedetta Bertolotti
University of Chemistry and Technology Prague
(UCT Prague), Czech Republic

Bhoomendra Bhongade
Department of Pharmaceutical Chemistry,
RAK College of Pharmaceutical Sciences,
RAK Medical and Health Sciences University,
Ras Al Khaimah, United Arab Emirates

Gordon Jacob Boehlich
Universität Hamburg, Institut für Pharmazie,
Hamburg, Germany

Anikó Borbás
Department of Pharmaceutical Chemistry,
University of Debrecen,
Debrecen, Hungary

Marc Bouillon
School of Natural Sciences, Bangor University,
Bangor, UK

Lei Cai
Hubei Key Laboratory of Natural Medicinal Chemistry
and Resource Evaluation, School of Pharmacy, Huazhong
University of Science and Technology,
Wuhan, P. R. China

María Emilia Cano
Universidad de Buenos Aires,Facultad de Ciencias
Exactas y Naturales, Departamento de Química Orgánica,
Pabellón 2, Ciudad Universitaria,
Buenos Aires, Argentina
Consejo Nacional de Investigaciones Científicas y

Técnicas (CONICET)-UBA, Centro de
Investigación en
Hidratos de Carbono
(CIHIDECAR),
Buenos Aires, Argentina

Antoine Carpentier
Département de Chimie, PROTEO,
RQRM,
Université Laval,
Québec City, Canada

Juan M. Casas-Solvas
Department of Chemistry and Physics,
University of Almería,
Almería, Spain

Thomas A. Charlton
University of Ottawa,
Ottawa, Canada

Aisling Ní Cheallaigh
School of Chemistry, University
College Dublin,
Belfield, Dublin 4, Ireland

Jose Luis Chiara
Instituto de Química Orgánica General,
IQOG-CSIC
Madrid, Spain

Laura Cipolla
Department of Biotechnology and
Bioscience,
University of Milano Bicocca,
Milano, Italy

Juan P. Colomer
Universidad Nacional de Córdoba,
Instituto de
Investigaciones en Físico-Química de
Córdoba –
(UNC – INFIQC – CONICET),
Facultad de Ciencias
Químicas, Departamento de Química
Orgánica,

Edificio de Ciencias 2, Ciudad
Universitaria –
Córdoba, Argentina

Conor J. Crawford
Department of Chemistry,
University College Dublin, Belfield,
Dublin 4, Ireland

Alejandro E. Cristófalo
Universidad de Buenos Aires,Facultad
de Ciencias
Exactas y Naturales, Departamento de
Química Orgánica,
Pabellón 2, Ciudad Universitaria,
Buenos Aires, Argentina
Consejo Nacional de Investigaciones
Científicas y
Técnicas (CONICET)-UBA, Centro de
Investigación en
Hidratos de Carbono (CIHIDECAR),
Buenos Aires, Argentina

Alexei V. Demchenko
University of Missouri – St. Louis, One
University Boulevard,
St. Louis, Missouri United States

Fruzsina Demeter
Department of Pharmaceutical
Chemistry, University of
Debrecen
Debrecen, Hungary,
MTA-DE Molecular Recognition and
Interaction Research Group,
University of Debrecen, Debrecen,
Hungary,
Doctoral School of Chemistry,
University of Debrecen,
Debrecen, Hungary

Vincent Denavit
Département de Chimie, PROTEO,
RQRM, Université Laval,
Québec City, Canada

Christian Denner
Institute of Organic Chemistry,
  University of Vienna,
Währingerstrasse 38, A-1090 Vienna,
  Austria

Guillaume Despras
Otto Diels Institute of Organic Chemistry,
Christiana Albertina University of Kiel,
Kiel, Germany

Perry Devo
Department of Pharmaceutical,
  Chemical and
Environmental Sciences, University of
  Greenwich Chemistry,
Greenwich, England

J. Hessel M. van Dijk
Leiden Institute of Chemistry, Leiden
  University,
Leiden, The Netherlands

Han Ding
Key Laboratory of Marine Medicine,
  Chinese Ministry
of Education, School of Medicine and
  Pharmacy,
Oceans University of China,
Qingdao, P. R. China

Lei Dong
Institut de Chimie et Biochimie
  Moléculaires
et Supramoléculaires, Laboratoire de
  Chimie
Organique 2—Glycochimie, UMR
  5246, CNRS,
Université Claude Bernard Lyon 1,
Villeurbanne, France

Barbara La Ferla
Department of Biotechnology and
  Bioscience,
University of Milano Bicocca,
Milano, Italy

Roland Fischer
Institute of Inorganic Chemistry,
Graz University of Technology,
Graz, Austria

Charles Gauthier
INRS-Institut Armand-Frappier,
Université du Québec
Laval, Canada

Marine Gavel
Normandie University,
  INSA Rouen,
UNIROUEN, CNRS
Rouen, France

Denis Giguère
Département de Chimie, PROTEO,
  RQRM,
Université Laval,
Québec City, Canada

Rita Gonçalves-Pereira
Centro de Química e strutural,
Faculdade de Ciências, Universidade de
  Lisboa,
Lisboa, Portugal

Takaaki Goto
Graduate School of Life and
  Environmental Sciences,
Kyoto Prefectural University,
Kyoto, Japan

Mieke Guinan
Lennard-Jones Laboratory,
  School of Chemical
  and Physical
Sciences, Keele University,
Keele, United Kingdom

Roberto Guizzardi
Department of Biotechnology and
  Bioscience,
University of Milano Bicocca,
Milano, Italy

Vojtěch Hamala
Institute of Chemical
   Process Fundamentals
   of the CAS,
Praha, Czech Republic

Dirk Hauck
Helmholtz Institute for Pharmaceutical
   Research Saarland
(HIPS), Helmholtz Centre for Infection
   Research (HZI),
Saarbrücken, Germany

Floriane Heis
Sorbonne Université, CNRS,
Institut Parisien de Chimie Moléculaire,
   IPCM
Paris, France

Mihály Herczeg
University of Debrecen,
Debrecen, Hungary,
Research Group for Oligosaccharide
   Chemistry of the
Hungarian Academy of Sciences,
Debrecen, Hungary

Daniela Imperio
Dipartimento di Scienze
   del Farmaco, Università
   degli Studi
del Piemonte Orientale "Amedeo
   Avogadro",
Novara, Italy

Nigel K. Jalsa
The University of the West Indies,
St. Augustine, Trinidad and Tobago

Antoine Joosten
Normandie University, INSA Rouen,
UNIROUEN, CNRS
Rouen, France

Varsha R. Jumde
Helmholtz Institute for Pharmaceutical
   Research Saarland

(HIPS), Helmholtz Centre for Infection
   Research (HZI),
Saarbrücken, Germany

Nándor Kánya
Department of Organic Chemistry,
University of Debrecen,
Debrecen, Hungary

Toshinari Kawada
Graduate School of Life and
   Environmental Sciences,
Kyoto Prefectural University,
Kyoto, Japan

Jindřich Karban
Institute of Chemical
   Process Fundamentals
   of the CAS,
Praha, Czech Republic

Martina Kašáková
University of Chemistry and
   Technology Prague
(UCT Prague), Czech Republic

Katharina Kettelhoit
Institute of Organic Chemistry,
Technische Universität
   Braunschweig,
Braunschweig, Germany

Victoria Kohout
Indiana University,
Bloomington, Indiana United States

Paul Kosma
Department of Organic Chemistry,
University of Natural Resources and
   Life Sciences,
Vienna, Austria

Pavol Kováč
NIDDK, LBC, National Institutes
   of Health,
Bethesda, Maryland, United States

Viktors Kumpinš
Faculty of Materials Science and
    Applied Chemistry,
Riga Technical University,
Riga, Latvia

Martin Kurfiřt
Institute of Chemical Process
    Fundamentals of the CAS,
Praha, Czech Republic

Martina Lahmann
School of Natural Sciences, Bangor
    University,
Bangor, United Kingdom

Danny Lainé
Département de Chimie,
    PROTEO, RQRM,
    Université Laval,
Québec City, Canada

Thomas Lecourt
Normandie University,
    INSA Rouen,
UNIROUEN, CNRS
Rouen, France

Yan Liu
Hubei Key Laboratory of Natural
    Medicinal Chemistry
and Resource Evaluation, School of
    Pharmacy, Huazhong
University of Science
    and Technology,
Wuhan, P. R. China

Jevgenija Luginina
Faculty of Materials Science and
    Applied Chemistry,
Riga Technical University,
Riga, Latvia

Dylan Lynch
School of Chemistry, Trinity College
    Dublin,

University of Dublin,
Dublin, Ireland

Milo Malanga
Cyclolab Ltd,
Budapest, Hungary

Enrique Mann
Instituto de Química Orgánica
    General, IQOG-CSIC
Madrid, Spain

Michael P. Mannino
University of Missouri – St. Louis, One
    University Boulevard,
St. Louis, Missouri United States

Nittert Marinus
Stratingh Institute for Chemistry,
    University of Groningen,
Groningen, The Netherlands

Kenya Matsushita
Graduate School of Engineering, Tottori
    University,
Tottori, Japan

Orla McCabe
School of Chemistry, University
    College Dublin
Belfield, Dublin 4, Ireland

Kévin Mébarki
Normandie University,
    INSA Rouen,
UNIROUEN, CNRS
Rouen, France

Yuichi Mikota
Faculty of Agriculture, Shizuoka
    University,
Shizuoka, Japan

Gavin J. Miller
Lennard-Jones Laboratory, School of
    Chemical and

Physical Sciences, Keele University,
Keele, United Kingdom

Adriaan J. Minnaard
Stratingh Institute for Chemistry,
    University of Groningen,
Groningen, The Netherlands

Hugo O. Montenegro
Universidad de Buenos Aires,Facultad
    de Ciencias
Exactas y Naturales, Departamento de
    Química Orgánica,
Pabellón 2, Ciudad Universitaria,
Buenos Aires, Argentina
Consejo Nacional de Investigaciones
    Científicas y
Técnicas (CONICET)-UBA, Centro de
    Investigación en
Hidratos de Carbono (CIHIDECAR),
Buenos Aires, Argentina

Jitka Moravcová
University of Chemistry and
    Technology Prague
(UCT Prague), Czech Republic

Leila Mousavifar
Université du Québec à Montréal,
Montréal, Canada

Mana Mohan Mukherjee
NIDDK, LBC, National Institutes of
    Health,
Bethesda, Maryland, United States

Kevin Muru
INRS-Institut Armand-Frappier,
    Université du Québec,
Laval, Canada

Stefan Oscarson
Centre for Synthesis and Chemical
    Biology,
University College Dublin,
Belfield, Dublin, Ireland

Matteo Panza
University of Missouri – St. Louis, One
    University Boulevard,
St. Louis, Missouri United States

Kamil Parkan
Department of Chemistry of Natural
    Compounds,
University of Chemistry and
    Technology (UCT, Prague),
Czech Republic

Hélène B. Pfister
NIDDK, LBC, National Institutes of
    Health,
Bethesda, Maryland United States

Mark Reihill
Centre for Synthesis and Chemical
    Biology,
University College Dublin,
Belfield, Dublin 4, Ireland

Bettina Riedl
University of Vienna, Institute of
    Organic Chemistry,
Vienna, Austria

Audric Rousset
Institut de Chimie et Biochimie
    Moléculaires
et Supramoléculaires, Laboratoire de
    Chimie
Organique 2—Glycochimie, UMR
    5246, CNRS,
Université Claude Bernard
    Lyon 1,
Villeurbanne, France

René Roy
Université du Québec à Montréal,
Montréal, Canada

Tze Chieh Shiao
Université du Québec à Montréal,
Montréal, Canada

Ondřej Šimák
Department of Chemistry
  of Natural Compounds,
University of Chemistry
  and Technology (UCT, Prague),
Czech Republic

Mark Smith
Rioscience LLC,
Sunnyvale, California United States

Lucie Červenková Šťastná
Institute of Chemical Process
  Fundamentals of the CAS,
Praha, Czech Republic

Jacob St-Gelais
Département de Chimie, PROTEO,
  RQRM, Université Laval,
Québec City, Canada

Arnold E. Stütz
Glycogroup, Institute of Organic
  Chemistry,
Graz University of Technology,
Graz, Austria

Afraz Subratti
The University of the West Indies,
St. Augustine, Trinidad and Tobago

Jun-ichi Tamura
Graduate School of Engineering, Tottori
  University,
Tottori, Japan
Faculty of Agriculture, Tottori University,
Tottori, Japan

Martin Thonhofer
Institute of Organic Chemistry,
Graz University of Technology,
Graz, Austria

Alexander Titz
Helmholtz Institute for Pharmaceutical
  Research Saarland

(HIPS), Helmholtz Centre for Infection
  Research (HZI),
Saarbrücken, Germany
Deutsches Zentrum für
  Infektionsforschung (DZIF),
Standort Hannover-Braunschweig,
Germany Saarland University,
Saarbrücken, Germany

Serena Traboni
Dipartimento di Scienze Chimiche,
Università degli Studi di Napoli
  Federico II,
Napoli, Italy

Nino Trattnig
Department of Organic Chemistry,
University of Natural Resources and
  Life Sciences,
Vienna, Austria

Thomas Tremblay
Département de Chimie, PROTEO,
  RQRM,
Université Laval,
Québec City, Canada

Māris Turks
Faculty of Materials Science and
  Applied Chemistry,
Riga Technical University,
Riga, Latvia

María Laura Uhrig
Universidad de Buenos Aires, Facultad
  de Ciencias
Exactas y Naturales, Departamento de
  Química Orgánica,
Pabellón 2, Ciudad Universitaria,
Buenos Aires, Argentina
Consejo Nacional de Investigaciones
  Científicas y
Técnicas (CONICET)-UBA, Centro de
  Investigación en
Hidratos de Carbono (CIHIDECAR),
Buenos Aires, Argentina

Mattia Vacchini
Department of Biotechnology and
    Bioscience, University of Milano
    Bicocca,
Milano, Italy

Karolína Vaňková
Department of Chemistry of Natural
    Compounds,
University of Chemistry and
    Technology (UCT, Prague),
Czech Republic

Antonio Vargas-Berenguel
Department of Chemistry
    and Physics,
University of Almería,
Almería, Spain

Stella Verkhnyatskaya
Stratingh Institute of Chemistry,
    University of Groningen,
Groningen, The Netherlands

Sébastien Vidal
Institut de Chimie et Biochimie
    Moléculaires
et Supramoléculaires, Laboratoire de
    Chimie
Organique 2—Glycochimie, UMR
    5246, CNRS,
Université Claude Bernard Lyon 1,
Villeurbanne, France

Marthe T. C. Walvoort
Stratingh Institute for Chemistry,
    University of Groningen,
Groningen, The Netherlands

Qian Wan
Hubei Key Laboratory
    of Natural Medicinal
    Chemistry
and Resource Evaluation, School of
    Pharmacy, Huazhong

University of Science and Technology,
Wuhan, P. R. China

Patrick Weber
Glycogroup, Institute of Organic
    Chemistry,
Graz University of Technology,
Graz, Austria

Rhodri Mir Williams
School of Natural Sciences, Bangor
    University,
Bangor, UK

Martin D. Witte
Stratingh Institute for Chemistry,
    University of Groningen,
Groningen, The Netherlands

Nuno M. Xavier
Centro de Química e strutural,
Faculdade de Ciências, Universidade de
    Lisboa,
Lisboa, Portugal

Xiong Xiao
Hubei Key Laboratory of Natural
    Medicinal Chemistry
and Resource Evaluation, School of
    Pharmacy, Huazhong
University of Science and Technology,
Wuhan, P. R. China

Yuko Yoneda
Faculty of Agriculture, Shizuoka
    University,
Shizuoka, Japan

Jing Zeng
Hubei Key Laboratory of Natural
    Medicinal Chemistry
and Resource Evaluation, School of
    Pharmacy, Huazhong
University of Science and Technology,
Wuhan, P. R. China

Part I

Synthetic Methods

1 Synthesis of Glycosyl Thiols via 1,4-Dithiothreitol-Mediated Selective Anomeric S-Deacetylation

Lei Cai, Yan Liu, Xiong Xiao, Jing Zeng, and Qian Wan*
Hubei Key Laboratory of Natural Medicinal Chemistry
and Resource Evaluation, School of Pharmacy, Huazhong
University of Science and Technology,
Wuhan, P. R. China

Han Ding[a]
Key Laboratory of Marine Medicine, Chinese Ministry
of Education, School of Medicine and Pharmacy,
Oceans University of China,
Qingdao, P. R. China

CONTENTS

We describe a practical method for the synthesis of glycosyl thiols (1-thiosugars) by 1,4-dithiothreitol (DTT)-mediated selective S-deacetylation. Glycosyl thiols are common building blocks for construction of S-linked glycosides, such as S-linked oligosaccharides, glycopeptides, and glycolipids.[1] The stability of S-glycosidic bond in these compounds allows for their wide application in the investigation of biological

---

* Corresponding author: wanqian@hust.edu.cn
[a] Checker: under supervision of Prof. Ming Li: lmsnouc@ouc.edu.cn

DTT (1.5 equiv)
NaHCO$_3$ (0.1 equiv)
SAc $\xrightarrow{\hspace{1cm}}$ SH
N,N-dimethylacetamide

processes and the development of new therapeutics.[2] A convenient and practical synthesis of glycosyl thiols is based on selective S-deacylation of peracylated thioglycosides.[3] However, most of the reported methods involve the use of a strong base, such as NaOH or NaOMe, and require careful control of the reaction conditions.[4] Weaker bases, for example, Et$_3$N or NaSMe, have also been applied to the selective anomeric S-deacylation, but only moderate yields were reported.[5] Recently, Zhu et al. reported an improved method applying aqueous NaSMe as base.[6] Inspired by native chemical ligation, we have developed a cysteine mediated S-deacetylation reaction.[3] This method has been further expanded to a DTT-mediated selective S-deacetylation reaction under weakly basic conditions through transthioesterification. The advantages of this method lie in mild reaction conditions, high efficiency on large scale, simple purification without column chromatography, and the ease of recovery of DTT. Herein, we present the detailed procedures of this DTT-mediated selective anomeric S-deacetylation as applied to 6-deoxy-D-pyranoses.

Fully acylated thiosugars (**1a, 1b**) reacted with DTT smoothly to produce glycosyl thiols **2a** and **2b** in good yields in the presence of 0.1 equiv of NaHCO$_3$ at room temperature (25°C) (Figure 1.1). Notably, the reaction must be quenched immediately once the starting material is completely consumed (TLC), as extended reaction times lead to formation of side products. In addition, the absence of moisture is essential, as its presence causes O-deacetylation. Furthermore, the use of DMA (N,N-dimethylacetamide) as reaction solvent is important: It allows for the isolation of the glycosyl thiols from the water-diluted reaction mixture by extraction with

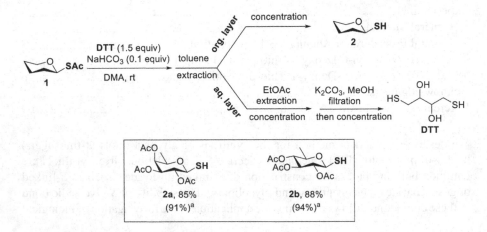

FIGURE 1.1   Selective anomeric S-deacetylation and the recovery and regeneration of DTT.
[a]The yields of recovered DTT in parentheses.

toluene. Normally, the products obtained by these processes are sufficiently pure for further transformations. The aqueous phase containing excess of DTT and acetylated DTT is saturated with NaCl followed by extraction with ethyl acetate. Further evaporation of ethyl acetate and treatment of the mixture of DTT and acetylated DTT with K_2CO_3 in MeOH regenerates a considerable amount of DTT. The regenerated DTT could be used in the next selective anomeric S-deacetylation without affecting the reaction efficiency.[3]

EXPERIMENTAL

GENERAL METHODS

All reactions were monitored by thin-layer chromatography on silica gel-coated TLC plates (HSGF 254, Yantai Chemical Industry Research Institute). The spots were visualized by charring with 10% H_2SO_4 in ethanol. Column chromatography was performed using silica gel (Qingdao Marine Chemical, Inc., China). NMR spectra were recorded with a Bruker AM-400 spectrometer, and the ^{1}H and ^{13}C NMR chemical shifts were referenced to the residual nondeuterated solvent peaks ($CDCl_3$ at δ_H 7.24 and δ_C 77.23). Optical rotations were measured with a PerkinElmer 341LC polarimeter using a quartz cell with 1-mL capacity and a 1-dm path length. Concentrations (c) are given in g/100 mL. High-resolution mass spectra were recorded with a Bruker micrOTOF II spectrometer using electrospray ionization . Commercially available AR grades reagents were purchased from Sinopharm Chemical Reagent Co., Ltd and used without further purification, unless otherwise specified. DMA was dried over 4 Å MS before use. DTT was purchased from Adamas as a racemic mixture.

GENERAL PROCEDURE FOR ANOMERIC S-DEACETYLATION

$NaHCO_3$ (0.1 equiv) was added to a 0.15 M solution of peracetylated glycosyl thiols **1** (1.0 equiv) and DTT (1.5 equiv) in DMA.* The mixture was stirred at room temperature (~25°C) until TLC showed that the starting materials were consumed.† The mixture was poured into water and extracted with toluene (3×). The combined organic layers were consecutively washed with water and brine, and then concentrated to furnish the glycosyl thiols **2**.‡ The aqueous phase was saturated with NaCl§ and extracted three times with ethyl acetate. The organic layers were combined and concentrated. The residue was dissolved in MeOH (containing 1% w/w K_2CO_3) and stirred at room temperature until all the acetylated DTT was consumed (2 h). The mixture was then filtered through Celite and washed with MeOH. The filtrate was concentrated to give DTT. These oxidation-prone glycosyl thiols **2a, 2b**

---

* The reagents addition sequence should be strictly followed because, at basic conditions in presence of air, DTT forms disulfides.
† ~4 h, depending on the reaction temperature and scale. It should be noted that progress of the reaction should be monitored often (every 20 min after the reaction proceeded 2 h) and terminated accordingly, because side reactions occur when the reaction time is extended unnecessarily.
‡ The purity is deemed sufficient for further transformations.
§ Solid NaCl was used instead of aqueous NaCl solution to make sure the aqueous phase was saturated.

and DTT could be stored in refrigerator at −20°C for 2 months without noticeable decomposition.

2,3,4-Tri-*O*-Acetyl-6-Deoxy-1-Thio-β-D-Galactopyranose (2A)

The reaction with 2,3,4-tri-*O*-acetyl-1-*S*-acetyl-6-deoxy-1-thio-β-D-galactopyranose (**1a**) (450 mg, 1.29 mmol, 1.0 equiv), DTT (0.30 g, 1.94 mmol, 1.5 equiv), and NaHCO$_3$ (10.9 mg, 0.13 mmol, 0.1 equiv) in dry DMA at rt according to the General Procedure gave crude **2a** as white amorphous solid (335 mg, 85%). Chromatography (3:1 petroleum ether–EtOAc) gave pure **2a**, R$_f$ = 0.30 (2:1 petroleum ether–EtOAc); [α]$_D^{20}$ + 34.7 (*c* 1.0, CHCl$_3$); ^{1}H NMR (400 MHz, CDCl$_3$) δ 5.26 (d, 1*H*, *J*$_{3,4}$ 3.2 Hz, H-4), 5.14 (t, 1*H*, *J*$_{1,2}$ = *J*$_{2,3}$ 10.0 Hz, H-2), 4.99 (dd, 1*H*, *J*$_{3,4}$ 3.2 Hz, *J*$_{2,3}$ 10.0 Hz, H-3), 4.47 (t, 1*H*, *J*$_{1,2}$ = *J*$_{1,SH}$ 10.0 Hz, H-1), 3.82 (q, 1*H*, *J*$_{5,6}$ 6.4 Hz, H-5), 2.31 (d, 1*H*, *J*$_{1,SH}$ 10.0 Hz, SH), 2.17 (s, 3*H*, COCH$_3$), 2.06 (s, 3*H*, COCH$_3$), 1.96 (s, 3*H*, COCH$_3$), 1.20 (d, 3*H*, *J*$_{5,6}$ 6.4 Hz, H-6). ^{13}C NMR (100 MHz, CDCl$_3$) δ 170.8 (CO), 170.3 (CO), 170.2 (CO), 79.1 (C-1), 74.0 (C-5), 72.2 (C-3), 71.2 (C-2), 70.5 (C-4), 21.1 (COCH$_3$), 20.9 (COCH$_3$), 20.8 (COCH$_3$), and 16.6 (CH$_3$). HRMS [M + Na]$^+$ calcd for C$_{12}$H$_{18}$O$_7$NaS, 329.0671; found, 329.0672.

The recovery of DTT was ~91%.

2,3,4-Tri-*O*-Acetyl-6-Deoxy-1-Thio-β-D-Glucopyranose (2B)

The reaction with 2,3,4-tri-*O*-acetyl-1-*S*-acetyl-6-deoxy-1-thio-β-D-glucopyranose (**1b**) (471 mg, 1.35 mmol, 1.0 equiv), DTT (0.31 g, 2.03 mmol, 1.5 equiv), and NaHCO$_3$ (11.4 mg, 0.14 mmol, 0.1 equiv) in DMA at rt according to the General Procedure gave **2b** as white amorphous solid (362 mg, 88%). Chromatography (2:1 petroleum ether–EtOAc). R$_f$ = 0.5 (1:1 petroleum ether–EtOAc) gave pure **2b**, [α]$_D^{20}$ + 9.1 (*c* 1.0, CHCl$_3$); ^{1}H NMR (400 MHz, CDCl$_3$) δ 5.12 (t, 1*H*, *J*$_{1,2}$ = *J*$_{2,3}$ 9.6 Hz, H-2), 4.92 (t, 1*H*, *J*$_{2,3}$ = *J*$_{3,4}$ 9.6 Hz, H-3), 4.82 (t, 1*H*, *J*$_{3,4}$ = *J*$_{4,5}$ 9.6 Hz, H-4), 4.48 (d, 1*H*, *J*$_{1,2}$ 9.6 Hz, H-1), 3.57 (dq, 1*H*, *J*$_{4,5}$ 9.6 Hz, *J*$_{5,6}$ 6.0 Hz, H-5), 2.23 (s, 1*H*, SH), 2.05 (s, 3*H*, COCH$_3$), 2.01 (s, 3*H*, COCH$_3$), 1.98 (s, 3*H*, COCH$_3$), 1.23 (d, 3*H*, *J*$_{5,6}$ 6.0 Hz, H-6). ^{13}C NMR (100 MHz, CDCl$_3$) δ 170.4 (CO), 169.9 (CO), 169.8 (CO), 78.5 (C-1), 75.0 (C-5), 74.2 (C-3), 73.7 (C-2), 73.3 (C-4), 21.0 (COCH$_3$), 20.9 (COCH$_3$), 20.8 (COCH$_3$), and 17.8 (CH$_3$). HRMS [M+Na]$^+$ calcd for C$_{12}$H$_{18}$NaO$_7$, 329.0671; found, 329.0678.

The recovery of DTT was ~94%.

ACKNOWLEDGMENTS

We thank the National Natural Science Foundation of China (21672077, 21472054, 21761132014, 21772050), the Specialized Research Fund for the Doctoral Program of Higher Education (20120142120092), the Recruitment Program of Global Youth Experts of China, and the State Key Laboratory of Bio-organic and Natural Products Chemistry (SKLBNPC13425) for support.

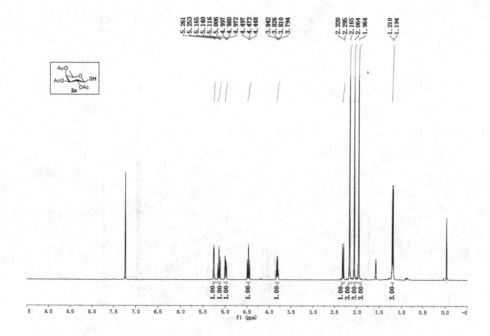

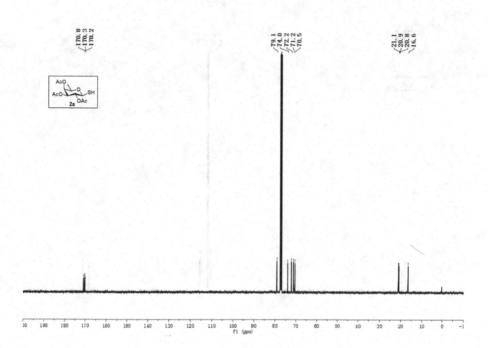

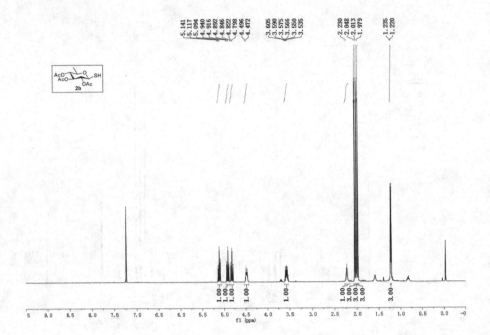

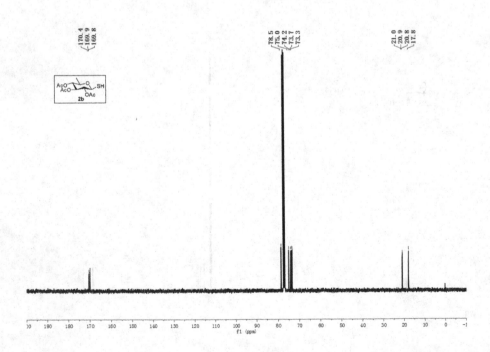

REFERENCES

1. Pachamuthu, K.; Schmidt, R. *Chem. Rev.* **2006**, *106*, 160–187.
2. Gingras, M.; Chabre, Y. M.; Roy, M. *Chem. Soc. Rev.* **2013**, *42*, 4823–4841.
3. Shu, P.; Zeng, J.; Tao, J.; Zhao, Y.; Yao, G.; Wan, Q. *Green Chem.* **2015**, *17*, 2545–2551.
4. (a) Blauvelt, M. L.; Khalili, M. K.; Jaung, W.; Paulsen, J.; Anderson, A. C.; Wilson, S. B.; Howell, A. R. *Bioorg. Med. Chem. Lett.* **2008**, *18*, 6374–6376; (b) Yamamoto, K.; Watanabe, N.; Matsuda, H.; Oohara, K.; Araya, T.; Hashimoto, M.; Miyairi, K.; Okazaki, I.; Saito, M.; Shimiza, T.; Kato, H.; Okuno, T. *Bioorg. Med. Chem. Lett.* **2005**, *15*, 4932–4935; (c) Stanetty, C.; Wolkerstorfer, A.; Amer, H.; Hofinger, A.; Jordis, U.; Claßen-Houben, D.; Kosma, P.; Beilstein, J. *Org. Chem.* **2012**, *8*, 705–711; (d) Wang, H.; Zhu, X. *Org. Biomol. Chem.* **2014**, *12*, 7119–7126.
5. (a) MacDougall, J. M.; Zhang, X.; Polgar, W. E.; Khroyan, T. V.; Toll, L.; Cashman, J. R. *J. Med. Chem.* **2004**, *47*, 5809–5815; (b) Gorska, K.; Huang, K.; Chaloin, O.; Winssinger, N. *Angew. Chem. Int. Ed.*, **2009**, *48*, 7695–7700; (c) Pilgrim, W.; Murphy, P. V. *J. Org. Chem.* **2010**, *75*, 6747–6755; (d) Mischnick, P.; Evers, B.; Thiem, J. *Carbohydr. Res.* **1994**, *264*, 293–299; (e) Moreno-Vargas, A. J.; Molina, L.; Carmona, A. T.; Ferrali, A.; Lambelet, M.; Spertini, O.; Robina, I. *Eur. J. Org. Chem.* **2008**, 2973–2982; (f) Gottschaldt, M.; Koth, D.; Mueller, D.; Klette, I.; Rau, S.; Goerls, H.; Schaefer, B.; Baum, R. P.; Yano, S. *Chem. Eur. J.* **2007**, *13*, 10273–10280.
6. Rao, J.; Zhang, G.; Zeng, X.; Zhu, X. In *Carbohydrate Chemistry: Proven Synthetic Methods*; Roy, R.; Vidal, S. Eds; CRC/Taylor & Francis: Boca Raton, FL, **2015**, *3*, pp. 89–96.

2 One-Step Transformation of Glycals into 1-Iodo Glycals

Karolína Vaňková, Ondřej Šimák, and Kamil Parkan*
Department of Chemistry of Natural Compounds,
University of Chemistry and Technology
Prague (UCT Prague), Czech Republic

Katharina Kettelhoit[a]
Technische Universität Braunschweig,
Braunschweig, Germany

CONTENTS

At present, 1-iodo glycals are valuable intermediates for the synthesis of various *C*-glycosyl compounds[1-13] using transition-metal catalyzed cross-coupling reactions such as Sonogashira, Stille, and Suzuki reaction or *ortho*-C–H activation.

Preparation and the utilization of 1-iodo-3,4,6-tri-*O*-TIPS-D-glucal and 1-iodo-3, 4,6-tri-*O*-TIPS-D-galactal were already published,[5,6] but some inconveniences continue to exist in their synthesis. Namely, the starting 3,4,6-tri-*O*-TIPS D-galactal has to be prepared from D-galactal by a time-consuming, low-yielding (~37%), two-step silylation. Subsequent lithiation followed by iodination afford 3,4,6-tri-*O*-TIPS-1-iodo-D-galactal in 65% yield only.[5,6] In the case of 1-iodo-3,4,6-tri-*O*-TIPS-D-glucal, the overall yield is high, but a direct stereoselective transformation of cross-coupling products into α- or β-*C*-glycosyl compounds failed,[11,14] which is in contrast to compound **3** having a *trans*-fused bicyclic system.[14,15]

---

* Corresponding author: parkank@vscht.cz
[a] Checker: under supervision of Daniel B. Werz: d.werz@tu-braunschweig.de

(t-Bu)₂Si–O … TIPSO … 1   →   3.5 equiv. t-BuLi, THF, -78 → 0 °C, 1 h   →   [Li]   →   2.5 equiv. CH₂I₂, -78 °C, 3 h   →   (t-Bu)₂Si–O … TIPSO … I   **3, 76 %**

TBSO OTIPS … TBSO … 2   →   3.5 equiv. t-BuLi, THF, -78 → 0 °C, 30 min   →   [Li]   →   2.5 equiv. CH₂I₂, -78 °C, 3 h   →   TBSO OTIPS … TBSO … I   **4, 87 %**

Here, we present an optimized one-pot synthesis of two new 1-iodo glycals, **3** and **4**, starting from easily accessible silylated glycals **1** and **2**, respectively.[15] In the first step, only 3.5 equiv of *t*-BuLi are used for lithiation, instead of the reported 4.5 equiv.[5,6] The reaction times are crucial, especially in the lithiation of galactal **2** where the reaction time may not exceed 30 min. Immediately after the appropriate lithiation reaction time, excess of diiodomethane (2.5 equiv) is added. For the isolation of **3** and **4** in the purity sufficient for the subsequent cross-coupling reactions, only partition of the reaction mixture between toluene and water is needed, followed by extraction of the organic layer with sodium thiosulfate solution. Pure (thin-layer chromatography [TLC]) 1-iodo-glycals* **3** and **4** are obtained in 76% and 87% yields, respectively.

EXPERIMENTAL

GENERAL METHODS

All reactions were performed under argon using oven-dried glassware. Solutions in organic solvents were dried with anhydrous $MgSO_4$ and concentrated at specified pressure/15–20°C. ^1H NMR spectra were recorded at 400.0 MHz for ^1H and 100.6 MHz for ^{13}C with a Bruker Avance III™ HD spectrometer. ^1H and ^{13}C resonances were fully assigned using H,H-COSY, H,C-HSQC, and H,C-HMBC techniques. Chemical shifts are referenced to acetone-d₆, δ 2.05, 206.26 ppm, respectively. Coupling constants (J) are reported in Hz with the following splitting abbreviations: s = singlet, d = doublet, t = triplet, and q = quartet. Infrared (IR) spectra were recorded with a Bruker Alpha FT-IR spectrometer using ATR technique. Absorption maxima (υ_{max}) are reported in wavenumbers (cm⁻¹). High-resolution mass spectra (HRMS) were measured on an LTQ Orbitrap XL (Thermo Fischer Scientific) spectrometer using ESI ionization technique. Nominal and exact *m/z* values are reported in Daltons. Optical rotations were measured on with AUTOPOL IV (Rudolph Research Analytical, United States) polarimeter with a path length of 1.0 dm and are reported with implied units of 10^{-1} deg·cm²·g⁻¹. Concentrations (*c*) are given in g/100 mL. TLC was carried out using Merck aluminum backed sheets coated with 60F 254 silica gel. Visualization of the spots was achieved using a UV-lamp Spectroline-ENF-240/F (Spectronics Corporation Westbury, United States) (λ_{max} = 254 nm) and/or

---

* 1-Iodo-glycals **3** and **4** are unstable. These compounds should be used directly in the next reaction or stored *without* exposure to atmospheric moisture in a dark and dry place in a refrigerator.

by spraying with cerium (IV) sulfate (1% in 10% H_2SO_4) solution. Flash column chromatography was carried out using neutral silica gel (Merck, 100–160 μm or MP Siltech 32–63 μm). Solvent ratios are reported in v/v.

1,5-ANHYDRO-4,6-O-(DI-TERT-BUTYLSILYLENE)-2-DEOXY-1-IODO-3-O-TRIISOPROPYLSILYL-D-ARABINO-HEX-1-ENITOL (3)

The persilylated D-glucal 1 (1.08 g, 2.44 mmol)[15] was codistilled with dry toluene (2 × 20 mL), and the reaction flask was kept under high vacuum overnight and filled with argon. Anhydrous THF (10 mL) was added and, under argon, t-BuLi[†] (5.03 mL of 1.7 M solution in pentane, 8.54 mmol) was added dropwise through a septum over 15 min with stirring at −78°C. The stirring of the bright yellow solution was continued at −78°C for 10 min, and the mixture was allowed to warm to 0°C and stirred at that temperature for 1 h. The reaction system (whole flask) was covered with aluminum foil, the mixture was cooled to −78°C, and diiodomethane (493 μL, 6.1 mmol) was added dropwise over 5 min. When the reaction was complete (~3 h, TLC: 12:1 hexane–CH_2Cl_2), the mixture was poured into a separatory funnel containing toluene (100 mL) and water (100 mL) and partitioned. The separated aqueous layer was washed with toluene (2 × 50 mL), the combined organic layers were washed with saturated $Na_2S_2O_3$ (80 mL) and water (100 mL), dried, and concentrated. Chromatography (100:1 hexane–Et_3N) afforded compound 3 (1.05 g, 76%[‡]) as a colorless oil, which solidified upon refrigeration. The residue, a white solid, was crystallized[§] (acetone–H_2O) to give compound 3 as white crystals. R_f = 0.55 (12:1 hexane–CH_2Cl_2); $[\alpha]_D^{25}$ −20.6 (c 0.2, acetone); mp 77–78°C (acetone/H_2O); ^{1}H NMR (acetone-d_6) δ 5.26 (d, 1H, $J_{2,3}$ 2.3 Hz, H-2), 4.49 (dd, 1H, $J_{3,4}$ 6.8, $J_{3,2}$ 2.3 Hz, H-3), 4.20–4.13 (m, 2H, H-6a, H-5), 4.05–3.96 (m, 2H, H-4, H-6b), 1.19–1.10 (m, 21H, ($CH_3)_2CHSi$)), 1.09 (s, 9H, ($CH_3)_3CSi$), 1.02 (s, 9H, ($CH_3)_3CSi$); ^{13}C NMR (acetone-d_6) δ 117.2 (C-2), 104.8 (C-1), 77.8 (C-4), 76.6 (C-5), 73.9 (C-3), 65.9 (C-6), 27.8, 27.32 (($CH_3)CSi$), 23.27, 20.4 (SiC(CH_3)), 18.57, 18.53 (($CH_3)_2CHSi$), 13.1 (SiCH($CH_3)_2$)); FT-IR ($CHCl_3$): $\tilde{v}$ 2962, 2865, 1627, 1472, 1462, 1388, 1374, 1365, 1122, 1107, 1058, 994, 938, 881, 828, 691, 682; MS [M + Na]+ 591.1; HRMS (ESI): m/z [M + Na]+ calcd for $C_{23}H_{45}O_4INaSi_2$, 591.1793; found, 591.1794. Anal. calcd for $C_{23}H_{45}IO_4Si_2$, C, 48.58; H, 7.98; I, 22.32. Found: C, 48.78, H, 7.75, I, 22.47.

1,5-ANHYDRO-3,4-O-DI-(TERT-BUTYLDIMETHYLSILYL)-2-DEOXY-6-O-TRIISOPROPYLSILYL-1-IODO-D-LYXO-HEX-1-ENITOL (3,4-DI-O-TERT-BUTYLDIMETHYLSILYL)-6-O-TRIISOPROPYLSILYL-1-IODO-D-GALACTAL) (4)

The persilylated D-galactal 2 (0.998 g, 1.88 mmol) was treated with t-BuLi (3.87 mL of 1.7 M solution in pentane, 6.58 mmol) as described for 3. The bright yellow

---

[‡] It is important to note that compound 3 was obtained also in higher 82% yield. The yield depends on the quality of 1.7 M solution of t-BuLi in pentane and on the reaction time of lithiation.

[§] Solubilize in minimum amount of acetone and then slowly add water until slight turbidity. Keep overnight at 0°C.

solution was stirred at −78°C for 10 min, allowed to warm to 0°C and stirred at that temperature for 30 min. The reaction system (whole flask) was covered with aluminum foil. After addition of diiodomethane (379 μL, 4.7 mmol), as described previously, TLC showed that the reaction was complete after ~3 h. After further processing, as described previously, chromatography (100:1 hexane–Et$_3$N) afforded compound **4** (1.07 g, 87%) as a colorless oil. R_f = 0.55 (12:1 hexane–CH$_2$Cl$_2$); $[\alpha]_D^{25}$ −18.3 (c 0.6, acetone); ^{1}H NMR (acetone-d$_6$) δ 5.24 (d, 1H, $J_{2,3}$ 4.1 Hz, H-2), 4.30–4.23 (m, 2H, H-5, H-3), 4.22–4.15 (m, 2H, H-4, H-6a), 4.01 (dd, 1H, J_{gem} 11.3, $J_{6b,5}$ 3.0 Hz, H-6b), 1.14–1.06 (m, 21H, (CH$_3$)$_2$CHSi), 0.95, 0.94 (2 × s, 2 × 9 H, (CH$_3$)$_3$CSi), 0.18, 0.17, 0.13, 0.12 (4 × s, 4 × 3 H, CH$_3$Si); ^{13}C NMR (acetone-d$_6$) δ 114.7 (C-2), 106.4 (C-1), 84.8 (C-5), 68.5, 68.3 (C-3,4), 62.1 (C-6), 26.4 ((CH$_3$)$_3$C), 18.8 ((CH$_3$)$_3$C), 18.4 ((CH$_3$)$_2$CH), 12.8 ((CH$_3$)$_2$CH); −3.79, −4.37, −4.48, −4.67 (CH$_3$Si); FT-IR (CHCl$_3$): ṽ 2957, 2930, 2891, 2865, 1620, 1472, 1463, 1406, 1389, 1384, 1362, 1256, 1135, 1095, 1069, 996, 939, 883, 838, 684, 659, 569, 512; MS [M + Na]$^+$ 679.2; HRMS (ESI): m/z [M + Na]$^+$ calcd for C$_{27}$H$_{57}$O$_4$INaSi$_3$ 679.2502; found, 679.2503. Anal. calcd for C$_{27}$H$_{57}$IO$_4$Si$_3$, C, 49.37; H, 8.75; I, 19.32. Found: C, 49.56; H, 8.64; I, 19.47.

ACKNOWLEDGMENTS

The work was supported by the Czech Science Foundation (15-17572S) and in part by financial support from specific university research (MSMT No. 20-SVV/2017).

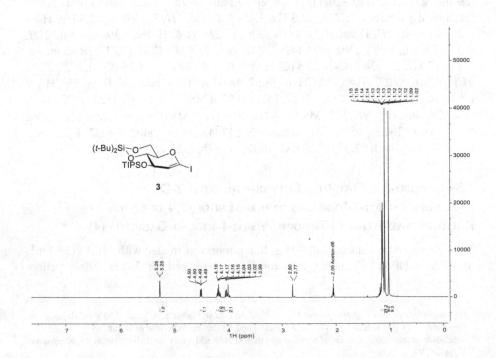

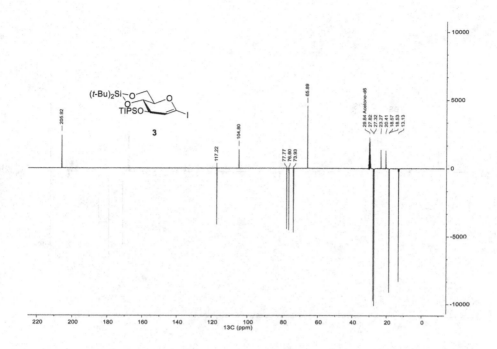

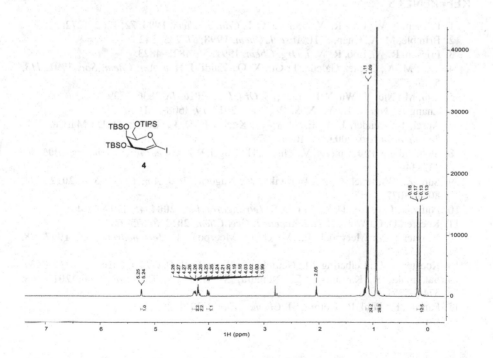

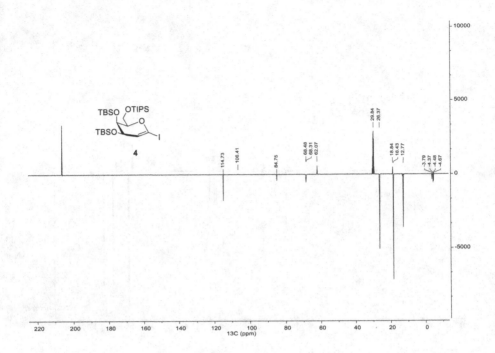

REFERENCES

1. Friesen, R. W.; Loo, R. W.; Sturino, C. F. *Can. J. Chem.* **1994**, *72*, 1262–1272.
2. Brimble, M. A.; Chan, S. H. *Aust. J. Chem.* **1998**, *51*, 235–242.
3. Friesen, R. W.; Loo, R. W. *J. Org. Chem.* **1991**, *56*, 4821–4823.
4. Tius, M. A.; Gomez-Galeno, J.; Gu, X. Q.; Zaidi, J. H. *J. Am. Chem. Soc.* **1991**, *113*, 5775–5783.
5. Liu, M.; Niu, Y.; Wu, Y. F.; Ye, X. S. *Org. Lett.* **2016**, *18*, 1836–1839.
6. Zhang, S.; Niu, Y. H.; Ye, X. S. *Org. Lett.* **2017**, *19*, 3608–3611.
7. Apsel, B.; Bender, J. A.; Escobar, M.; Kaelin Jr, D. E.; Lopez, O. D.; Martin, S. F. *Tetrahedron Lett.* **2003**, *44*, 1075–1077.
8. Steunenberg, P.; Jeanneret, V.; Zhu, Y. H.; Vogel, P. *Tetrahedron: Asymmetry,* **2005**, *16*, 337–346.
9. Sun, Z.; Winschel, G. A.; Borovika, A.; Nagorny, P. *J. Am. Chem. Soc.* **2012**, *134*, 8074–8077.
10. Potuzak, J. S.; Tan, D. S.; Tan, D. S. *Tetrahedron Lett.* **2004**, *45*, 1797–1801.
11. Koester, D. C.; Werz, D. B. *Beilstein J. Org. Chem.* **2012**, *8*, 675–682.
12. Jeanneret, V.; Meerpoel, L.; Vogel, P.; Meerpoel, L. *Tetrahedron Lett.* **1997**, *38*, 543–546.
13. Koester, D. C.; Leibeling, M.; Neufeld, R.; Werz, D. B. *Org. Lett.* **2010**, *12*, 3934–3937.
14. Sakamaki, S.; Kawanishi, E.; Nomura, S.; Ishikawa, T. *Tetrahedron,* **2012**, *68*, 5744–5753.
15. Parkan, K.; Pohl, R.; Kotora, M. *Chem.—Eur. J.* **2014**, *20*, 4414–4419.

3 Optimized Henry Reaction Conditions for the Synthesis of an L-Fucose C-Glycosyl Derivative

Dirk Hauck, and Varsha R. Jumde
Helmholtz Institute for Pharmaceutical Research Saarland
(HIPS), Helmholtz Centre for Infection Research (HZI),
Saarbrücken, Germany

Conor J. Crawford[a]
University College Dublin, Belfield,
Dublin 4, Ireland

Alexander Titz*
Helmholtz Institute for Pharmaceutical Research Saarland
(HIPS), Helmholtz Centre for Infection Research (HZI),
Saarbrücken, Germany
Deutsches Zentrum für Infektionsforschung (DZIF)
Standort Hannover-Braunschweig,
Germany Saarland University,
Saarbrücken, Germany

CONTENTS

---

* Corresponding author: alexander.titz@helmholtz-hzi.de
[a] Checker: under supervision of Stefan Oscarson: stefan.oscarson@ucd.ie

1   2   3   4, 53%

C-Glycosyl compounds are found in many natural products and are also often introduced into synthetic glycoconjugates to increase stability toward acid hydrolysis and glycosidase degradation of their parent *O*-glycosides.[1] D-Mannose- and L-fucose-derived *C*-glycosyl substances have been developed as potent inhibitors of the bacterial lectin LecB.[2,3] Their synthesis starts with a Henry addition/condensation sequence of nitromethane with L-fucose, followed by reduction of the nitro group and further derivatization.

The Henry addition of nitromethane to various carbohydrates followed by water elimination and intramolecular cyclization has been developed many decades ago mainly by the lab of Sowden[4–6] and later the labs of Köll[7,8] and Petruš.[9] The specific Henry reaction on L-fucose as substrate was first developed by Phiasivongsa et al. using DBU as base in 1,4-dioxane in the presence of molecular sieves. It gives products in reasonable yields with the formation of side products.[10] Using that protocol, the side product formation was generally extensive in our hands, rendering purification by chromatography cumbersome.

Inspired by the early work by Sowden[4–6], we report here a modification of the known[10] procedure with L-fucose (1, performing the Henry addition in DMSO with catalytic amounts of sodium methoxide as base). Crude intermediate diastereomeric mixture of nitroalditols 2 were used without further purification, water was eliminated (→3) and intramolecularly cyclized in situ in refluxing acidified water to give the pyranosyl *C*-glycosyl derivative 4.

The first reaction step, addition of nitromethane to fucose, proceeds with full conversion to nitroalditols 2, as judged by thin-layer chromatography (TLC), supported by nuclear magnetic resonance (NMR) and LC–MS. Subsequently, water was eliminated from addition product 2 by heating in aqueous HCl at pH 4, and nitroolefin 3 was cyclized in situ, to give crude 4. One chromatography followed by crystallization of 4 from EtOH yielded pure derivative 4 (53%, in two crops).

EXPERIMENTAL

GENERAL METHODS

L-Fucose was purchased from Jennewein Biotechnologie GmbH (Rheinbreitbach, Germany). All other commercially available chemicals and solvents were used without further purification.

TLC was performed on silica gel 60–coated aluminum sheets containing fluorescence indicator (Merck KGaA, Darmstadt, Germany) and developed using aqueous $KMnO_4$ solution (1%) or a molybdate solution (a 0.02 M solution of ammonium cerium sulfate dihydrate and 0.02 M ammonium molybdate tetrahydrate in aqueous

10% H_2SO_4). Prepacked silica gel 60 columns and a Teledyne Isco Combiflash Rf200 system were used for preparative medium pressure liquid chromatography (MPLC). Melting points were determined on a Stuart SMP30 apparatus. NMR spectroscopy was performed with a Bruker Avance III 500 UltraShield spectrometer at 500 (^{1}H) and 126 MHz (^{13}C). Chemical shifts are given in parts per million (ppm) and were calibrated on residual solvent peaks as internal standard. Multiplicities were specified as d (doublet), t (triplet), or q (quartet). The signals were assigned following the fucose numbering with the help of ^{1}H,^{1}H-COSY, and DEPT-135-edited ^{1}H,^{13}C-HSQC experiments. Analytical HPLC–MS was performed on a Thermo Dionex Ultimate 3000 HPLC coupled to a Bruker amaZon SL for low-resolution mass spectra, and the data were analyzed using DataAnalysis (Bruker Daltonics, Bremen, Germany). HPLC was performed on an RP-18 column (100/2 Nucleoshell RP18plus, 2.7 µm from Macherey Nagel, Germany) as stationary phase and LCMS-grade-distilled MeCN and ddH_2O were used for the preparation of mobile phases. The gradient used was 5 → 95% MeCN in H_2O containing 0.1% formic acid, at a flow rate 600 µL min^{-1}.

β-L-FUCOPYRANOSYL NITROMETHANE (4)

Dry (r.t., 5×10^{-3} mbar, 2–12 h) L-fucose (1, 1.0 g, 6.1 mmol) was dissolved in dry DMSO (6 mL) under a nitrogen atmosphere. Nitromethane (4.0 mL, 12 equiv) was added with stirring at room temperature, followed by dropwise addition of sodium methoxide (1.2 mL, 1 M in MeOH, 0.2 equiv) upon which the homogeneous reaction turned intense golden-orange. The progress of the reaction was monitored by TLC (2: R_f = 0.25; 15% MeOH in CH_2Cl_2, molybdate staining). When all fucose was consumed (~4–6 h), the mixture was poured onto cold (0°C) 10 mM aqueous HCl (120 mL), and the pH was adjusted to 4 using 1 M aqueous HCl. This solution was then heated to reflux over night when full conversion was observed by TLC (4: R_f = 0.50; 15% MeOH in CH_2Cl_2, potassium permanganate staining) accompanied by the formation of minor amounts of the slightly faster migrating furanosyl side product (NMR data for the furanoside: ^{1}H NMR [500 MHz, MeOH-d$_4$] δ 4.77 [dd, 1 H, J 13.0, 3.1 Hz, 1H of CH_2NO_2], 4.67 [dt, 1 H J 8.7, 3.3 Hz,], 4.62 [dd, 1 H, J 13.0, 8.5 Hz, 1H of CH_2NO_2], 4.06–4.01 [m, 1 H], 4.01–3.98 [m, 1 H], 3.87 [dd, 1 H J 6.6, 4.6 Hz,], 3.60 [dd, 1 H, J 4.5, 3.0 Hz, H-anomeric], 1.22 [d, 3 H, J 6.7 Hz, CH_3]). ^{13}C NMR (126 MHz, MeOH-d$_4$) δ 91.32 (C-anomer), 80.39, 79.02, 78.73, 76.23 (CH_2NO_2), 68.73, 19.68 (CH_3). After cooling to r.t., the pH of the clear solution was adjusted to 6 with aqueous NaOH (1 M) and the solution was first lyophilized followed by further overnight drying (r.t., 5×10^{-3} mbar) to remove residual DMSO and nitromethane. Purification by column chromatography on silica with CH_2Cl_2– MeOH (gradient 1 → 15%) gave the product as off-white foam (1.13 g, 90%) containing the largely inseparable furanosyl impurity. Crystallization from EtOH gave pure 4 (669 mg, 53%) as white needles. mp 180–181°C (Lit.[10]: 182°C from THF–iPr$_2$O). ^{1}H NMR (500 MHz, MeOH-d$_4$) δ 4.82 (dd, 1H, $J_{CH_2NO_2,CH_2NO_2}$ 13.0 Hz, J_{1,CH_2NO_2} 2.3 Hz, –CH_2NO_2), 4.49 (dd, 1H, $J_{CH_2NO_2,CH_2NO_2}$ 13.1 Hz, J_{1,CH_2NO_2} 9.7 Hz, –CH_2NO_2), 3.90 (td, 1H, $J_{1,2} = J_{1,CH_2NO_2}$ 9.6 Hz, J_{1,CH_2NO_2} 2.3 Hz, H-1), 3.65 (d, 1H, $J_{3,4}$ 3.4 Hz, H-4), 3.62

(q, 1H, $J_{5,6}$ 6.5 Hz, H-5), 3.50 (ddd, 1H, $J_{2,3}$ 9.4 Hz, $J_{3,4}$ 3.2 Hz, J 0.9 Hz, H-3), 3.43 (t, 1H, $J_{1,2} = J_{2,3}$ 9.5 Hz, H-2), 1.20 (d, 3H, $J_{5,6}$ 6.4 Hz, H-6). ^{13}C NMR (126 MHz, MeOH-d$_4$) δ 78.46 (CH_2NO_2), 78.29 (C-1), 76.25 (C-3), 75.73 (C-5), 73.43 (C-4), 69.27 (C-2), 16.93 (C-6). LC–MS m/z [$C_7H_{13}NO_6$ + Na]$^+$: calcd. 230.1, found: 229.9.

ACKNOWLEDGMENTS

We thank the Helmholtz Association (grant no. VH-NG-934) and the European Research Council (ERC Starting Grant SWEETBULLETS, grant no. 716311) for financial support.

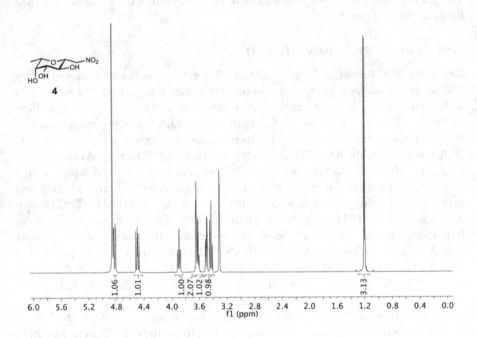

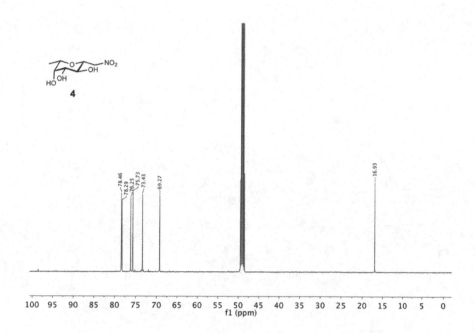

REFERENCES

1. Yang, Y.; Yu, B. *Chem. Rev.* **2017**, *117*, 12281–12356.
2. Sommer, R.; Exner, T. E.; Titz, A. *PLOS ONE*, **2014**, *9*, e112822.
3. Sommer, R.; Wagner, S.; Rox, K.; Varrot, A.; Hauck, D.; Wamhoff, E.-C.; Schreiber, J.; Ryckmans, T.; Brunner, T.; Rademacher, C.; Hartmann, R. W.; Brönstrup, M.; Imberty, A.; Titz, A. *J. Am. Chem. Soc.* **2018**, *140*, 2537–2545.
4. Sowden, J. C.; Fischer, H. O. L. *J. Am. Chem. Soc.* **1946**, *68*, 1511–1513.
5. Sowden, J. C.; Oftedahl, M. L. *J. Org. Chem.* **1961**, *26*, 1974–1977.
6. Sowden, J. C.; Bowers, C. H.; Lloyd, K. O. *J. Org. Chem.* **1964**, *29*, 130–132.
7. Köll, P.; Stenns, C.; Seelhorst, W.; Brandenburg, H. *Liebigs Ann. Chem.* **1991**, 201–206.
8. Förtsch, A.; Kogelberg, H.; Köll, P. *Carbohydr. Res.* **1987**, *164*, 391–402.
9. Petrušová, M.; Fedoroňko, M.; Petruš, L. *Chem. Papers*, **1990**, *44*, 267–271.
10. Phiasivongsa, P.; Samoshin, V.; Gross, P. *Tetrahedron Lett.* **2003**, *44*, 5495–5498.

4 Microwave-Assisted Synthesis of N-Substituted 1-Azido Glucuronamides

Nuno M. Xavier*, and Rita Gonçalves-Pereira
Centro de Química e Strutural,
Faculdade de Ciências, Universidade de Lisboa,
Lisboa, Portugal

Rosalino Balo[a]
Centro Singular de Investigación en Química Biolóxica e
Materiais Moleculares (CiQUS),
Departamento de Química Orgánica,
Universidade de Santiago de Compostela
Santiago de Compostela, Spain

CONTENTS

Glycosyl azides are useful synthetic intermediates in carbohydrate chemistry, since the azide functionality may undergo a diversity of reactions and enables accessing a variety of N-glycosyl derivatives.[1,2] Various methodologies and protocols have been described for the synthesis of anomeric azides.[1–11] The most practical and commonly used one is the reaction of glycosyl acetates, which are readily prepared and rather

---

* Corresponding author: nmxavier@fc.ul.pt.
[a] Checker: supervised by Ramón J. Estévez: ramon.estevez@usc.es.

1 R = $C_{12}H_{25}$ **3** R = $C_{12}H_{25}$

2 R = H_2C **4** R = H_2C

stable glycosyl donors, with trimethylsilyl azide ($TMSN_3$) in the presence of a Lewis acid such as boron trifluoride etherate ($BF_3 \cdot Et_2O$) or $SnCl_4$.

Glycosyl azides from uronic acids/esters constitute bifunctional building blocks for new glycoderivatives, including 1,6-linked oligosaccharide mimetics, glycosyl amino acids, or glycopeptide mimetics.[12–14] One of the methods for the synthesis of these types of anomeric azides is the selective oxidation of the primary hydroxyl group of unprotected glycosyl azides, to give the corresponding 1-azido uronic acids,[15,16] which can be further esterified.[15] The treatment of peracetylated glucuronic acid,[17] glucuronic esters,[18,19] and glucuronamides[19,20] with $TMSN_3$ in the presence of $SnCl_4$ as catalyst at room temperature was reported to give the corresponding glycosyl azides in moderate to very good yields. Long reaction times are usually required due to the low reactivity of glucuronyl donors arising from the presence of an electron-withdrawing carboxylic, carboxylate, or carboxamide group at C-5.

Herein, we report a procedure for fast anomeric azidation of peracetylated N-substituted glucuronamides effected by microwave (MW) irradiation.[21,22] We have recently described the MW-assisted anomeric azidation of penta-O-acetyl-D-glucopyranose with $TMSN_3$ in acetonitrile using $BF_3 \cdot Et_2O$ as Lewis acid.[23] The most effective operative conditions were power = 150 W, pressure = 250 Psi, and $T = 65°C$, enabling complete conversion within 2 h, which is a considerably shorter reaction time than those reported previously, e.g., 12–24 h,[24,25] for $TMSN_3$-based azidation.

When peracetylated N-benzyl glucopyranuronamide was subjected to these MW conditions, it underwent de-O-acetylation at C-4. With trimethylsilyl triflate (TMSOTf) as promoter, instead of $BF_3 \cdot Et_2O$, the expected glycosyl azide was formed after 50 min and isolated in 71% yield. The optimal conditions for anomeric azidation, i.e., using 9 equiv. of $TMSN_3$ and 8 equiv. of TMSOTf in CH_3CN at 65°C, were also effective for N-propargyl- and N-dodecyl peracetylated glucuronamides, leading to the corresponding glycosyl azides in comparable yields.[23]

We further report herein the application of this methodology to a glucuronamide derivative containing a nitrogenous heteroaromatic system at the amide moiety.

The peracetylated N-dodecyl and N-(benzyltriazolyl)methyl glucuronamidyl donors 1–2 were synthesized in a well-established pathway previously reported by us.[23,26]

Treatment of 1 and 2 with TMSN$_3$ in the presence of TMSOTf, employing our MW-assisted method, led to the corresponding glycosyl azides 3 and 4 as inseparable mixtures of α/β anomers in 1:0.6 (62%) and 1:0.7 (60%) ratios, respectively. The formation of anomeric mixtures reflects the participation of both the acetate group at C-2 and of the amide group at C-6 in assisting the intermediate glycosyl cation and directing the stereochemistry of the nucleophilic attack by TMSN$_3$. The remote participation of the C-6 group has been previously ascribed as playing a determining role in the stereochemical outcome of O- and N-glycosylation reactions using glucuronyl donors and derivatives.[19,23]

In conclusion, a reliable MW-assisted protocol for the synthesis of glucuronamide-based anomeric azides was developed, using the TMSN$_3$-based azidation of peracetylated glucuronamides. The products are obtained in good yields and considerably shorter reaction times (50 min), when compared to those reported for conventional procedures (12–16 h).[19-20]

EXPERIMENTAL

GENERAL METHODS

Chemicals were purchased from Sigma-Aldrich. HPLC grade acetonitrile (from CARLO ERBA reagents) was used as supplied. MW experiments were carried out using a CEM Discover SP Microwave Synthesizer. The dynamic method was used and the operating conditions were as follows: power = 150 W, pressure = 250 Psi, and T = 65°C with high-speed stirring. The reactions were monitored by TLC using Merck 60 F$_{254}$ silica gel aluminum plates with detection under UV light (254 nm) and/or by charring with 10% H$_2$SO$_4$ in EtOH. Flash column chromatography was performed on silica gel 60 G (0.040–0.063 mm, E. Merck). NMR spectra were acquired with a BRUKER Avance 400 spectrometer operating at 400.13 MHz for ^{1}H or 100.62 MHz for ^{13}C and using CDCl$_3$ as solvent. Chemical shifts are given in parts per million (ppm) and are reported relative to internal TMS or to the residual undeuterated solvent peak (7.26 ppm for ^{1}H and 77.16 for ^{13}C). High-resolution mass spectra were measured with a high-resolution QqTOF Impact II mass spectrometer equipped with an ESI ion source (Bruker Daltonics). Spectra were recorded in positive ESI mode with external calibration. Solutions in organic solvents were dried with anhydrous MgSO$_4$ and concentrated at reduced pressure and <40°C.

GENERAL PROCEDURE FOR THE SYNTHESIS OF N-SUBSTITUTED 1-AZIDO GLUCOPYRANURONAMIDES

To a solution of N-substituted glucopyranuronamide (1 mmol) in acetonitrile (26 mL) in a MW reaction vial, trimethylsilyl azide* (9 equiv) was added, followed

---

* TMSN$_3$ should be handled with care and under inert atmosphere, since it is moisture sensitive and releases toxic and explosive hydrogen azide in contact with water.

by TMSOTf (8 equiv). The mixture was subjected to MW irradiation (150 W, P_{max} = 250 Psi) with stirring at 65°C for 50 min. DCM and sat. aq. NaHCO$_3$ were added. When the mixture became neutral, it was extracted with dichloromethane (3×). The combined organic phases were washed with water, dried, concentrated, and chromatographed.

N-DODECYL 2,3,4-TRI-O-ACETYL-1-AZIDO-1-DEOXY-α,β-D-GLUCOPYRANURONAMIDE (3)

According to the general procedure, N-dodecyl 1,2,3,4-tetra-O-acetyl-α,β-D-glucopyranuronamide[23] (**1**, 534 mg, 1.01 mmol) was treated with TMSN$_3$ (95%, 1.3 mL, 9.1 mmol) in the presence of TMSOTf (1.5 mL, 8.1 mmol). When the reaction was complete (TLC, 3:7 EtOAc–hexane), chromatography (3:7 → 1:1 EtOAc–hexane) gave **3** (anomeric mixture, α/β 1:0.6) as colorless oil (320 mg, 62%); R_f = 0.3 (3:7 EtOAc–hexane,). ^{1}H NMR (CDCl$_3$) δ 6.51 (t, 0.6 H, J 5.5 Hz, NH, β), 6.44 (t, 1H, J 5.5 Hz, NH, α), 5.66 (d, 1H, $J_{1,2(\alpha)}$ 4.2 Hz, H-1 α), 5.43 (t, 1H, $J_{2,3(\alpha)} = J_{3,4(\alpha)}$ 10 Hz, H-3 α), 5.30 (t, 0.6H, $J_{2,3(\beta)} = J_{3,4(\beta)}$ 9.5 Hz, H-3 β), 5.19–5.06 (m, 1.6H, H-4 α, H-4 β), 4.98–4.89 (m, 1.6H, H-2 α, H-2 β), 4.82 (d, 0.6H, $J_{1,2(\beta)}$ 8.9 Hz, H-1 β), 4.37 (d, 1H, $J_{4,5(\alpha)}$ 10.1 Hz, H-5 α), 4.05 (d, 0.6H, $J_{4,5(\beta)}$ 9.9 Hz, H-5 β), 3.28–3.17 (m, 3.2H, CH$_2$-7, α, β), 2.11, 2.08, 2.06, 2.06, 2.02, 2.01 (6 s, 14.4H, CH$_3$, OAc, α, β), 1.58–1.45 (m, 3.2H, CH$_2$-8, α, β), 1.38–1.19 (m, 28.8H, CH$_2$-9 to CH$_2$-17), 0.88 (t, 4.8H, CH$_3$-18); ^{13}C NMR (CDCl$_3$) δ 170.0, 169.8, 169.6, 169.6, 169.5, 169.2 (CO, Ac, α, β), 166.3 (CO, amide, α), 165.7 (CO, amide, β), 87.8 (C-1 β), 86.0 (C-1 α), 74.4 (C-5 β), 71.9 (C-3 β), 70.6 (C-2 β), 70.1 (C-2 α), 70.1 (C-5 α), 69.2 (C-4 α), 69.2 (C-4 β), 68.8 (C-3 α), 39.3, 39.3 (C-7, α, β), 31.9, 29.6, 29.6, 29.5, 29.5, 29.3, 29.3, 29.2, 26.8, 26.8, 22.6 (C-8 to C-17, α, β), 20.6, 20.6, 20.5, 20.5, 20.5, (CH$_3$, Ac, α, β), 14.1 (C-18); HRMS: [M + H]$^+$ calcd for C$_{24}$H$_{40}$N$_4$O$_8$, 513.2919; found, 513.2923. Anal. calcd for C$_{24}$H$_{40}$N$_4$O$_8$: C, 56.24; H, 7.87; N, 10.93. Found: C, 56.24; H, 7.98; N, 10.80.

N-(1'-BENZYL-1'H-1',2',3'-TRIAZOL-4'-YL)METHYL 2,3,4-TRI-O-ACETYL-1-AZIDO-1-DEOXY-α,β-D-GLUCOPYRANURONAMIDE (4)

According to the general procedure, N-(1'-benzyl-1H-1',2',3'-triazol-4'-yl)methyl 1,2,3,4-tetra-O-acetyl-α,β-D-glucopyranuronamide[23] (**2**, 545 mg, 1.02 mmol) was treated with TMSN$_3$ (1.3 mL, 9.2 mmol) in the presence of TMSOTf (1.5 mL, 8.2 mmol). When the reaction was complete (TLC, EtOAc), chromatography (1:2 → 2:1 EtOAc–cyclohexane) gave **4** (anomeric mixture, α/β 1:0.7) as colorless oil (315 mg, 60%); R_f = 0.7 (EtOAc). ^{1}H NMR (CDCl$_3$) δ 7.50, 7.48 (2 s, 1.7H, H-9, α, β), 7.42–7.25 (m, 8.5H, CH, Ph), 7.21–7.11 (m, NH, α, β), 5.65 (d, 1H, $J_{1,2(\alpha)}$ 4.2 Hz, H-1 α), 5.56–5.45

(m, 3.4*H*, CH$_2$, Bn, α, β), 5.42 (t, 1*H*, $J_{2,3(\alpha)} = J_{3,4(\alpha)}$ 9.8 Hz, H-3 α), 5.27 (t, 0.7*H*, $J_{2,3(\beta)}$ = $J_{3,4(\beta)}$ 9.5 Hz, H-3 β), 5.16–5.02 (m, 1.7*H*, H-4 α, H-4 β), 4.95–4.86 (m, 1.7*H*, H-2 α, H-2 β), 4.73 (d, 0.7*H*, $J_{1,2(\beta)}$ 8.9 Hz, H-1 β), 4.60–4.41 (m, 3.4*H*, CH$_2$-7, α, β), 4.36 (d, 1*H*, $J_{4,5(\alpha)}$ 10.3 Hz, H-5 α), 4.00 (d, 0.7*H*, $J_{4,5(\beta)}$ 9.9 Hz, H-5 β), 2.09, 2.07, 2.03, 2.01, 2.01 (5 s, 15.3*H*, CH$_3$, OAc, α, β); ^{13}C NMR (CDCl$_3$) δ 170.1, 170.0, 169.9, 169.8, 169.7, 169.3 (CO, Ac, α, β), 166.7, 166.0 (CO, amide, α, β), 144.6, 144.5 (C-8, α, β), 134.6, 134.5 (Cq, Ph, α, β), 129.3, 128.9, 128.9, 128.4 (CH, Ph, α, β), 122.6, 122.6 (C-9, α, β), 88.1 (C-1 β), 86.1 (C-1 α), 74.3 (C-5 β), 71.9 (C-3 β), 70.6 (C-2 β), 70.1, 70.0 (C-2 α, C-5 α), 69.2 (C-4 α), 69.1 (C-4 β), 68.8 (C-3 α), 54.4 (CH$_2$Ph, α, β), 34.8, 34.8 (CH$_2$-7, α, β), 20.8, 20.8, 20.7, 20.7 (CH$_3$, OAc, α, β); HRMS: [M + H]$^+$ calcd for C$_{22}$H$_{25}$N$_7$O$_8$, 516.1836; found 516.1837. Anal. calcd for C$_{22}$H$_{25}$N$_7$O$_8$: C, 51.26; H, 4.89; N, 19.02. Found: C, 51.69; H, 5.29; N, 16.30.

ACKNOWLEDGMENTS

"Fundação para a Ciência e Tecnologia" (FCT) is acknowledged for funding through the FCT Investigator Program (IF/01488/2013) the exploratory project IF/01488/2013/CP1159/CT0006 and the projects UID/MULTI/00612/2013, UID/MULTI/00612/2019, UIDB/00100/2020 and UIDP/00100/2020 (CQE).

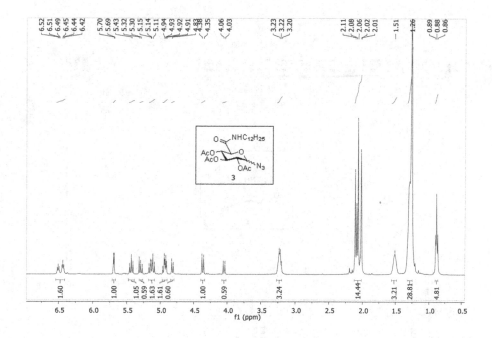

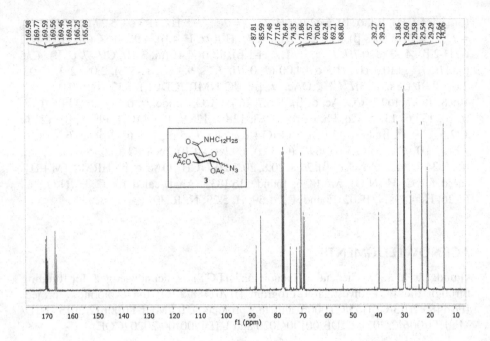

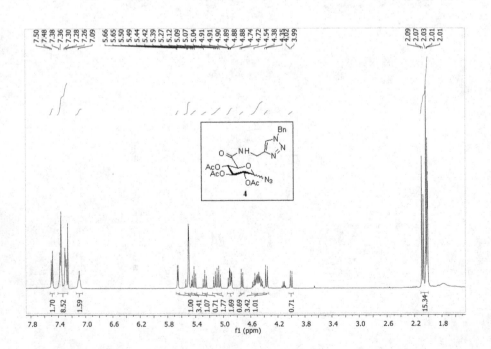

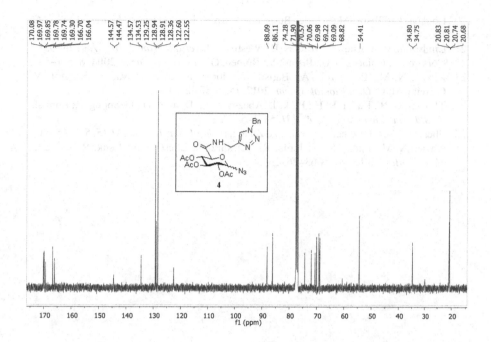

REFERENCES

1. Györgydeák, Z.; Thiem, J. *Adv. Carbohydr. Chem. Biochem.* **2006**, *60*, 103–182.
2. Witczak. Z. J. Recent advances in the synthesis of functionalized carbohydrate azides. In *Specialist Periodical Reports: Carbohydrate Chemistry*; Rauter, A. P., Lindhorst, T., Eds; Royal Society of Chemistry, London, **2010**, Vol. 36, pp. 176–193.
3. Giuliano, R. M.; Davis, R. S.; Boyko, W. J. *J. Carbohydr. Chem.* **1994**, *13*, 1135–1143.
4. Kirschning, A.; Jesberger, M.; Monenschein, H. *Tetrahedron Lett.* **1999**, *40*, 8999–9002.
5. Kawabata, H.; Kubo, S.; Hayashi, M. *Carbohydr. Res.* **2001**, *333*, 153–158.
6. Reddy, B. G.; Madhusudanan, K. P.; Vankar, Y. D. *J. Org. Chem.* **2004**, *69*, 2630–2633.
7. Kumar, R.; Tiwari, P.; Maulik, P. R.; Misra, A. K. *Eur. J. Org. Chem.* **2006**, *2006*, 74–79.
8. Tanaka, T.; Nagai, H.; Noguchi, M.; Kobayashi, A.; Shoda, S.-I. *Chem. Commun.* **2009**, 3378–3379.
9. Sabesan, S.; Neira S. *Carbohydr. Res.* **1992**, *223*, 169–185.
10. Lepage, M. L.; Bodlenner, A.; Compain, P. *Eur. J. Org. Chem.* **2013**, *2013*, 1963–1972.
11. Cui, T.; Smith, R.; Zhu, X. *Carbohydr. Res.* **2015**, *416*, 14–20.
12. Tian, G.-Z. ; Wang, X.-L.; Hu, J.; Wang, X.-B.; Guo, X.-Q.; Yin, J. *Chinese Chem. Lett.* **2015**, *26*, 922–930.
13. Cara, C. J.; Skropeta, D. *Tetrahedron*, **2015**, *71*, 9357–9365.
14. Röckendorf, N.; Lindhorst, T. K. *J. Org. Chem.* **2004**, *69*, 4441–4445.
15. Györgydeák, Z.; Thiem, J. *Carbohydr. Res.* **1995**, *268*, 85–92.
16. Ying, L.; Gervay-Hague, J. *Carbohydr. Res.* **2003**, *338*, 835–841.
17. Malkinson, J. P.; Falconer, R. A.; Toth, I. *J. Org. Chem.* **2000**, *65*, 5249–5252.
18. Cara; C. J.; Skropeta, D. *Tetrahedron*, **2015**, *71*, 9357–9365.
19. Tosin, M.; Murphy, P. V. *Org. Lett.* **2002**, *4*, 3675–3678.

20. Loukou, C.; Tosin, M.; Müller-Bunz H.; Murphy, P. V. *Carbohydr. Res.* **2007**, *342*, 1953–1959.
21. Lindström, P.; Tierney, J.; Wathey, B.; Westman, J. *Tetrahedron*, **2001**, *57*, 9225–9283.
22. Corsaro, A.; Chiacchio, U.; Pistarà V.; Romeo, G. *Curr. Org. Chem.* **2004**, *8*, 511–538.
23. Xavier, N. M.; Porcheron, A.; Batista, D.; Jorda, R.; Řezníčková, E.; Kryštof, V.; Oliveira, M. C. *Org. Biomol. Chem.* **2017**, *15*, 4667–4680.
24. Thomas, G. B.; Rader, L. H.; Park, J.; Abezgauz, L.; Danino, D.; DeShong, P.; English, D. S. *J. Am. Chem. Soc.* **2009**, *131*, 5471–5477.
25. Shen, H.; Shen, C.; Chen,C.; Wang, A. Zhang, P. *Catal. Sci. Technol.* **2015**, *5*, 2065–207.
26. Xavier, N. M.; Lucas, S. D.; Jorda, R.; Schwarz, S.; Loesche, A., Csuk, R.; Oliveira, M. C. *Synlett*, **2015**, *26*, 2663–2672.

5 Click Approach to Lipoic Acid Glycoconjugates

Thomas Tremblay, Antoine Carpentier,
and Denis Giguère*
Département de Chimie, PROTEO, RQRM,
Université Laval,
Québec City, Canada

Charles Gauthier[a]
INRS-Institut Armand-Frappier,
Université du Québec
Laval, Canada

CONTENTS

Carbohydrate-modified nanomaterials are models for investigating medically relevant binding events.[1] Preparation of glyco-nanoparticles is quite challenging, and various strategies have been employed for the synthesis of these molecular probes.[2] Saccharide-functionalized gold nanoparticles possess unique physical properties. The affinity of gold for thiol groups is well known, and, consequently, the reduced form of lipoic acid (a disulfide) is a valuable candidate for chemical ligation. Thus, gold glyco-nanoparticles conjugated through lipoic acid moiety have been used as tools to study lectin-carbohydrate interactions,[3] as biosensor for the plant toxin, ricin,[4] as *in vitro* imaging platform[5] and as agents for controlling nonspecific adsorption of blood serum.[6]

Sugar-containing lipoic acid ligands are usually prepared as glycosylamide conjugates with unprotected carbohydrates. They are highly polar compounds which are

---

* Corresponding author: denis.giguere@chm.ulaval.ca
[a] Checker: charles.gauthier@iaf.inrs.ca

difficult to purify. We present here a click approach to such a lipoic acid glycoconjugate.[7] This strategy for preparing carbohydrate conjugates has several advantages. First, the disulfide bond of the lipoic acid moiety is compatible with the click reaction conditions. Second, this approach avoids multistep synthesis, and the readily accessible product lacking the amide moiety can be easily purified using conventional methods. Finally, the resulting 1,2,3-triazole linkage has proven to be particularly stable toward various reaction conditions, especially the reductive conditions necessary for efficient anchoring onto gold nanoparticles.

Hepta-O-acetyl-β-lactosyl azide **1**[8] underwent facile Cu(I)-catalyzed azide-alkyne 1,3-dipolar cycloaddition with known alkyne **2**.[9] The 1,2,3-triazole intermediate **3** was easily purified using standard silica gel flash chromatography and was isolated in 84% yield. Finally, de-O-acetylation with methanolic sodium methoxide provided pure product **4** in quantitative yield.

EXPERIMENTAL

GENERAL METHODS

Reactions in organic media were carried out under nitrogen, and ACS-grade solvents were used without further purification. Reactions were monitored by thin-layer chromatography (TLC) using silica gel F_{254}-coated aluminum plates (Silicycle). Visualization of the spots was affected by exposure to UV light and charring with phenol-sulfuric acid spray (phenol [3.0 g] in 5% H_2SO_4 in EtOH [100 mL]). Optical rotations were measured with a JASCO DIP-360 digital polarimeter. Nuclear magnetic resonance (NMR) spectra were recorded with an Agilent DD2 500 MHz spectrometer. Proton (^{1}H) and carbon (^{13}C) chemical shifts (δ) are reported in ppm relative to the chemical shift of residual chloroform (7.26 ppm for ^{1}H spectrum and 77.16 ppm for ^{13}C spectrum) or residual methanol-d_4 (3.31 ppm for ^{1}H spectrum and 49.00 ppm for ^{13}C spectrum). Coupling constants (J) are reported in Hertz (Hz), and the following abbreviations are used: singlet (s), doublet (d), doublet of doublets (dd), triplet (t), multiplet (m), and broad (br). Assignments of NMR signals were made by homonuclear correlation spectroscopy and heteronuclear single quantum coherence two-dimensional spectroscopy. High-resolution mass spectra (HRMS) were measured with an Agilent 6210 LC time of flight mass spectrometer in electrospray mode. Elemental analysis was performed with a FLASH 2000 Analyzer (Thermo Scientific). Copper(I) iodide, diisopropylethylamine, tetrahydrofuran (THF), and resin Dowex H$^+$ were purchased from Sigma-Aldrich Chemical Co., Inc. and methanol was purchased from Fisher Scientific.

N-{[1-(2,3,4,6-Tetra-O-Acetyl-β-ᴅ-Galactopyranosyl-(1→4)-2,3,6-tri-O-Acetyl-β-ᴅ-Glucopyranosyl)-1,2,3-Triazol-4-yl]methyl}-5-(1,2-Dithiolan-3-yl)pentanamide (3)

To a stirred solution of 2,3,4,6-tetra-O-acetyl-β-ᴅ-galactopyranosyl-(1→4)-2,3,6-tri-O-acetyl-β-ᴅ-glucopyranosyl azide $\mathbf{1}^8$ (1.303 g, 1.995 mmol, 1.0 equiv) and 5-(1,2-dithiolane-3-yl)-N-(prop-2-yn-1-yl)pentanamide $\mathbf{2}^9$ (499.9 mg, 1.995 mmol, 1.0 equiv) in THF (20 mL) was added copper(I) iodide (39 mg, 0.200 mmol, 0.1 equiv) and diisopropylethylamine (0.7 mL, 4,00 mmol, 2.0 equiv), and the mixture was stirred at room temperature for 16 h. Ethyl acetate (50 mL) was added and the mixture was washed with aqueous 5% EDTA (50 mL).* The aqueous phase was extracted with EtOAc (3 × 50 mL), and the combined organic extracts were washed successively with an aqueous HCl 1 M (200 mL), dried over Na_2SO_4, and chromatography (4:1 EtOAc–Et₂O) gave compound **3** as a yellow foam (1.525 g, 1.685 mmol, 84%).† R_f 0.3 (4:1 EtOAc–Et₂O); $[\alpha]_D$ −13.4 (c 0.5, CHCl₃); ¹H NMR (CDCl₃) δ 7.71 (s, 1H, H_a), 6.21 (t, 1H, J 5.7 Hz, NH), 5.78 (d, 1H $J_{1,2}$ 8.6 Hz, H-1^I), 5.41–5.33 (m, 3H, H-2^I, H-4^I, H-4^{II}), 5.12 (dd, 1H, $J_{2,3}$ 10.4 Hz, $J_{2,1}$ 7.9 Hz, H-2^{II}), 4.96 (dd, 1H, $J_{3,2}$ 10.4 Hz, $J_{3,4}$ 3.5 Hz, H-3^{II}), 4.52 (d, 1H, $J_{1,2}$ 7.9 Hz, H-1^{II}), 4.51–4.45 (m, 3H, H_c, H-6a^I), 4.16–4.06 (m, 3H, H-6b^I, H-6a^{II}, H-6b^{II}), 3.96 (t, 1H, $J_{3,2} = J_{3,4}$ 9.2 Hz, H-3^I), 3.92–3.87 (m, 2H, H-5^I, H-5^{II}), 3.56 (m, 1H, H_h), 3.19–3.08 (m, 2H, H_j), 2.48–2.41 (m, 1H, H_i), 2.20 (t, 2H, J 7.5 Hz, H_d), 2.15, 2.10, 2.07, 2.05, 2.05, 1.96, 1.86 (7s, 21 H, COCH₃), 1.93–1.87 (m, 1H, H_i), 1.74–1.59 (m, 4H, H_e, H_g), 1.51–1.36 (m, 2H, H_f); ¹³C NMR (CDCl₃) δ 172.8 (CONH), 172.9, 170.5, 170.4, 170.2, 169.6, 169.2, 169.2 (COCH₃), 145.4 (C_b) 121.0 (C_a), 101.2 (C-1^{II}), 85.7 (C-1^I), 76.0 (C-5^I), 75.7 (C-3^I), 72.6 (C-2^I), 71.0 (C-3^{II}), 70.9 (C-5^{II}), 70.7 (C-4^I), 69.1 (C-2^{II}), 66.7 (C-4^{II}), 61.8 (C-6^{II}), 60.9 (C-6^I), 56.6 (C_h), 56.5 (C_h) 40.4 (2 C_i), 38.6 (C_j) 36.3 (2 C_d) 34.9 (C_c), 34.7 (2 C_g), 29.0 (2 C_f), 25.3 (2 C_e), 21.0, 20.84, 20.82, 20.79, 20.77, 20.7, and 20.4 (COCH₃). ESI⁺–HRMS: m/z [M + Na]⁺ calcd for $C_{37}H_{52}N_4NaO_{18}S_2$, 927.2599; found, 927.2610. Anal. calcd for $C_{37}H_{52}N_4O_{18}S_2$: C, 49.11; H, 5.79; N, 6.19. Found: C, 49.23; H, 5.80; N, 6.00.

N-{[1-(β-ᴅ-Galactopyranosyl-(1→4)-β-ᴅ-Glucopyranosyl)-1,2,3-Triazol-4-yl]methyl}-5-(1,2-Dithiolan-3-yl)pentanamide (4)

To a stirred solution of compound **3** (1.502 g, 1.660 mmol, 1.0 equiv) in methanol (50 mL) was added a solution of 1M NaOMe in methanol (3.7 mL, 3.7 mmol, 2.3 equiv) until pH 8–9 is reached. The mixture was stirred at room temperature for 2 h. Acidic resin Dowex H⁺ (unwashed) was added until pH 7 and the solution was filtered and concentrated under reduced pressure to give compound **4** as a pure yellow foam (0.903 g, 1.477 mmol, 89%).‡ R_f 0.2 (8:2 CH₂Cl₂–MeOH); $[\alpha]_D^{25}$ + 3.6 (c 0.5, MeOH); ¹H NMR (methanol-d₄) δ 8.06 (s, 1H, H_a), 5.63 (d, 1H, $J_{1,2}$ 9.3 Hz, H-1^I), 4.45 (s, 2H, H_c), 4.42 (d, 1H, $J_{1,2}$ 7.7 Hz, H-1^{II}), 3.96 (t, 1H, $J_{2,1} = J_{2,3}$ 9.1 Hz, H-2^I), 3.91–3.88 (m, 2H, H-6a^I, H-6b^I), 3.84 (dd, 1H, $J_{4,3}$ 3.3 Hz, $J_{4,5}$ 1.0 Hz, H-4^{II}),

---

* This step is necessary to removed trace amount of copper.
† Attempt to crystallize compound **3** was unsuccessful using various solvents.
‡ Attempt to crystallize compound **4** was unsuccessful using various solvents.

3.82–3.77 (m, 2H, H-6aᴵᴵ, H-4ᴵ), 3.76–3.70 (m, 3H, H-3ᴵ, H-5ᴵ, H-6bᴵᴵ), 3.62 (ddd, 1H, $J_{5,4}$ 1.0 Hz, H-5ᴵᴵ), 3.61–3.54 (m, 2H, H-2ᴵᴵ, H_h), 3.51 (dd, 1H, $J_{3,2}$ 9.7 Hz, $J_{3,4}$ 3.3 Hz, H-3ᴵᴵ), 3.20–3.07 (m, 2H, H_j), 2.49–2.42 (m, 1H, H_i), 2.24 (t, 2H, J 7.4 Hz, H_d), 1.93–1.84 (m, 1H, H_i), 1.76–1.59 (m, 4H, H_e, H_g), 1.52–1.39 (m, 2H, H_f); ¹³C NMR (methanol-d₄) δ 176.0 (CONH), 146.4 (C_b), 123.4 (C_a), 105.1 (C-1ᴵᴵ), 89.3 (C-1ᴵ), 79.7 (C-4ᴵ), 79.5 (C-5ᴵ), 77.1 (C-5ᴵᴵ), 76.8 (C-3ᴵ), 74.8 (C-3ᴵᴵ), 73.6 (C-2ᴵ), 72.5 (C-2ᴵᴵ), 70.1 (C-4ᴵᴵ), 62.5 (C-6ᴵᴵ), 61.5 (C-6ᴵ), 57.5 (C_h), 41.3 (C_i), 39.4 (C_j), 36.7 (C_d), 35.7 (C_g), 35.6 (C_e), 29.9 (C_f), and 26.6 (C_e). ESI⁺–HRMS: m/z [M + H]⁺ calcd for $C_{23}H_{39}N_4O_{11}S_2$, 611, 2051; found, 611, 2052. Anal. calcd for $C_{23}H_{38}N_4O_{11}S_2$: C, 45.23; H, 6.27; N, 9.17. Found: C, 43.71; H, 6.23; N, 8.80.

ACKNOWLEDGMENTS

The work was supported by the Natural Sciences and Engineering Research Council of Canada (NSERC), the Fonds de Recherche du Québec-Nature et Technologies, and the Université Laval. T.T. acknowledges PROTEO and Fonds Arthur-Labrie for a postgraduate fellowship.

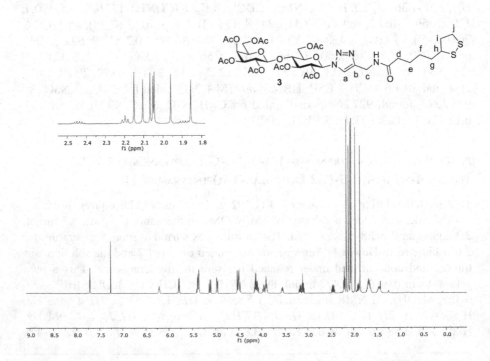

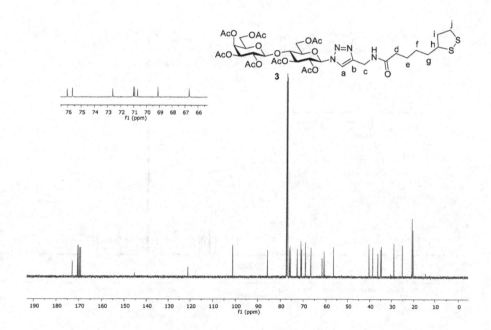

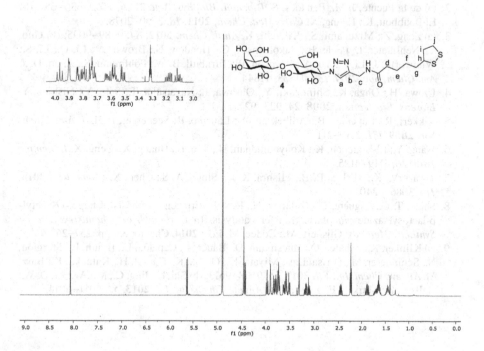

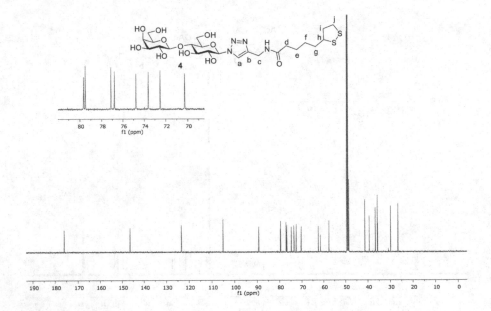

REFERENCES

1. Adak, A. K.; Lin, H.-J.; Lin, C.-C. *Org. Biomol. Chem.* **2014**, *12*, 5563–5573.
2. (a) de la Fuente, J. M.; Penadés, S. *Biochim. Biophys. Acta* **2006**, *1760*, 636–651; (b) El-Boubbou, K.; Huang, X. *Curr. Med. Chem.* **2011**, *18*, 2060–2078.
3. (a) Zeng, Z.; Mizukami, S.; Kikuchi, K. *Anal. Chem.* **2012**, *84*, 9089–9095; (b) Guo, Y.; Nehlmeier, I.; Poole, E.; Sakonsinsiri, C.; Hondow, N.; Brown, A.; Li, Q.; Li, S.; Whitworth, J.; Li, Z.; Yu, A.; Brydson, R.; Turnbull, B. W.; Pöhlmann, S.; Zhou, D. *J. Am. Chem. Soc.* **2017**, *139*, 11833–11844.
4. Uzawa, H.; Ohga, K.; Shinozaki, Y.; Ohsawa, I.; Nagatsuka, T.; Seto, Y.; Nishida, Y. *Biosens. Bioelectron.* **2008**, *24*, 923–927.
5. Kikkeri, R.; Lepenies, B.; Adibekian, A.; Laurino, P.; Seeberger, P. H. *J. Am. Chem. Soc.* **2009**, *131*, 2110–2112.
6. Wang, Y.; El-Boubbou, K.; Kouyoumdjian, H.; Sun, B.; Huang, X.; Zeng, X. *Langmuir* **2010**, *26*, 4119–4125.
7. Tiwari, V.; K.; Mishra, B. B.; Mishra, K. B.; Singh, A. S.; Chen, X. *Chem. Rev.* **2016**, *116*, 3086–3240.
8. Shiao, T. C.; Giguère, D.; Galanos, N.; Roy, R. Efficient synthesis of hepta-*O*-acetyl-β-lactosyl azide *via* phase transfer catalysis. In *Carbohydrate Chemistry: Proven Synthetic Methods*; Gijsbert, M.; Codee, M. Eds; **2014**, Chapter 33, pp. 257–262.
9. (a) Klinker, K.; Schäfer, O.; Huesmann, D.; Bauer, T.; Capelôa, L.; Braun, L.; Stergiou, N.; Schinnerer, M.; Dirisala, A.; Miyata, K.; Osada, K.; Cabral, H.; Kataoka, K.; Barz, M. *Angew. Chem. Int. Ed.* **2017**, *56*, 9608–9613; (b) Shi, L.; Jing, C.; Ma, W.; Li, D.-W.; Halls, J. E.; Marken, F.; Long, Y.-T. *Angew. Chem. Int. Ed.* **2013**, *52*, 6011–6014.

6 Synthesis of *N*-Glucosyl Ethyl and Butyl Phosphoramidates

Afraz Subratti, and Nigel K. Jalsa*
The University of the West Indies,
St. Augustine, Trinidad and Tobago

Perry Devo[a]
Department of Pharmaceutical, Chemical and
Environmental Sciences, University of Greenwich Chemistry,
Greenwich, England

CONTENTS

The Staudinger-phosphite reaction has found extensive use in diverse biological applications.[1–4] A subset of this reaction has been employed in the synthesis of glycosyl phosphoramidates, either conjugated to peptides[5,6] and proteins[1] or as standalone derivatives, which have been proposed as useful building blocks for drug development.[7]

Herein, we report the syntheses of diethyl phosphoramidate (**2**) and dibutyl phosphoramidate (**3**) derivatives of glucose in excellent yields.

* Corresponding author: nigel.jalsa@sta.uwi.edu
[a] Checker: under supervision of Adrian Dobbs and Babur Chowdhry. Babur Chowdhry's e-mail: b.z.chowdhry@greenwich.ac.uk

1

P(OEt)₃
THF, rt, 24 h

P(OBu)₃
THF, rt, 24 h

2
91 %

3
93 %

EXPERIMENTAL

GENERAL METHODS

All chemicals used were reagent grade and used as purchased, unless otherwise stated. Reaction mixtures were stirred magnetically. Gravity column chromatography was carried out using high-purity silica gel (porosity: 60 Å, particle size: 63–200 μm, bulk density: 0.5 g/mL, pH range: 6.0–8.0, residual water: <7.0%). Silica gel plates on aluminum backing (Silica G TLC Plates, w/UV 254) were used for thin layer chromatography, with acidified ammonium molybdate (ammonium molybdate [VI] tetrahydrate [25 g] in 1 M H_2SO_4 [500 mL]) for visualization. Characterization of compounds and confirmation of the structure was done by (1) high-resolution mass spectrometry (HRMS) by Bruker Daltonics micrOTOF-Q instrument via electron spray ionization; (2) optical rotation ($[\alpha]_D$) measurement (Bellingham & Stanley ADP 220 polarimeter at 25.0°C); (3) melting point determination (Mel-Temp® Digital Melting Point Apparatus: 1101D Mel-Temp®); (4) elemental analysis (PerkinElmer 2400 Series II CHNS/O); and (5) nuclear magnetic resonance (NMR) spectroscopy with Bruker 300 or 600 MHz spectrometers. Chemical shifts are reported in ppm, and coupling constants in Hertz multiplicities are stated as follows: s (singlet), *appt*s (apparent singlet), d (doublet), dd (doublet of doublets), *appt*dd (apparent doublet of doublets), ddd (doublet of doublet of doublets), dt (doublet of triplets), *appt*dt (apparent doublet of triplets), t (triplet), *p*t (pseudo triplet), td (triplet of doublets), q (quartet), and m (multiplet), etc. Combustion analysis was performed with a PerkinElmer 2400 Series II CHNS/O. The glycosyl azide **1** was synthesized as described.[8]

Solutions in organic solvents were dried with anhydrous Na_2SO_4 and concentrated at reduced pressure and <40°C.

GENERAL SYNTHETIC PROTOCOL

Glucosyl azide **1** (1000 mg, 2.679 mmol) was dissolved in freshly distilled tetrahydrofuran (THF) (10 mL) and triethyl phosphite (1.15 mL, 6.706 mmol, 2.5 equiv) or tributyl phosphite (1.81 mL, 6.700 mmol, 2.5 equiv) was added. The solution was stirred at room temperature (25.0°C) and the reaction was monitored by TLC (EtOAc). When the starting material was consumed (~24 h), the mixture was concentrated *in vacuo*, diluted with CH_2Cl_2 (50 mL) and washed with water (3 × 50 mL). The organic layer was dried, filtered, and concentrated. Chromatography (EtOAc) afforded the phosphoramidate **2** (from triethyl phosphite) as a white solid (1171 mg, 91%), R_f 0.4 (EtOAc) and **3** (from tributyl phosphite) as a white solid (1340 mg, 93%), R_f 0.6 (EtOAc).

N-(2,3,4,6-TETRA-*O*-ACETYL-β-D-GLUCOPYRANOSYL)-BISETHOXYPHOSPHORAMIDATE (2)

$[\alpha]_D$ + 3.2 (*c* 1.0, $CDCl_3$); mp 141–144°C (EtOH). ^1H NMR ($CDCl_3$) δ 5.26 (t, 1*H*, $J_{2,3}$ 9.5 Hz, $J_{3,4}$ 9.5 Hz, H-3), 5.03 (t, 1*H*, $J_{3,4}$ 9.5 Hz, $J_{4,5}$ 9.5 Hz, H-4), 4.88 (t, 1*H*, $J_{1,2}$ 9.5 Hz, $J_{2,3}$ 9.5 Hz, H-2), 4.57 (m, 1*H*, H-1), 4.23 (dd, 1*H*, $J_{5,6}$ 5.0 Hz, $J_{6,6'}$ 12.4 Hz, H-6), 4.12 (dd, 1*H*, $J_{5,6'}$ 2.1 Hz, $J_{6,6'}$ 12.4 Hz, H-6'), 4.14–4.10 (m, 4*H*, CH_2CH_3), 3.79 (ddd, 1*H*, $J_{4,5}$ 9.5 Hz, $J_{5,6}$ 5.0 Hz, $J_{5,6'}$ 2.1 Hz, H-5), 3.65 (*app*t, 1*H*, N–H), 2.08 (s, 3*H*, $COCH_3$), 2.07 (s, 3*H*, $COCH_3$), 2.04 (s, 3*H*, $COCH_3$), 2.02 (s, 3*H*, $COCH_3$), 1.33 (t, 3*H*, *J* 4.9 Hz, CH_2CH_3), 1.31 (t, 3*H*, *J* 5.1 Hz, $-CH_2CH_3$); ^{13}C NMR ($CDCl_3$) δ 170.6 (1*C*, $COCH_3$), 170.5 (1*C*, $COCH_3$), 170.0 (1*C*, $COCH_3$), 169.6 (1*C*, $COCH_3$), 82.7 (1*C*, J_{C1-H1} 157.0 Hz, C-1), 73.2 (1*C*, C-5), 72.8 (1*C*, C-3), 71.5 (1*C*, C-2), 68.5 (1*C*, C-4), 62.71 (1*C*, CH_2CH_3), 62.74 (1*C*, CH_2CH_3), 62.1 (1*C*, C-6), 20.71 (1*C*, $COCH_3$), 20.66 (1*C*, $COCH_3$), 20.61 (1*C*, $COCH_3$), 20.60 (1*C*, $COCH_3$), 16.2 (1*C*, CH_2CH_3), 16.1 (1*C*, CH_2CH_3); HRMS: *m/z* $[M + Na]^+$ calcd for $C_{18}H_{30}NO_{12}PNa$, 506.1403; found, 506.1407. Anal. calcd for $C_{18}H_{30}NO_{12}P$: C, 44.72; H, 6.26; N, 2.90; O, 39.72; P, 6.41. Found: C, 44.80; H, 6.28; N, 2.93; O, 39.78.

N-(2,3,4,6-TETRA-*O*-ACETYL-β-D-GLUCOPYRANOSYL)-BISBUTOXYPHOSPHORAMIDATE, 3

$[\alpha]_D$ + 4.7 (*c* 0.9, $CDCl_3$); mp 144–145°C (EtOH). ^1H NMR ($CDCl_3$) δ 5.26 (t, 1*H*, $J_{2,3}$ 9.6 Hz, $J_{3,4}$ 9.6 Hz, H-3), 5.04 (t, 1*H*, $J_{3,4}$ 9.6 Hz, $J_{4,5}$ 9.6 Hz, H-4), 4.88 (t, 1*H*, $J_{1,2}$ 9.6 Hz, $J_{2,3}$ 9.6 Hz, H-2), 4.57 (dd, 1*H*, $J_{1,2}$ 9.6 Hz, $J_{1,NH}$ 11.2 Hz, H-1), 4.24 (dd, 1*H*, $J_{5,6}$ 4.8 Hz, $J_{6,6'}$ 12.4 Hz, H-6), 4.12 (dd, 1*H*, $J_{5,6'}$ 2.2 Hz, $J_{6,6'}$ 12.4 Hz, H-6'), 4.03–3.92 (m, 4*H*, 2× $-CH_2CH_3$), 3.76 (ddd, 1*H*, $J_{4,5}$ 9.6 Hz, $J_{5,6}$ 4.8 Hz, $J_{5,6'}$ 2.2 Hz, H-5), 3.62

(*pseud*t, 1*H*, N–H), 2.08 (s, 3*H*, COCH$_3$), 2.06 (s, 3*H*, COCH$_3$), 2.03 (s, 3*H*, COCH$_3$), 2.02 (s, 3*H*, COCH$_3$), 1.66 (m, 4*H*, 2× CH$_2$CH$_2$CH$_2$CH$_3$), 1.39 (sext, 4*H*, J_{H-H} 7.5 Hz, 2× CH$_2$CH$_2$CH$_2$CH$_3$), 0.94 (*app*dt, 6*H*, J_{H-H} 7.5 Hz, 2× CH$_2$CH$_2$CH$_2$CH$_3$); ^{13}C NMR (CDCl$_3$) δ 170.6 (1*C*, COCH$_3$), 170.5 (1*C*, COCH$_3$), 170.0 (1*C*, COCH$_3$), 169.6 (1*C*, COCH$_3$), 82.8 (1*C*, J_{C1-H1} 155.4 Hz, C-1), 73.2 (1*C*, C-5), 72.8 (1*C*, C-3), 71.5 (1*C*, C-2), 68.4 (1*C*, C-4), 66.6 (1*C*, CH$_2$CH$_2$CH$_2$CH$_3$), 66.4 (1*C*, CH$_2$CH$_2$CH$_2$CH$_3$), 62.0 (1*C*, C-6), 32.3 (1*C*, CH$_2$CH$_2$CH$_2$CH$_3$), 32.2 (1*C*, CH$_2$CH$_2$CH$_2$CH$_3$), 20.71 (1*C*, COCH$_3$), 20.66 (1*C*, COCH$_3$), 20.60 (2*C*, COCH$_3$), 18.72 (1*C*, CH$_2$CH$_2$CH$_2$CH$_3$), 18.71 (1*C*, CH$_2$CH$_2$CH$_2$CH$_3$), 13.61 (1*C*, CH$_2$CH$_2$CH$_2$CH$_3$), 13.55 (1*C*, CH$_2$CH$_2$CH$_2$CH$_3$); HRMS: *m/z* [M + Na]$^+$ calcd for C$_{22}$H$_{38}$NO$_{12}$PNa, 562.2029; found, 562.2036. Anal. calcd for C$_{22}$H$_{38}$NO$_{12}$P: C, 48.98; H, 7.10; N, 2.60; O, 35.59; P, 5.74. Found: C, 49.03; H, 7.19; N, 2.61; O, 35.60.

ACKNOWLEDGMENTS

The authors thank The University of the West Indies for financial support.

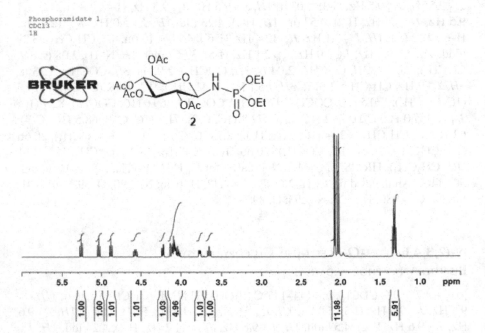

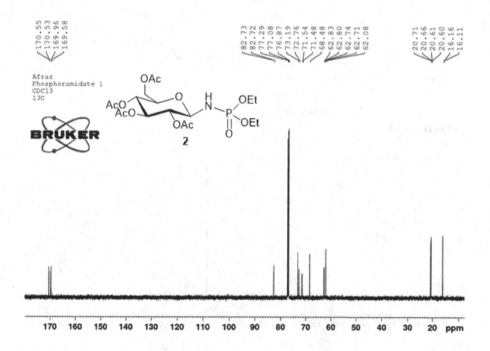

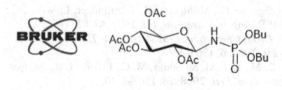

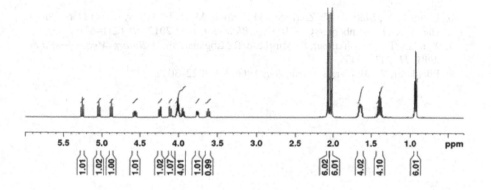

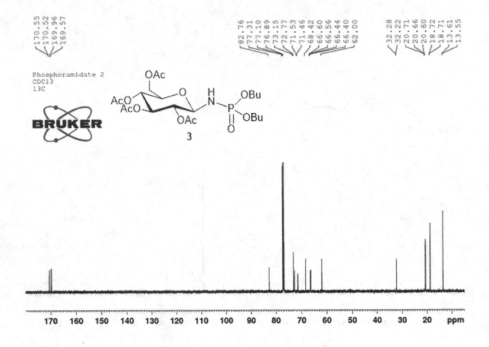

REFERENCES

1. Serwa, R.; Wilkening, I.; Del Signore, G.; Mühlberg, M.; Claußnitzer, I.; Weise, C.; Gerrits, M.; Hackenberger, C. P. R. *Angew. Chem. Int. Ed.* **2009**, *48*, 8234–8239.
2. Letsinger, R. L.; Schott, M. E. *J. Am. Chem. Soc.* **1981**, *103*, 7394–7396.
3. Xue, J.; Wu, J.; Guo, Z. *Org. Lett.* **2004**, *6*, 1365–1368.
4. Kline, T.; Trent, M. S.; Stead, C. M.; Lee, M. S.; Sousa, M. C.; Felise, H. B.; Nguyen, H. V.; Miller, S. I. *Bioorg. Med. Chem. Lett.* **2008**, *18*, 1507–1510.
5. Jaradat, D. M. M.; Hamouda, H.; Hackenberger, C. P. R. *Eur. J. Org. Chem.* **2010**, 5004–5009.
6. Böhrsch, V.; Mathew, T.; Zieringer, M.; Vallée, M. R. J.; Artner, L. M.; Dernedde, J.; Haag, R.; Hackenberger, C. P. R. *Org. Biomol. Chem.* **2012**, *10*, 6211–6216.
7. Kannan, T.; Vinodhkumar, S.; Varghese, B.; Loganathan, D. *Bioorg. Med. Chem. Lett.* **2001**, *11*, 2433–2435.
8. Pfleiderer, W.; Buhler, E. *Chem. Ber.* **1966**, *99*, 3022–3039.

7 p-Tolyl 2,3,4,6-Tetra-O-Benzoyl-1-Thio-β-D-Galactopyranoside: Direct Synthesis from the Readily Available α-per-O-Benzoyl Derivative

*Mana Mohan Mukherjee, and Pavol Kováč**
NIDDK, LBC, National Institutes of Health,
Bethesda, Maryland, United States

Marek Baráth[a]
Institute of Chemistry, Slovak Academy of Sciences,
Bratislava, Slovakia

CONTENTS

Thioglycosides are versatile intermediates in synthetic carbohydrate chemistry.[1-3] Preparation of the fully acetylated thioglycosides from per-O-acetylated carbohydrates is a simple conversion. However, the disadvantage of per-O-acetylated glycosyl donors, as with other 2-O-acetylated glycosyl donors, is the frequent formation of orthoesters, and transfer of the acetyl group from position 2 to the free hydroxyl group of the glycosyl acceptor,[4-6] which occurs during glycosidation/glycosylation. Such transacylation occurs to a much lesser extent with glycosyl donors having the hydroxyl group at O-2 benzoylated.[6]

* Corresponding author: kpn@helix.nih.gov
[a] Checker: chemmbar@savba.sk

Per-O-benzoylated thioglycosides are normally prepared indirectly, by sequential deacetylation of fully acetylated thioglycosides and benzoylation.[7-9]

We have previously reported the direct preparation of thioglycosides[10] from a series of fully benzoylated mono- and disaccharides. Benzoylated thioglycosides were only rarely prepared in this way.[11] The first synthesis and characterization of p-tolyl 2,3,4,6-tetra-O-benzoyl-1-thio-β-D-galactopyranoside (3), described elsewhere in this volume,[12] also involved deacetylation of its fully acetylated counterpart followed by benzoylation. Here we describe the direct synthesis of thioglycoside 3 from 1,2,3,4,6-penta-O-benzoyl-α-D-galactopyranose[13] (1), whose reaction with p-tolylthiol was promoted with trimethylsilyl trifluoromethanesulfonate (TMSOTf). The first characterization of the crystalline α-anomer 2, formed along with 3, is also described.

EXPERIMENTAL

GENERAL METHODS

Unless specified otherwise, all reagents and solvents were purchased from Sigma Chemical Company and used as supplied. Reactions were monitored by thin-layer chromatography (TLC) on silica gel 60 glass slides. Spots were visualized by charring with H_2SO_4 in EtOH (5% v/v) and/or UV light. Melting points were determined with a Kofler hot stage. Optical rotations were measured at ambient temperature with a Jasco P-2000 digital polarimeter. NMR spectra were measured at 25°C for solutions in benzene-d_6 or $CDCl_3$, at 600 MHz for ^{1}H, and at 150 MHz for ^{13}C with a Bruker Avance Spectrometer. Assignments of NMR signals were aided by 1D and 2D experiments (^{1}H–^{1}H homonuclear decoupling, APT, COSY, HSQC) run with the software supplied with the spectrometer. Chemical shifts were referenced to that of tetramethylsilane (0 ppm) or signals of residual non-deuterated benzene (7.16 for ^{1}H) and for ^{13}C, signal of the solvents (benzene, 128.39 ppm; $CHCl_3$, 77.00 ppm). Solutions in organic solvents were dried with anhydrous $MgSO_4$ and concentrated at reduced pressure at <40°C.

P-TOLYL 2,3,4,6-TETRA-O-BENZOYL-1-THIO-α-D-GALACTOPYRANOSIDE (2) AND P-TOLYL 2,3,4,6-TETRA-O-BENZOYL-1-THIO-β-D-GALACTOPYRANOSIDE (3)

p-Thiocresol (0.25 g, 2 mmol) was added to a solution of 1,2,3,4,6-penta-O-benzoyl-α-D-galactopyranose[13] (1, 0.7 g, 1 mmol) in CH_2Cl_2 (10 mL), followed by dropwise addition of TMSOTf (36 μL, 0.2 mmol). The mixture was kept at room temperature for 30 h, when TLC (20:1 toluene–EtOAc) showed complete consumption of the starting material and formation of two close, faster moving products (R_f ~0.5–0.6). The product with slower mobility largely predominated.* The reaction was quenched with triethylamine (0.01 mL) and the mixture was concentrated. A solution of the crude product in DCM was washed successively with 1 M NaOH (150 mL), to remove excess of EtSH, and brine. The aqueous phases were backwashed with DCM (3 × 15 mL), and combined organic phases were dried and concentrated. At this stage, NMR confirmed the absence of the starting material and large preponderance of the β-glycoside formed. Chromatography of the crude mixture (50 g silica gel, 97:3→ 15:1 toluene–EtOAc) gave first a small amount of a mixture of p-tolyl 2,3,4,6-tetra-O-benzoyl-1-thio-α-D-galactopyranoside (2) and p-tolyl 2,3,4,6-tetra-O-benzoyl-1-thio-β-D-galactopyranoside (3), followed by pure compound 3 (combined yield, 0.56 g, 80%).† The mixed fraction was resolved via preparative TLC using 15:1 CCl₄– EtOAc to give pure p-tolyl 2,3,4,6-tetra-O-benzoyl-1-thio-α-D-galactopyranoside (2), mp 143–144°C (MeOH, twice); [α]$_D^{25}$ +162.7 (c 1.0, CHCl₃); ¹H NMR (C₆D₆) δ 8.20–8.19 (m, 2H, ArH), 8.18–8.10 (m, 4H, ArH), 7.99–7.97 (m, 2H, ArH), 7.38– 7.36 (d, 2H, J 8.4 Hz, ArH), 7.11 (t, 1H, J 7.5 Hz, ArH), 7.08–7.02 (m, 3H, ArH), 6.98–6.93 (m, 3H, ArH), 6.92–6.87 (m, 3H, ArH), 6.76 (t, 2H, J 7.8 Hz, ArH), 6.72 (t, 2H, J 7.8 Hz, ArH), 6.47 (d, 1H, $J_{1,2}$ 5.5 Hz, H-1), 6.37 (dd, 1H, $J_{2,3}$ 10.8 Hz, $J_{2,1}$ 5.5 Hz, H-2), 6.29 (dd, 1H, $J_{3,2}$ 10.8 Hz, $J_{3,4}$ 3.2 Hz, H-3), 6.16 (br. d, 1H, J 2.2 Hz, H-4), 4.93 (m, 1H, H-5), 4.70 (dd, 1H, $J_{6,5}$ 11.4 Hz, $J_{6,6}$ 7.8 Hz, H-6), 4.25 (dd, 1H, $J_{6,5}$ 11.4 Hz, $J_{6,6}$ 4.7 Hz, H-6), 1.93 (s, 3H, CH₃); ¹³C NMR (C₆D₆) δ, 166.3 (C=O), 166.1 (2C=O), 165.9 (C=O), 138.1, 133.8, 133.7, 133.5, 133.3, 130.8, 130.6, 130.55, 130.5, 130.4, 130.1, 130.0, 129.9, 129.8, 129.2, 129.1, 128.8, 128.7, 87.3 (C-1), 70.3 (C-2), 70.2 (C-4), 69.9 (C-3), 68.8 (C-5), 63.2 (C-6), 21.3 (CH₃); ESI-MS: m/z [M+NH₄⁺] calcd for C₄₁H₃₈O₉SN⁺, 720.2267; found, 720.2271. Anal. calcd for C₄₁H₃₄O₉S: C, 70.07; H, 4.88. Found: C, 69.93; H, 5.03.

p-Tolyl 2,3,4,6-tetra-O-benzoyl-1-thio-β-D-galactopyranoside (3, colorless foam), showed [α]$_D^{25}$ +66.7 (c 1.0, CHCl₃); Ref. 12 [α]$_D$ +66 (c 1.0, CHCl₃). ¹H NMR (CDCl₃) δ 8.04–8.02 (dd, 2H, J 8.4 Hz, J 1.2 Hz, ArH), 7.99–7.97 (dd, 2H, J 8.4 Hz, J 1.2 Hz, ArH), 7.93–7.89 (dd, 2H, J 8.2 Hz, J 1.1 Hz, ArH), 7.77–7.73 (dd, 2H, J 8.3 Hz, J 1.1 Hz, ArH), 7.61 (t, 1H, J 7.5 Hz, ArH), 7.57 (t, 1H, J 7.5 Hz, ArH), 7.53 (t, 1H, J 7.5 Hz, ArH), 7.50–7.46 (br. d, 2H, J 8.0 Hz, ArH), 7.46–7.42 (m, 4H, ArH), 7.42–7.38 (m, 3H, ArH), 7.24–7.21 (t, 2H, J 7.9 Hz, ArH), 7.09–7.06 (d, 2H, J 7.9 Hz, ArH), 5.99 (br. d, 1H, J 3.1 Hz, H-4), 5.75 (t, 1H, $J_{2,3}$ 9.9 Hz, H-2), 5.59 (dd, 1H, $J_{3,2}$ 9.9 Hz, $J_{3,4}$ 3.3 Hz, H-3), 4.98 (d, 1H, $J_{1,2}$ 10.0 Hz, H-1), 4.65 (dd, 1H, $J_{6,5}$ 11.4 Hz, $J_{6,6}$ 6.8 Hz, H-6), 4.44 (dd, 1H, $J_{6,5}$ 11.4 Hz, $J_{6,6}$ 5.8 Hz, H-6), 4.37 (app t, 1H, J 6.4 Hz, H-5), 2.37 (s, 3H, CH₃); ¹³C NMR (CDCl₃) δ, 166.0 (C=O), 165.5 (C=O), 165.4 (C=O), 165.1

---

* The starting compound 1 moves only marginally slower than the major product 3, but the former chars also on heating without acid spray. This property of 1 is the easiest means to monitor the complete consumption of 1. The best separation of the two products can be achieved using 15:1 CCl₄–EtOAc as eluent, but in that case, 1 and 3 show virtually the same chromatographic mobility, and use of this solvent is environmentally and cost prohibitive.

† While the combined yield of the formed 2 and 3 was consistent (~80%), the ratio of the formed α and β thioglycosides in the original authors' and the Checker's experiments varied from 1:7 to 1:10.

(C=O), 138.6, 134.4, 133.5, 133.3, 133.2, 130.0, 129.81, 129.80, 129.7, 129.4, 129.3, 128.9, 128.7, 128.5, 128.4, 128.2, 127.3, 86.1 (C-1), 75.0 (C-5), 72.9 (C-3), 68.3 (C-4), 67.9 (C-2), 62.5 (C-6), 21.3 (CH$_3$); ESI–MS: m/z [M+NH$_4^+$] calcd for C$_{41}$H$_{38}$O$_9$SN$^+$, 720.2267; found, 720.2280. Anal. calcd for C$_{41}$H$_{34}$O$_9$S: C, 70.07; H, 4.88. Found: C, 69.99; H, 5.02.

a)

b)

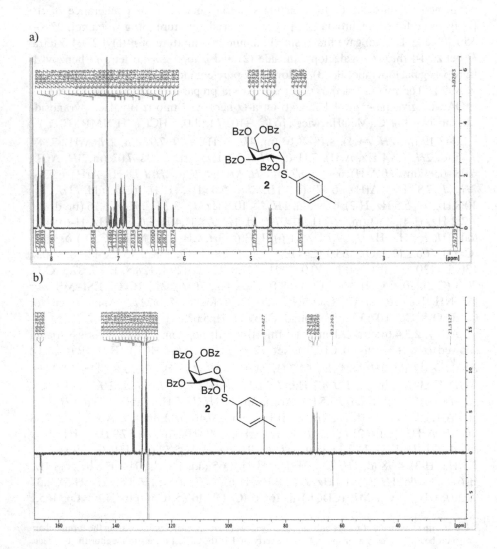

a)

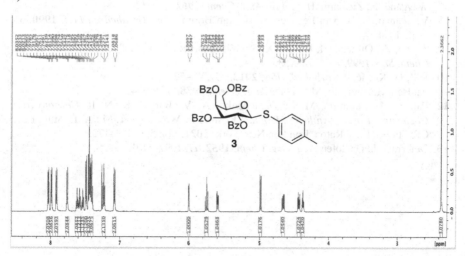

b)

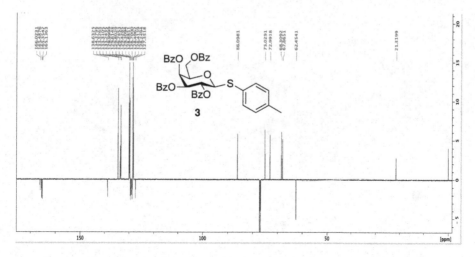

REFERENCES

1. Fügedi, P.; Garegg, P. J.; Lonn, H.; Norberg, T. *Glycoconjugate J.* **1987**, *4*, 97–108.
2. Garegg, P. *Adv. Carbohydr. Chem. Biochem.* **1997**, *52*, 179–205.
3. Lian, G.; Zhang, X.; Yu, B. *Carbohydr. Res.* **2015**, *403*, 13–22.
4. Ziegler, T.; Kováč, P.; Glaudemans, C. P. J. *Liebigs Ann. Chem.* **1990**, *6*, 613–615.
5. Nilsson, M.; Westman, J.; Svahn, C. M. *J. Carbohydr. Chem.* **1992**, *12*, 23–37.

6. Kováč, P. *Carbohydr. Res.* **1986**, *153*, 237–251.
7. Weygand, F.; Ziemann, H. *Liebigs Ann. Chem.* **1962**, *657*, 179–198.
8. Veeneman, G. H.; van Leeuven, S. H.; van Boom, J. H. *Tetrahedron Lett.* **1990**, *31*, 1331–1334.
9. Zhang, Z.; Ollmann, I. R.; Ye, X. S.; Wischnat, R.; Baasov, T.; Wong, C. H. *J. Am. Chem. Soc.* **1999**, *121*, 734–753.
10. Sail, D.; Kováč, P. *Carbohydr. Res.* **2012**, *357*, 47–52.
11. Paulsen, H.; Brenken, M. *Liebigs Ann. Chem.* **1988**, 649–654.
12. Panza, M.; Mannino, M. P.; Demchenko, A. V., Baryal, K. N. In *Carbohydrate Chemistry: Proven Synthetic Methods*; Kosma, P.; Wrodnigg, T. M.; A. E. Stütz Eds.; CRC Press, Boca Raton, London, New York, **2021**, Vol. 5, p. 163–172.
13. Deferrari, J.; Deulofeu, V. *J. Org. Chem.* **1952**, *17*, 1097–1101.

8 One-Pot Chemoselective S- vs O-Deacetylation and Subsequent Thioetherification

*Leila Mousavifar, Tze Chieh Shiao, and René Roy**
Université du Québec à Montréal,
Montréal, Canada

Thomas A. Charlton[a]
University of Ottawa,
Ottawa, Canada

CONTENTS

In our ongoing program aimed at preparing antagonists against the adhesion of uropathogenic *Escherichia coli* (UPEC) infections,[1-7] we needed α-D-mannopyranosides possessing hydrophobic aglycones to satisfy the binding site requirements of the so-called tyrosine-gate.[5-7] Amongst the potent ligands recently identified,[7] a series of hydrophobic aglycones linked by thioethers showed high promises with K_d in the low nM range.[7] Our previous syntheses used photolytic thiol–ene reaction with allylic mannopyranosides,[7] a process extensively used in glycomimetic syntheses,[8,9] but the limited access, stringent odors of thiols, and their limited commercial availability prompted us to investigate an alternative approach. We choose to investigate the previously described chemoselective de-S-acetylation of sugar thioacetates in the presence of O-acetyl groups followed by the one-pot thioetherification of the resulting thiolates.[10,11]

---

* Corresponding author: roy.rene@uqam.ca
[a] Checker: under supervision of R. N. Ben: Robert.Ben@uottawa.ca

To achieve this goal, the known allyl α-D-mannopyranoside (1)[12–17] was treated with thioacetic acid under more recent photocatalyzed conditions that traditionally depended on using 2,2′-azobisisobutyronitrile (AIBN).[18,19] The newer conditions are considered more appealing than those using AIBN which has become less commercially available due to its explosive potential. Hence, allyl α-D-mannoside 1 was treated with a solution of thioacetic acid and 2,2-dimethoxy-2-phenylacetophenone (DMPA) as photocatalyst in dry tetrahydrofuran (THF) to give the thioacetate precursor 2 in 94% yield.

Thioacetate 2 was then chemoselectively de-S-acetylated using hydrazinium acetate in degassed dimethylformamide (DMF) at room temperature to afford intermediate 3, which was used for the next step without further characterization, due to its propensity to form the corresponding disulfide. Thioetherification of 3 was achieved in a one-pot process using triethylamine and 4-bromobenzyl bromide (4) to provide thioethers 5 in overall yields ranging from 42 to 58%, the yield greatly depending on the level of oxidation into disulfide. Compound 5 is of interest on its own given the possibility to be further modified by palladium-catalyzed Heck, Suzuki, or Sonogashira reactions.

EXPERIMENTAL

GENERAL METHODS

Reactions were carried out under nitrogen using commercially available ACS-grade solvents which were stored over 4 Å molecular sieves. Solutions in organic solvents were dried over anhydrous Na_2SO_4, filtered, and concentrated under reduced pressure. Reagents were obtained from Sigma Aldrich Canada LTD and used without further purification, except for compound 4, which showed m.p. of 61–62° (lit. m.p. 60.5–61°,[20] lit. m.p. 61–62°)[21] after recrystallization from EtOH. Melting points were measured on a Fisher Jones apparatus and are uncorrected. Optical rotations were measured with a JASCO P-1010 polarimeter. The photolysis reactions were run in quartz cuvette of 3 mL (12.5 × 45 mm) (Fisher Scientific, cat. no. Fisherbrand™ 14958130, Fishersci.ca); the checker used Hellma 101-QS fluorescence cuvette of 3.5 mL (12.5 × 45 mm) (cat. no. 101-10-40) (Hellma-Analytics.com). Reactions were monitored by thin-layer chromatography (TLC) using silica gel 60 F_{254} coated plates (E. Merck). ^1H and ^{13}C NMR spectra were recorded at 300 or 600 MHz and 75 or 150 MHz, respectively, with Varian Gemini 2000 (300 MHz) and Varian Inova (600 MHz) spectrometers. All NMR spectra were measured at 25°C in indicated deuterated solvents. Proton and carbon chemical shifts (δ) are reported in ppm relative to the chemical shift of residual $CHCl_3$, which was set at 7.28 ppm (^1H) and 77.16 ppm (^{13}C). Coupling constants (J) are reported in Hertz (Hz), and the following abbreviations are used for peak multiplicities: singlet (s), doublet (d), doublet of doublets (dd), doublet of doublet with equal coupling constants (t_{ap}), triplet (t), and multiplet (m). Assignments were made using COSY (correlated spectroscopy) and HSQC (heteronuclear single quantum coherence) experiments. High-resolution mass spectra (HRMS) were measured with an LC–MS–TOF (liquid chromatography–mass spectrometry–time of flight) instrument (Agilent Technologies) in positive and/or negative electrospray mode by the analytical platform of UQAM.

3-(ACETYLTHIO)PROPYL 2,3,4,6-TETRA-O-ACETYL-α-D-MANNOPYRANOSIDE (2)

To a solution of allyl 2,3,4,6-tetra-O-acetyl-α-D-mannopyranoside[11–16] (1, 194 mg, 0.50 mmol) in dry THF (2 mL) was added a solution of thioacetic acid (107 µL, 1.50 mmol, 3.0 equiv) and DMPA (5 mg, 0.04 mmol) in THF (100 µL, quartz cuvette). The mixture was degassed with N_2 and irradiated for 1–2 h at room temperature using a 365-nm UV lamp (UVGL-58 Mineralight). When TLC monitoring showed that the reaction was complete (TLC, 3:2 hexane–EtOAc), the mixture was diluted with EtOAc and washed with H_2O. The organic layer was separated and dried over Na_2SO_4, filtered, concentrated, and chromatography (hexane → 7:3 hexane–EtOAc) afforded compounds 2 as a colorless oil (218 mg, 0.47 mmol, 94%). R_f 0.3, 3:2 hexane–EtOAc; $[\alpha]_D^{20}$ +48 (c 1.0, CHCl₃). ¹H NMR (600 MHz, CDCl₃) δ 5.27 (dd, 1H, $J_{2,3}$ 3.4 Hz, $J_{3,4}$ 10.0 Hz, H-3), 5.22 (dd, 1H, $J_{3,4} = J_{4,5}$ 10.0 Hz, H-4), 5.18 (dd, 1H, $J_{1,2}$ 1.8 Hz, $J_{2,3}$ 3.4 Hz, H-2), 4.75 (d, 1H, $J_{1,2}$ 1.8 Hz, H-1), 4.22 (dd, 1H, $J_{6a,6b}$ 12.2 Hz, $J_{5,6a}$ 5.4 Hz, H-6$_a$), 4.06 (dd, 1H, $J_{6a,6b}$ 12.2 Hz, $J_{5,6b}$ 2.4 Hz, H-6$_b$), 3.92 (ddd, 1H, H-5), 3.70 (ddd, 1H, $J_{1'a,1'b}$ 9.9 Hz, $J_{1'a,2'a}$ 6.9 Hz, $J_{1'a,2'b}$ 5.7 Hz, H1'a), 3.46 (ddd, 1H, H$_{1'b}$), 2.91 (t, 2H, $J_{2',3'}$ 7.0 Hz, CH₂SAc), 2.28 (s, 3H, SCOCH₃), 2.10, 2.06, 1.99, 1.94 (4×s, 12H, 4 × COCH₃), 1.86 (m, 2H, H2'a, H2'b); ¹³C NMR (CDCl₃) δ 195.3 (SCO), 170.4, 169.8, 169.7, 169.5 (4×CO), 97.5 (C-1), 69.4 (C-2), 68.9 (C-3), 68.4 (C-5), 66.6 (OCH₂), 66.0 (C-4), 62.3 (C-6), 30.4 (SAc), 29.1 (OCH₂CH₂), 25.6 (CH₂S), 20.7, 20.6, 20.5, 20.5 (4 × CH₃). ESI⁺–HRMS: m/z [M + Na]⁺ calcd for $C_{19}H_{28}O_{11}S$ + Na+: 487.1245; found, 487.1251. Anal. calcd for $C_{19}H_{28}O_{11}S$: C, 49.13; H, 6.08. Found: C, 49.24; H, 6.13.

3-[(4-BROMOBENZYL)THIOPROPYL] 2,3,4,6-TETRA-O-ACETYL-α-D-MANNOPYRANOSIDE (5)

To a solution of compound 2 (160 mg, 0.34 mmol) in dry and degassed DMF (3 mL) was added hydrazinium acetate (79 mg, 0.86 mmol, 2.5 equiv). The mixture was stirred at 50°C until hydrazinium acetate solubilized (~5 min), and then at room temperature or with heating as needed to maintain the solution clear. When monitoring (TLC, 3:2 hexane–EtOAc) showed the conversion of all starting material (~20 min) into intermediate 3 (R_f = 0.34, 3:2 hexane–EtOAc), which is charring but UV transparent, the mixture was diluted with EtOAc and washed with 1 N HCl, saturated solution of NaHCO₃, H_2O (4×), brine, dried and concentrated. To the residue was added a solution of 4-bromobenzyl bromide (4) (170 mg, 0.68 mmol, 2.0 equiv) in dry DMF (3 mL), followed by triethylamine (57 µL, 0.41 mmol). The mixture was stirred until TLC (7:3 hexane/EtOAc) showed complete transformation (~6 h) into the final thioether **5**. The mixture was diluted with EtOAc and washed successively with water, brine and dried, concentrated, and chromatography (hexane $\rightarrow$ 7:3 hexane–EtOAc) afforded the desired compound **5** as colorless oil (85 mg, 0.14 mmol, 42%). $R_f = 0.21$, 7:3 hexane/EtOAc. $[\alpha]_D^{20}$ +31 (*c* 1.1, CHCl$_3$). ^{1}H NMR (600 MHz, CDCl$_3$) δ 7.42 (d, 1H, $J_{H,H}$ 8.4 Hz, H-arom), 7.18 (d, 1H, $J_{H,H}$ 8.4 Hz, H-arom), 5.29 (dd, 1*H*, $J_{2,3}$ 3.4 Hz, $J_{3,4}$ 10.0 Hz, H-3), 5.26 (dd, 1*H*, $J_{3,4} = J_{4,5}$ 10.0 Hz, H-4), 5.20 (dd, 1*H*, $J_{1,2}$ 1.8 Hz, $J_{2,3}$ 3.4 Hz, H-2), 4.75 (d, 1*H*, $J_{1,2}$ 1.8 Hz, H-1), 4.25 (dd, 1*H*, $J_{6a,6b}$ 12.2 Hz, $J_{5,6a}$ 5.4 Hz, H-6$_a$), 4.09 (dd, 1*H*, $J_{6a,6b}$ 12.2 Hz, $J_{5,6b}$ 2.4 Hz, H-6$_b$), 3.93 (ddd, 1*H*, H-5), 3.75 (ddd, 1*H*, $J_{1'a,1'b}$ 9.9 Hz, $J_{1'a,2'a}$ 6.9 Hz, $J_{1'a,2'b}$ 5.7 Hz, H$_{1'a}$), 3.64 (s, 2*H*, SCH$_2$Ph), 3.48 (ddd, 1*H*, H$_{1'b}$), 2.48 (t, 2H, $J_{H,H}$ 7.0 Hz, H$_3$'S), 2.14, 2.08, 2.04, 1.99 (4 s, 12*H*, 4 × COCH$_3$), 1.81 (m, 2*H*, H$_2$'); ^{13}C NMR (CDCl$_3$) δ 170.5, 170.0, 169.8, 169.6 (4 × CO), 137.3 (C$_{arom-q}$), 131.5 (C$_{arom}$), 130.4 (C$_{arom}$), 120.7 (C$_{arom-q}$), 97.5 (C-1), 69.4 (C-2), 69.0 (C-3), 68.5 (C-5), 66.4 (C-1'), 66.0 (C-4), 62.3 (C-6), 35.6 (SCH$_2$Ph), 28.6 (C-2'), 27.8 (C-3'S), 20.8, 20.6, 20.6, 20.6 (4 × Ac). ESI$^+$–HRMS: *m/z* [M + Na]$^+$ calcd for C$_{24}$H$_{31}$BrO$_{10}$S + Na+: 613.0714; found, 613.0715. Anal. calcd for C$_{24}$H$_{31}$BrO$_{10}$S: C, 48.74; H, 5.28. Found: C, 48.73; H, 5.26.

ACKNOWLEDGMENTS

The work was supported from Natural Sciences and Engineering Research Council of Canada, for a Canadian Research Chair in Therapeutic Chemistry (NSERC), and from the Fonds de Recherche du Québec—Nature et technologies (FRQNT) team grant to R. R.

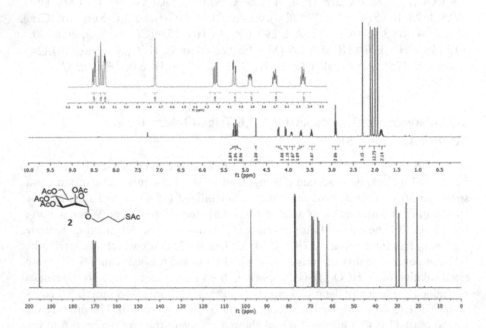

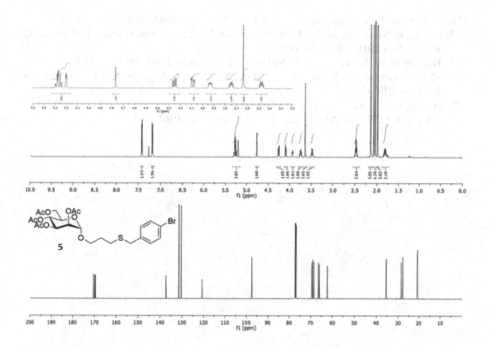

REFERENCES

1. Nagahori, N.; Lee, R. T.; Nishimura, S.-I.; Pagé, D.; Roy, R.; Lee, Y. C. *Chem. Bio. Chem.* **2002**, *3*, 836–844.
2. Touaibia, M.; Shiao, T. C.; Papadopoulos, A.; Vaucher, J.; Wang, Q.; Benhamioud, K.; Roy, R. *Chem. Commun.* **2007**, 380–382.
3. Touaibia, M.; Wellens, A.; Shiao, T. C.; Wang, Q.; Sirois, S.; Bouckaert, J.; Roy, R. *Chem. Med. Chem.* **2007**, *2*, 1190–1201.
4. Touaibia, M.; Roy, R. *Mini-Reviews Med. Chem.* **2007**, *7*, 1270–1283.
5. Wellens, A.; Lahmann, M.; Touaibia, M.; De Greve, H.; Oscarson, S.; Roy, R.; Remaut, H.; Bouckaert, J. *Biochem.* **2012**, *51*, 4790–4799.
6. Roos, G.; Wellens, A.; Touaibia, M.; Geerlings, P.; Roy, R.; Wyns, L.; Bouckaert, J. *ACS Med. Chem. Lett.* **2013**, *4*, 1085–1090.
7. Touaibia, M.; Krammer, E.-M.; Shiao, T. C.; Yamakawa, N.; Wang, Q.; Glinschert, A.; Papadopoulos, A.; Mousavifar, L.; Maes, E.; Oscarson, S.; Vergoten, G.; Lensink, M. F.; Roy, R.; Bouckaert, J. *Molecules* **2017**, *22*, 1101.
8. Dondoni, A.; Marra, A. *Chem. Soc. Rev.* **2012**, *41*, 573–586.
9. Conte, M. L.; Robb, M. J.; Hed, Y.; Marra, A.; Malkoch, M.; Hawker, C. J.; Dondoni, A. *J. Polymer Sci. A: Polym. Chem.* **2011**, *49*, 4468–4475.
10. Park, W. K.C.; Meunier, S.; Zanini, D.; Roy, R. *Carbohydr. Lett.* **1995**, *1*, 179–184.
11. Rao, J.; Zhang, G.; Zeng, X.; Zhu, X. *Carbohydr. Chem.: Proven Synth. Methods* **2015**, *3*, 89–96.
12. Winnik, F. M.; Carver, J. P.; Krepinsky, J. J. *J. Org. Chem.* **1982**, *47*, 2701–2707.
13. Takano, T.; Nakatsubo, F.; Murakami, K. *Carbohydr. Res.* **1990**, *203*, 341–342.
14. Pagé, D.; Roy, R. *Glycoconjugate J.* **1997**, *14*, 345–356.
15. Matsuoka, K.; Terabatake, M.; Esumi, Y.; Terunuma, D.; Kuzuhara, H. *Tetrahedron Lett.* **1999**, *40*, 7839–7842.

16. Wolfenden, M. L.; Cloninger, M. J. *J. Am. Chem. Soc.* **2005**, *127*, 12168–12169.
17. Ortega, P.; Serramia, M. J.; Munoz-Fernandez, M. A.; Javier de la Mata, F.; Gomez, R. *Tetrahedron* **2010**, *66*, 3326–3331.
18. Houseman, B. T.; Gawalt, E. S.; Mrksich, M. *Langmuir* **2003**, *19*, 1522–1531.
19. Mori, T.; Hatano, K.; Matsuoka, K.; Esumi, Y.; Toone, E. J.; Terunuma, D. *Tetrahedron* **2005**, *61*, 2751–2760.
20. Huyser, E. S. *J. Am. Soc. Chem.* **1960**, *82*, 391–393.
21. Addy, J. K.; Laird, R. M.; Parker, R. E. *J. Chem. Soc.* **1961**, 1708–1711.

9 Sakurai Anomeric Allylation of Methyl α-Pyranosides

Marc E Bouillon, Rhodri Mir Williams,
and Martina Lahmann**
School of Natural Sciences, Bangor University,
Bangor, UK

Orla McCabe[a]
School of Chemistry, University College Dublin,
Belfield, Dublin 4, Ireland

CONTENTS

Even though *O*- and *N*-glycosides are much more common, a wide variety of *C*-glycosyl structures is produced by nature. Typical representatives are aryl *C*-glycosyl flavones found in *Viola yedoensis*[1] and the fatty esters papulacandin A–E isolated from fermentation broth of *Papularia spherosperma*. These were described in 1976[2] and synthesized almost 30 years later.[3] The stability of these pseudo-glycosides makes this class of compounds interesting and particularly useful for the design of inhibitors and glycomimetics.[4] 3-(Glycopyranosyl)-1-propenes are also useful synthetic intermediates because their propenyl group can easily be further functionalized.

Numerous synthetic methods have been established over the years[5] for making *C*-glycosyl compounds but we find the method originally published by Hosomi et al.[6]

* Corresponding authors: m.bouillon@bangor.ac.uk; m.lahmann@bangor.ac.uk
[a] Checker: orla.mccabe@ucd.ie

BnO—OBn / BnO / BnO—OMe **1**

AllTMS, TMSOTf, MeCN, 0 °C, 82%

→

BnO—OBn / BnO / BnO **2α** + BnO—OBn / BnO / OBn **2β**

OMe / —OBn / BnO—OBn **3**

AllTMS, TMSOTf, MeCN, 0 °C, 78%

→

—OBn / BnO—OBn **4α** + —OBn / BnO—OBn **4β**

to be the most reliable one. Accordingly, a solution of the starting material in dry acetonitrile, a peracetylated glycoside or a benzylated methyl glycoside, is treated with the aglycone in the presence of Lewis acids at 0°C.[7] Here we describe conversion of methyl 2,3,4,6-tetra-O-benzyl-α-D-galactopyranoside (**1**) and methyl 2,3,4-tri-O-benzyl-α-D-fucopyranoside (**3**) into their corresponding D-galactosyl propenes **2α** and **2β** and L-fucosyl propenes **4α**, **4β**, respectively.[*]

EXPERIMENTAL

GENERAL METHODS

Per-benzylated methyl glycopyranosides were synthesized according to published procedures.[8] All reagents (Sigma-Aldrich, Acros Organics, and Fisher Scientific) were used as received. Reactions were monitored by thin-layer chromatography (TLC) and run in purified solvents using glassware which was dried using a heat gun, while being flushed by stream of dry nitrogen. Column chromatography was carried out on silica gel (VWR Chemicals, 40–63 μm) using mixtures of petroleum ether (bp 40–60°C) or toluene and diethyl ether. TLC was done using precoated Merck aluminum TLC-plates (silica gel 60 F254). Spots were visualized by UV light (254 nm) and by charring with a vanillin staining solution (1 g vanillin, 20 mL acetic acid, 200 mL MeOH, 10 mL conc. H_2SO_4). ^{1}H and ^{13}C NMR spectra were recorded on a Bruker Avance III 400 MHz UltraShield Plus spectrometer with tetramethylsilane (CDCl$_3$) as internal standard. NMR signal assignments are based on ^{1}H,^{1}H-COSY, DEPTQ, HSQC, and HMBC experiments. Elementary analyses were performed by OEA Laboratories Ltd, Cornwall, United Kingdom. Optical rotations were determined at a wavelength of 589.3 nm (sodium D line) using a 1-mL (0.25 dm) quartz cell with the Bellingham + Stanley ADP440 digital polarimeter at ambient temperature. Solutions in organic solvents were dried with anhydrous $MgSO_4$ and concentrated at reduced pressure at <45°C.

---

[*] The stability of the glycosyl propenes depends on the saccharide. The galactosyl derivatives should be converted as soon as possible because they degrade easily, even when stored in a sealed flask at −18°C, whereas for the fucosyl derivative we did not see any noticeable decomposition when stored at −18°C.

3-(2′,3′,4′,6′-Tetra-O-Benzyl-α-d-Galactopyranosyl)-1-Propene (2α) and
3-(2′,3′,4′,6′-Tetra-O-Benzyl-β-d-Galactopyranosyl)-1-Propene (2β)

Trimethylsilyl triflate (830 μL, 4.55 mmol, 0.5 equiv) was added dropwise, at 0°C under N_2, to a solution of methyl 2,3,4,6-tetra-O-benzyl-α-d-galactopyranoside **1** (5.05 g, 9.10 mmol) and allyltrimethylsilane (4.33 mL, 27.3 mmol, 3 equiv) in dry acetonitrile (45 mL). The mixture was stirred while allowed to warm slowly to 10–12°C, when TLC (4:1 petroleum ether–EtOAc) indicated full conversion (~8 h). The cooling to 0°C was resumed, and the reaction was quenched by slow addition of sat. aq. $NaHCO_3$ (25 mL). EtOAc (90 mL) was added, and the organic solvents were removed. The aqueous residue was extracted with Et_2O (1 × 125 mL). The organic phase was washed with brine (25 mL), dried, filtered, and concentrated, to yield the crude product **2** (4.72 g, α/β ~7:1, NMR) as a brown syrup. The crude product was taken up in DCM (5 mL), adsorbed on silica gel (9 g) and chromatographed (Ø = 6 cm, dry bed heights = 14 cm, $V_{Fraction}$ = 200 mL, 9:1 → 4:1 petroleum ether–Et_2O) to yield **2α/β** (4.22 g, 7.47 mmol, 82%) as colorless syrup. This material was rechromatographed (Ø = 8 cm, dry bed heights = 20 cm dry, $V_{Fraction}$ = 250 mL, toluene-Et_2O, 99:1 → 20:1), to give first **2β** (174 mg, 0.31 mmol, ~3%) as colorless syrup, followed by mixed fractions, and then **2α** (3.70 g, 6.55 mmol, 72%) as colorless syrup.

2β, $[\alpha]_D^{20}$ +8.6 (c 2.4, $CHCl_3$); $[\alpha]_D^{25}$ +8.8 (c 1.31, $CHCl_3$)[9]; ^{1}H NMR (400 MHz, $CDCl_3$, TMS): δ 7.39–7.24 (m, 20H, 20-Ph), 5.92 (ddt, 1H, J_{trans}1a,2 17.2 Hz, J_{cis}1b,2 10.2 Hz, $J_{2,3}$ 6.9 Hz, H-2), 5.08 (dd, 1H, $J_{1a,1b}$ 1.6 Hz, H-1a), 5.03 (dd ~ d, 1H, H-1b), 4.94 (d, 1 H J 11.9 Hz, H_{Bn}-4′), 4.74 (d, 1H, J 11.7 Hz, H_{Bn}-3′), 4.66 (d, 1H, H_{Bn}-3′), 4.64 (d, 1H, H_{Bn}-2′), 4.63 (d, 1H, H_{Bn}-4′), 4.47 (d, 1H, J 11.8 Hz, H_{Bn}-6′), 4.41 (d, 1H, H_{Bn}-6′), 3.98 (d, 1H, $J_{3',4'}$ 2.7 Hz, H-4′), 3.71 (dd ~ t, 1H, $J_{2',3'}$ 9.4 Hz, H-2′), 3.59 (dd, 1H, H-3′), 3.58–3.49 (m, 3H, H-5′, H-6a′, H-6b′), 3.30 (dt, 1H, 1H, $J_{1',2'}$ 9.0 Hz, $J_{1',3}$ 3.0 Hz, H-1′), 2.66–2.54 (m, 1H, CH_2-3a), 2.38–2.26 (m, 1H, CH_2-3b); ^{13}C NMR (100 MHz, $CDCl_3$, TMS): δ 138.8, 138.5, 138.4, 138.0 (4 Ph_q), 135.2 (C-2), 128.4–127.5 (20 Ph), 116.6 (C-1), 84.9 (C-3′), 79.4 (C-1′), 78.6 (C-2′), 77.1 (C-5′), 77.3, 74.4 (Bn), 73.7 (C-4′), 73.5, 72.2 (Bn), 69.0 (C-6′), 36.2 (C-3). m/z mass calcd. for $C_{37}H_{40}O_5Na$: 587.2773, found: 587.2748. The compound is unstable and correct analytical figures could not be obtained.

2α, $[\alpha]_D^{20}$+46.4 (c 4.0, $CHCl_3$); $[\alpha]_D^{20}$ +56.6 (c 5.0, CH_2Cl_2); $[\alpha]_D^{25}$ +55.7 (c 1.0, CH_2Cl_2)[10], ^{1}H NMR (400 MHz, $CDCl_3$, TMS): δ 7.38–7.20 (m, 20H, 20-Ph), 5.76 (ddt, 1H, J_{trans}1a,2 17.2 Hz, J_{cis}1b,2 10.2 Hz, $J_{2,3}$ 6.9 Hz, H-2), 5.07 (dd, 1H, $J_{1a,1b}$ 1.6 Hz, H-1a), 5.02 (dd ~ d_{broad}, 1H, H-1b), 4.70 (d, 1H, J 11.6 Hz, H_{Bn}-4′), 4.69 (d, 1H, J 12.1 Hz, H_{Bn}-3′), 4.59 (d, 2H, J 12.3 Hz, H_{Bn}-3′, H_{Bn}-2′), 4.57 (d, 1H, H_{Bn}-4′), 4.53 (d, 1H, J 12.1 Hz, H_{Bn}-6′), 4.51 (d, 1H, H_{Bn}-2′), 4.48 (d, 1H, H_{Bn}-6′), 4.11–4.03 (m, 1H, H-5′), 4.03–3.97 (m, 2H, H-1′, H-4′), 3.84 (dd, 1H, $J_{6a',6b'}$ 10.5 Hz, $J_{5',6a'}$ 7.3 Hz, H-6a′), 3.78–3.72 (m, 1H, H-2′), 3.71 (dd, 1H, $J_{2',3'}$ 6.8 Hz, $J_{3',4'}$ 2.6 Hz, H-3′), 3.67 (dd, 1H, $J_{5',6b'}$ 4.7 Hz, H-6b′), 2.48–2.28 (m, 2H, CH_2-3); ^{13}C NMR (100 MHz, $CDCl_3$, TMS): δ 138.6, 138.5, 138.5, 138.3 (4 Ph_q), 135.1 (C-2), 128.4–127.5 (20 Ph), 116.8 (C-1), 76.4 (C-2′, C-3′), 74.3 (C-4′), 73.2, 73.1, 73.0 (Bn), 72.6 (C-5′), 70.8 (C-1′), 67.2 (C-6′), 32.2 (C-3); m/z mass calcd for $C_{37}H_{39}O_5$ Na: 587.2773, found: 587.2750; the compound is unstable and correct analytical figures could not be obtained.

3-(2′,3′,4′-Tri-O-Benzyl-α-l-Fucopyranosyl)-1-Propene (4α) and
3-(2′,3′,4′-Tri-O-Benzyl-β-l-Fucopyranosyl)-1-Propene (4β)

Trimethylsilyl triflate (700 μL, 3.79 mmol, 0.5 equiv) was added dropwise at 0°C under N_2 to a solution of methyl 2,3,4-tri-O-benzyl-α-l-fucopyranoside **3** (3.40 g, 7.58 mmol) and allyltrimethylsilane (3.60 mL, 22.7 mmol, 3 equiv) in dry acetonitrile (38 mL). The mixture was stirred at 0°C until TLC indicated full conversion of the starting material (1 h) and quenched by slow addition of sat. aq. $NaHCO_3$ solution (17 mL). EtOAc was added (75 mL) and organic solvents were evaporated (200 mbar). The aqueous residue was extracted with Et_2O (85 mL), and the organic phase was washed with brine (20 mL), dried, filtered, and concentrated to yield the crude product **4α/β** as a yellow syrup (97:3 α/β ratio, NMR). A solution of the crude product in DCM (4 mL) was adsorbed on silica (8 g) and chromatographed (∅ = 5.5 cm, dry bed heights = 14 cm, $V_{Fraction}$ = 150 mL, petroleum ether–Et_2O, 9:1 → 17:3) to give **4α** and **4β** (2.78 g, 6.06 mmol, 80%) as a colorless syrup (α/β, ~97:3). Separation of the pseudo anomers was achieved by gravity column chromatography (same dimensions as above) with toluene–Et_2O (99:1), yielding first pure **4β** (56 mg, 0.122 mmol, ~2%) and then **4α** (2.62 g, 5.71 mmol, 75%) as colorless syrups.

4β, $[\alpha]_D^{20}$ −6.4 (c 2.8, CHCl₃), ¹H NMR (400 MHz, CDCl₃, TMS): δ 7.42–7.22 (m, 15H, 15-Ph), 5.94 (ddt, 1H, $J_{trans1,2}$ 17.2 Hz, $J_{cis1,2}$ 10.3 Hz, $J_{2,3}$ 6.9 Hz, H-2), 5.08 (dd, 1H, $J_{1a,1b}$ 2.0 Hz, H-1a), 5.03 (dd, 1H, H-1b), 4.99 (d, 1H, J 12.0 Hz, 1H, H_{Bn}-4′), 4.95 (d, 1H, J 10.8 Hz, H_{Bn}-2′), 4.76 (d, 1H, H_{Bn}-3′), 4.72 (d, 1H, H_{Bn}-4′), 4.70 (d, 1H, H_{Bn}-3′), 4.66 (d, 1H, H_{Bn}-2′), 3.70 (dd~t, 1H, $J_{2',3'}$ 9.4 Hz, H-2′), 3.64 (dd ~ d$_{broad}$, 1H, $J_{3',4'}$ 2.8 Hz, H-4′), 3.59 (dd, 1H, $J_{2',3'}$ 9.4 Hz, H-3′), 3.42 (dq ~ q$_{broad}$, 1H, $J_{5',6'}$ 6.4 Hz, $J_{4',5'}$ ~1 Hz, H-5′), 3.26 (ddd ~ dt, 1H, $J_{1',2'}$ 9.1 Hz, $J_{1',3b}$ 8.4 Hz, $J_{1',3a}$ 3.0 Hz, H-1′), 2.59 (dddd, 1H, J_{gem} 14.9 Hz, $^4J_{3a,1}$ 1.5 Hz, CH₂-3a), 2.32 (ddd, 1H, CH₂-3b), 1.16 (d, 3H, H-6′); ¹³C NMR (100 MHz, CDCl₃, TMS): δ 138.7, 138.5, 138.5 (3 Ph$_q$), 135.5 (C-2), 128.4–127.5 (15 Ph), 116.4 (C-1), 85.3 (C-3′), 79.3 (C-1′), 78.6 (C-2′), 76.5 (C-4′), 75.3 (Bn-2′), 74.5 (Bn-4′), 74.1 (C-5′), 72.4 (Bn-3′), 36.2 (C-3), 17.3 (C-6′). m/z mass calcd for $C_{30}H_{34}O_4Na$: 481.2355, found: 481.2341.

4α, $[\alpha]_D^{20}$ −39.4 (c 4.0, CHCl₃); $[\alpha]_D^{20}$ −16.6 (c 1.5, CHCl₃)[11]; ¹H NMR (400 MHz, CDCl₃, TMS): δ 7.38–7.23 (m, 15H, 15-Ph), 5.76 (ddt, 1H, $J_{trans1,2}$ 17.1 Hz, $J_{cis1,2}$ 10.2 Hz, $J_{2,3}$ 6.9 Hz, H-2), 5.05 (m, 2H, CH₂-1), 4.75 (d, 1H, J 12.1 Hz, H_{Bn}-3′), 4.73 (d, 1H, J 11.8 Hz, H_{Bn}-4′), 4.65 (d, 1H, H_{Bn}-3′), 4.60 (d, 2H, J 11.7 Hz, H_{Bn}-2′, H_{Bn}-4′), 4.52 (d, 1H, H_{Bn}-2′), 4.06 (m, 1H, H-1′), 3.95 (m, 1H, H-5′), 3.83–3.73 (m, 3H, H-2′, H-3′, H-4′), 2.45–2.26 (m, 2H, CH₂-3), 1.29 (d, 3H, $J_{5,6}$ 6.6 Hz, H₃-6′). ¹³C NMR (100 MHz, CDCl₃, TMS) δ 138.8, 138.6, 138.4 (3 Ph$_q$), 135.4 (C-2), 128.4–127.5 (Ph), 116.6 (C-1), 76.9 (C-3′), 76.6 (C-2′), 75.8 (C-4′), 73.1 (Bn-3′), 73.0 (Bn-2′, Bn-4′), 70.2 (C-1′), 68.7 (C-5′), 32.5 (C-3), 15.2 (C-6′); m/z mass calcd for $C_{30}H_{34}O_4Na$: 481.2355, found: 481.2369. Anal. calcd for $C_{30}H_{34}O_4$: C, 78.57; H, 7.47. Found: C, 78.65; H, 7.41.

ACKNOWLEDGMENTS

The Authors thank S.G. Jones at Bangor University for providing the MS data and BEACON + for financial support to M. Bouillon and M. Lahmann. BEACON+ is supported by the Welsh Government and the European Regional Development Fund.

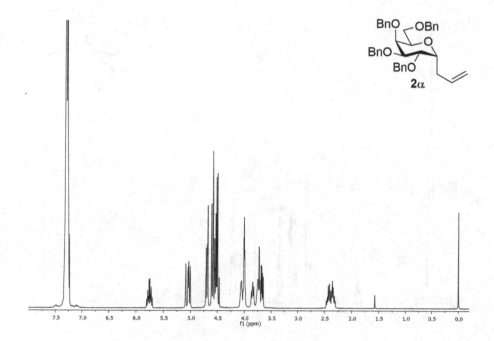

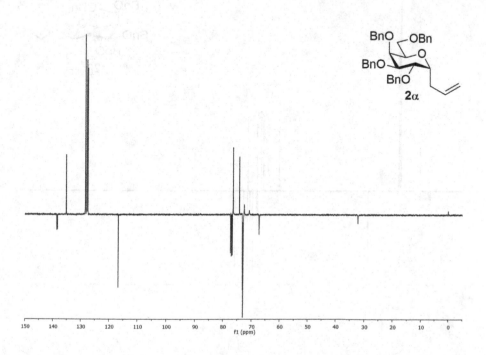

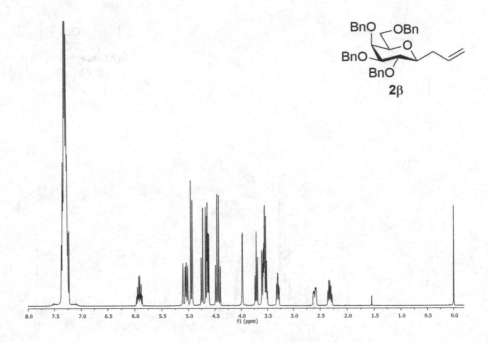

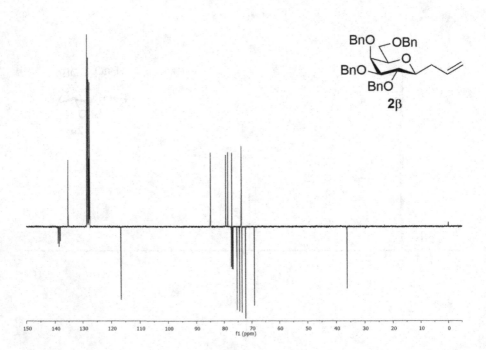

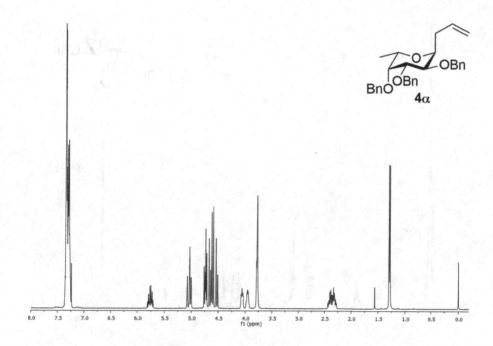

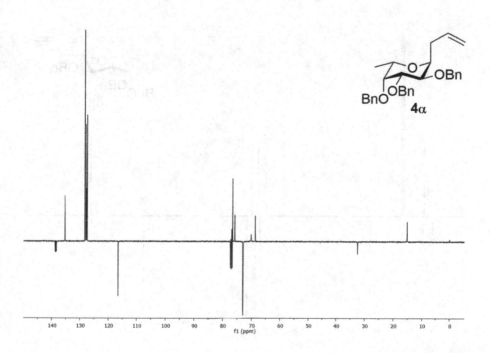

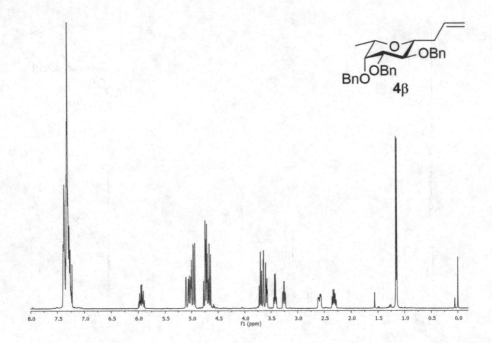

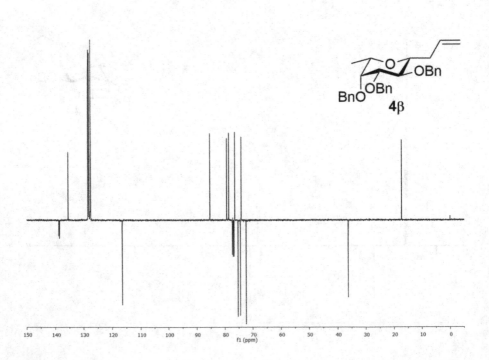

REFERENCES

1. Xie, C.; Veitch, N. C.; Houghton, P. J.; Simmonds, M. S. *J. Chem. Pharm. Bull.* **2003**, *51*, 1204–1207.
2. Traxler, P.; Gruner, J.; Auden, J. A. L. *J. Antibiot.* **1977**, *30*, 289–296.
3. Denmark, S. E.; Regens, C. S.; Kobayashi, T. *J. Am. Chem. Soc.* **2007**, *129*, 2774–2776.
4. Compain, P.; Martin, O. R. *Bioorg. Med. Chem.* **2001**, *9*, 3077–3092.
5. Yang, Y.; Yu, B. *Chem. Rev.* **2017**, *117*, 12281–12356.
6. Hosomi, A.; Sakata, Y.; Sakurai, H. *Tetrahedron Lett.* **1984**, *25*, 2383–2386.
7. Gaertzen, O.; Misske, A. M.; Wolbers, P.; Hoffmann, H. M. R. *Tetrahedron Lett.* **1999**, *40*, 6359–6363.
8. Matwiejuk, M.; Thiem, J. *Eur. J. Org. Chem.* **2011**, *2011*, 5860–5878.
9. Thota, V. N.; Gervay-Hague, J; Kulkarni, S. S. *Org. Biomol. Chem.* **2012**, *10*, 8132–8139.
10. Palomo, C.; Oiarbide, M.; Landa, A.; González-Rego; M. C., García, J. M.; González, A.; Odriozola, J. M.; Martín-Pastor, M.; Linden, A. *J. Am. Chem. Soc.* **2002**, *124*, 8637–8643.
11. La Ferla, B.; Russo, L.; Airoldi, C.; Nicotra, F. *Tetrahedron: Asymmetry* **2009**, *20*, 744–745.

10 Conversion of Simple Sugars to Highly Functionalized Fluorocyclopentanes

Patrick Weber*, and Arnold E. Stütz
Glycogroup, Institute of Organic Chemistry,
Graz University of Technology,
Graz, Austria

Roland Fischer
Institute of Inorganic Chemistry,
Graz University of Technology,
Graz, Austria

Christian Denner[a]
Institute of Organic Chemistry, University of Vienna,
Währingerstrasse 38, A-1090 Vienna, Austria

CONTENTS

Highly functionalized cyclopentanes have frequently been found as core scaffolds or partial structures of biologically active natural products, not only in well-known prostaglandins[1] but also in carbohydrate conjugates such as trehazolins[2,3] and allosamidin[4] as well as in carba nucleosides neplanocin[5] and aristeromycin.[6] Moreover, there are many examples of cyclopentane derivatives designed for medicinal purposes.[7] Based on early work by Padwa[8] and Oppolzer,[9] Vasella and Bernet[10] have thoroughly

* Corresponding Author; e-mail: patrick.weber@tugraz.at.
[a] Checker; e-mail: christian.denner@univie.ac.at.

explored an efficient route to C-branched aminocyclopentane-polyols relying on a (2 + 3)-cycloaddition reaction. Jäger's laboratory[11] as well as Reymond and his group[12] provided early examples of hydroxymethyl aminocyclopentane triols as efficient glycosidase inhibitors. In the same context, we have exploited[13] these guiding references for our modified approach to related compounds, including deoxyfluoro derivatives featuring β-D-*gluco* configuration.

For example, by reaction with *N*-benzylhydroxylamine·HCl in the presence of pyridine, the known 5-enofuranose[14] was converted into partially protected bicyclic isoxazolidine **2**. Methanol was found a superior solvent in this transformation, both in terms of yields (>90%) as well as stereoselectivity. The free hydroxyl group at C-2 allows for further regio- and stereoselective structural elaboration. For example, introduction of a fluorine substituent was achieved by conventional reaction employing Et₃N·3HF and Xtal-Fluor-E®, a crystalline deoxyfluorination reagent and comparably much safer replacement for diethylaminosulfur trifluoride and its analogs. Due to the structural features of the bicyclic isoxazolidine **2**, *endo*-approach of the reagents is sterically highly restricted forcing the system into forming a carbocation at C-2 which is subsequently quenched by fluoride approaching virtually exclusively from the *exo* face, thus providing fluorocyclopentane **3** in acceptable yield with retention of the "D-*gluco*" configuration (for proof of the retention, see below). Conventional hydrogenolysis employing Pd/C in presence of HCl gave polyfunctionalized fluorocyclopentylamine **4** as the hydrochloride. X-ray crystallography provided unambiguous structure proof for the configuration at all three newly formed chiral centers (**Figure 10.1**).

EXPERIMENTAL

GENERAL METHODS

Commercially available reagents and solvents were purchased from Sigma Aldrich, ABCR, or Fisher Scientific and were used without further purification, unless otherwise specified. Cyclohexane was distilled prior to use. Analytical thin-layer chromatography (TLC) was performed on precoated aluminum plates silica gel 60 F_{254}. Spots were detected by UV light (254 nm) and/or charring with a solution of vanillin (9 g) in a mixture of H_2O (950 mL)/EtOH (750 mL)/H_2SO_4 (120 mL) or ceric ammonium molybdate (100 g ammonium molybdate/8 g ceric sulfate in 1000 mL 10% aqueous H_2SO_4). Solutions in organic solvents were dried over Na_2SO_4 and concentrated *in vacuo* at <40°C. For column chromatography, silica gel 60 (230–400 mesh, E. Merck 9385) was used. Melting points were measured

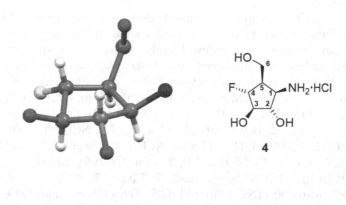

FIGURE 10.1 XRD structure of compound **4** (CCDC: 1948738). (For the sake of consistency, numbering follows the monosaccharide pattern starting with the amino substituted carbon [C-1] clockwise along the ring.)

with a Stuart SMP10. Optical rotations were measured at 20°C on a Perkin Elmer 341 polarimeter at 589 nm and a path length of 10 cm. NMR spectra were recorded for solutions in $CDCl_3$ or D_2O with a Varian INOVA 500 operating at 499.82 MHz (^{1}H), and at 470.33 MHz (^{19}F) or a Bruker Ultrashield spectrometer at 300.36 (^{1}H) and 75.53 MHz (^{13}C). Chemical shifts are listed in δ employing residual, nondeuterated solvent as the internal standard[15] ($CDCl_3$: 7.26 ppm [^{1}H], 77.16 ppm [^{13}C]; D_2O: 4.79 ppm [^{1}H]). Signals were assigned unambiguously by COSY and HSQC analysis. Aromatic protons and carbons were found in the expected regions and are not listed. For the sake of consistency, numbering follows the monosaccharide pattern starting with the amino substituted carbon (C-1) clockwise along the ring. MALDI-TOF mass spectrometry was performed on a Micromass TofSpec 2E time-of-flight mass spectrometer. For single crystal X-ray diffractometry (XRD), all suitable crystals were covered with a layer of silicone oil. A single crystal was selected, mounted on a glass rod on a copper pin, and placed in a cold N_2 stream provided by an Oxford Cryosystems cryometer (T = 100 K). XRD data collection was performed on a Bruker APEX II diffractometer with use of Mo $K\alpha$ radiation (λ = 0.71073 Å) from an IµS microsource and a charge coupled device (CCD) area detector.

(3A*R*,4*R*,5*R*,6*S*,6A*R*)-1-Benzyl-5,6-Bis(benzyloxy)-4-Fluorohexahydro-1*H*-Cyclopenta[*c*] Isoxazole (3)

To a solution of the known[14] 5-enofuranose **1** (410 mg, 1.26 mmol) in methanol (8 mL) was added pyridine (0.2 mL) and *N*-benzylhydroxylamine hydrochloride (250 mg, 1.57 mmol), and the mixture was stirred at rt for 16 h when complete conversion was observed by TLC (2:1 cyclohexane–EtOAc, R_f = 0.30). Solvents were removed and a solution of the residue in CH_2Cl_2 (20 mL) was washed with saturated aqueous $NaHCO_3$ (20 mL). The aqueous phase was twice extracted with CH_2Cl_2. The combined organic phases were dried, filtered, and the solvent was removed to furnish isoxazolidine **2** as colorless, slightly hygroscopic amorphous solid (509 mg, 94%). For analysis, a small amount was chromatographed (10:1 cyclohexane–EtOAc) to give **2**; mp 121-122°C (cyclohexane–EtOAc, 2×); $[\alpha]_D$ +11.7 (c 1.2, $CHCl_3$); ^1H NMR ($CDCl_3$) δ 4.76 (d, 1H, J 11.8 Hz, OCH_2Ph), 4.55 (d, 1H, $J_{1,2}$ 11.8 Hz, OCH_2Ph), 4.30 (d, 1H, J 11.5 Hz, OCH_2Ph), 4.17 (d, 1H, J 11.5 Hz, OCH_2Ph), 3.96 (m, 1H, $J_{5,6a}$ 6.5 Hz, $J_{6a,6b}$ 9.2 Hz, H-6a), 3.92 (d, 1H, J 12.5 Hz, NCH_2Ph), 3.80–3.66 (m, 3H, H-6b, H-2, H-4), 3.60 (d, 1H, J 12.5 Hz, NCH_2Ph), 3.59 (dd, 1H, J 5.1 Hz, H-3), 3.48 (dd, 1H, $J_{1,2}$ 9.6 Hz, $J_{1,5}$ 5.8 Hz, H-1), 2.91 (m, 1H, H-5), 2.15 (bs, 1H, 4-OH); 13C NMR ($CDCl_3$) δ 87.1 (C-3), 86.1 (C-2), 78.2 (C-4), 72.8, 72.2 (2 × OCH_2Ph), 71.1 (C-1), 69.5 (C-6), 60.4 (NCH_2Ph), 50.1 (C-5); TOF–MS: m/z mass $[M + H]^+$ calcd for $C_{27}H_{30}NO_4$, 432.2175; found, 432.2172.

To a stirred solution of crude isoxazolidine **2** (480 mg, 1.11 mmol) in CH_2Cl_2 (10 mL), $Et_3N\cdot3HF$ (0.897 mL, 5.56 mmol, 5 equiv) and Xtal-Fluor-E® (611 mg, 2.67 mmol, 2.4 equiv) were added at 0°C. When completed conversion of the starting material was observed (~3 h, 2:1 cyclohexane–EtOAc, R_f = 0.62), saturated aqueous $NaHCO_3$ (20 mL) was added and the phases were separated. The aqueous phase was washed twice with CH_2Cl_2, the organic phase, combined with the washings, were dried, filtered, and the solvent was removed. The residue was chromatographed (20:1 cyclohexane–EtOAc) to provide compound **3** as a yellow syrup (304 mg, 63%); $[\alpha]_D$ +12.8 (c 1.0, $CHCl_3$); ^1H NMR ($CDCl_3$) δ 4.67 (ddd, 1H, $J_{3,4} = J_{4,5}$ 6.6 Hz, $J_{4,F}$ 54.1 Hz, H-4), 4.67 (d, 1H, J 12.0 Hz, OCH_2Ph), 4.62 (d, 1H, J 12.0 Hz, OCH_2Ph), 4.43 (d, 1H, J 11.7 Hz, OCH_2Ph), 4.31 (d, 1H, J 11.7 Hz, OCH_2Ph), 4.05 (dd, 1H, $J_{5,6a}$ 7.3 Hz, $J_{6a,6b}$ 9.2 Hz, H-6a), 3.91 (d, 1H, J 12.4 Hz, NCH_2Ph), 3.90 (ddd, 1H, $J_{2,3} = J_{3,4}$ 8.0 Hz, $J_{3,F}$ 17.6 Hz, H-3), 3.81 (dd, 1H, $J_{5,6b}$ 2.6 Hz, H-6b), 3.76 (m, 1H, H-2), 3.59 (d, 1H, J 12.7 Hz, NCH_2Ph), 3.56 (dd, 1H, $J_{1,2}$ 9.7 Hz, $J_{1,5}$ 6.1 Hz, H-1), 3.12 (m, 1H, $J_{5,F}$ 31 Hz, H-5); ^{13}C NMR ($CDCl_3$) δ 100.3 ($J_{4,F}$ 187.1 Hz, C-4), 85.4 ($J_{3,F}$ 18.9 Hz, C-3), 84.2 (C-2), 72.5, 72.4 (2 × OCH_2Ph), 70.9 ($J_{1,F}$ 2.7 Hz, C-1), 69.5 ($J_{6,F}$ 1.3 Hz, C-6), 60.1 (NCH_2Ph), 48.6 ($J_{5,F}$ 21.9 Hz, C-5); ^{19}F NMR ($CDCl_3$) δ −182.9 (bm); TOF–MS: m/z mass $[M + H]^+$ calcd for $C_{27}H_{29}FNO_3$, 434.2131; found, 434.2128. Anal. calcd for $C_{27}H_{28}FNO_3$: C, 74.80; H, 6.51; N, 3.23. Found: C, 74.63; H, 6.48; N, 3.20.

(1R,2S,3R,4R,5R)-3-Amino-5-Fluoro-4-(Hydroxymethyl)cyclopentane-1,2-Diol Hydrochloride (4)

A solution of deoxyfluoro compound **3** (304 mg, 0.701 mmol) in methanol (5 mL) and THF (5 mL) was adjusted to pH 1 (2 M HCl, 0.5 mL). Pearlman's catalyst (770 mg, 20% $Pd(OH)_2/C$) was added and the suspension was stirred under H_2 at ambient pressure. After complete conversion (8:4:1 $CHCl_3$–MeOH-25% NH_4OH, R_f = 0.36), the catalyst was filtered off and the filtrate was concentrated. Crystallization from MeOH–EtOAc gave colorless polyol **4** (103 mg, 73%); mp 179–180°C (decomposition); $[\alpha]_D$ +6.7 (c 1.0, H_2O); ^1H NMR (D_2O) δ 4.87 (ddd, 1H, $J_{3,4} = J_{4,5}$ 6.2 Hz, $J_{4,F}$ 54.0 Hz, H-4), 4.18 (ddd, 1H, $J_{2,3}$ 8.5 Hz, $J_{3,F}$ 20.5 Hz, H-3), 4.07 (dd, 1H, $J_{1,2}$ 8.8 Hz, H-2), 3.96 (dd, 1H, $J_{5,6a}$ 5.6 Hz, $J_{6a,6b}$ 11.7 Hz, H-6a), 3.93 (dd, 1H, $J_{5,6b}$ 4.6 Hz, H-6b), 3.88 (dd, 1H, $J_{1,5}$ 10.4 Hz, H-1), 2.75 (ddddd, 1H, $J_{5,F}$ 24.9 Hz, H-5); ^{13}C NMR (D_2O) δ 96.0 ($J_{4,F}$ 184.8 Hz, C-4), 77.7 ($J_{3,F}$ 21.0 Hz, C-3), 75.3 ($J_{2,F}$ 9.2 Hz, C-2), 58.1 ($J_{6,F}$ 2.4 Hz, C-6), 53.8 ($J_{1,F}$ 4.6 Hz, C-1), 41.3 ($J_{5,F}$ 22.3 Hz, C-5); ^{19}F NMR (D_2O) δ -187.5 (ddd); TOF–MS: m/z mass [M + H]$^+$ calcd for $C_6H_{13}FNO_3$, 166.0880; found, 166.0880. Anal. calcd for $C_6H_{13}ClFNO_3$: C, 35.74; H, 6.50; N, 6.95. Found: C, 35.44; H, 6.44; N, 6.96.

ACKNOWLEDGMENTS

Patrick Weber is recipient of a DOC Fellowship of the Austrian Academy of Sciences at the Institute of Organic Chemistry, Graz University of Technology.

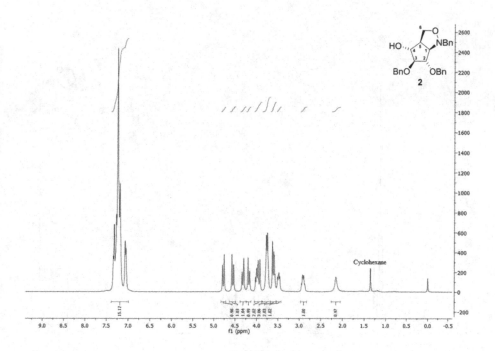

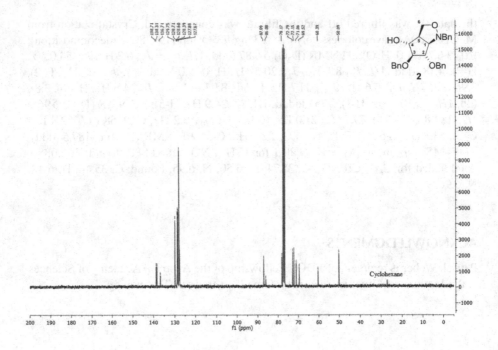

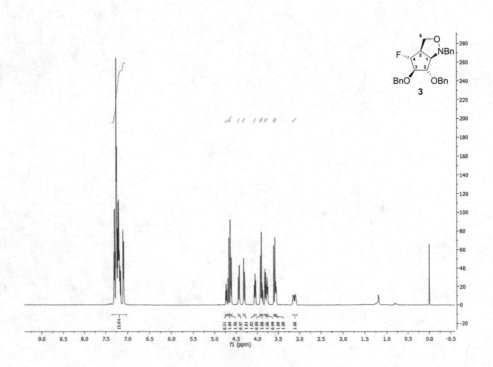

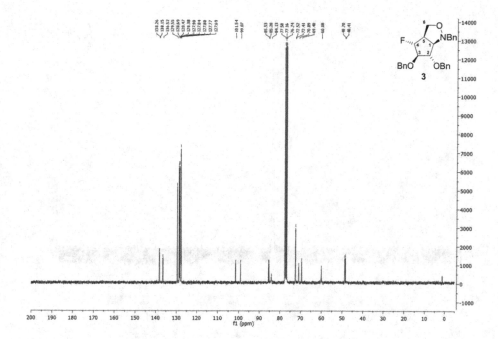

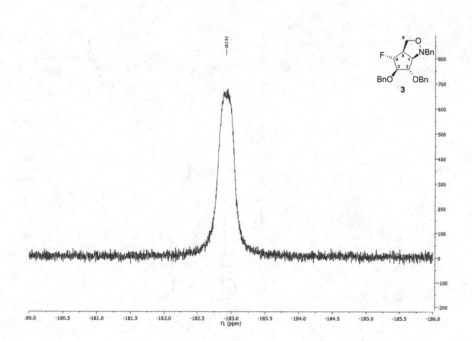

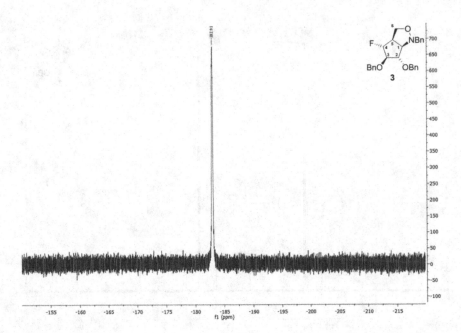

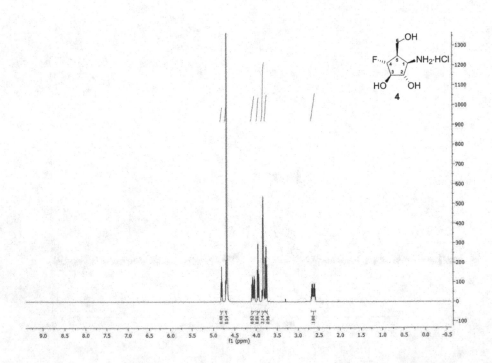

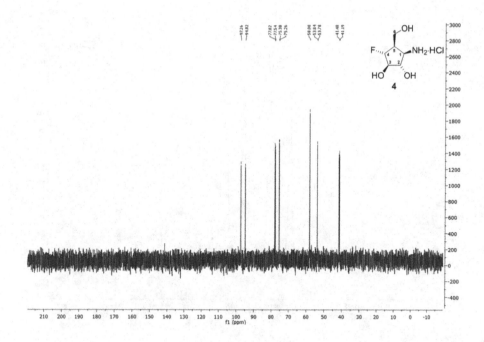

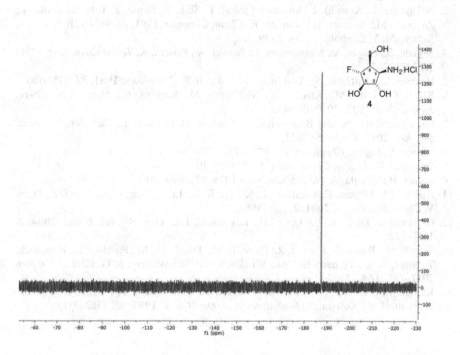

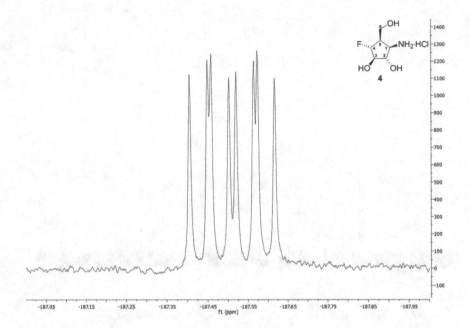

REFERENCES

1. Lands, W. E. M. *Ann. Rev. Physiol.* **1979**, *41*, 633–652.
2. Nakayama, T.; Amachi, T.; Murao, S.; Sakai, T.; Shin, T.; Kenny, P. T. M.; Iwashita, T.; Zagorski, M.; Komura, H.; Nomoto, K. *Chem. Commun.* **1991**, *14*, 919–921.
3. Kobayashi, Y. *Carbohydr. Res.* **1999**, *315*, 3–15.
4. Sakuda, S.; Isogai, A.; Matsumoto, S.; Suzuki, A.; Koseki, K. *Tetrahedron Lett.* **1986**, *27*, 2475–2478.
5. Hayashi, M.; Yainuma, S.; Yoshioka, H.; Nakatsu, K. *J. Antibiot.* **1981**, *34*, 675–680.
6. Kishi, T.; Muroi, M.; Kusaka, T.; Nishikawa, M.; Kamiya, K.; Mizuno, K. *Chem. Pharm. Bull.* **1972**, *20*, 940–946.
7. For a recent review see: Boutureira, O.; Matheu, M. I.; Diaz, Y.; Castillon, S. *Chem. Soc. Rev.* **2013**, *42*, 5056–5072.
8. Padwa, A. *Angew. Chem. Int. Ed.* **1976**, *15*, 123–136.
9. Oppolzer, W. *Angew. Chem. Int. Ed.* **1977**, *16*, 10–23.
10. Bernet, B.; Vasella, A. *Helv. Chim. Acta* **1979**, *62*, 1990–2016.
11. Kleban, M.; Hilgers, P.; Greul, J. N.; Kugler, R. D.; Li, J.; Picasso, S.; Vogel, P.; Jäger, V. *Chem. Bio. Chem.* **2001**, *2*, 356–368.
12. Gartenmann Dickson, L.; Leroy, E.; Reymonds, J.-L. *Org. Biomol. Chem.* **2004**, *2*, 1217–1226.
13. Schalli, M.; Tysoe, C.; Fischer, R.; Pabst, B. M.; Thonhofer, M.; Paschke, E.; Rappitsch, T.; Stütz, A. E.; Tschernutter, M.; Windischhofer, W.; Withers, S. G. *Carbohydr. Res.* **2017**, *443–444*, 15–22.
14. Kumar, V.; Gauniyal, H. M.; Shaw, A. K. *Tetrahedron Asymm*, **2007**, *18*, 2069–2078.
15. Gottlieb, H. E.; Kotlyar, V.; Nudelman, A. *J. Org. Chem.* **1997**, *62*, 7512–7515.

11 Carbene-Mediated Quaternarization of the Anomeric Position of Carbohydrates

*Kévin Mébarki, Marine Gavel, Antoine Joosten, and Thomas Lecourt**
Normandie University, INSA Rouen,
UNIROUEN, CNRS
Rouen, France

Floriane Heis[a]
Sorbonne Université, CNRS,
Institut Parisien de Chimie Moléculaire, IPCM
Paris, France

CONTENTS

---

* Corresponding author: Thomas.Lecourt@insa-rouen.fr
[a] Checker: floriane.heis@upmc.fr under supervision of Matthieu Sollogoub

Carbohydrates with a quaternary position are key players for designing nonnatural sugar derivatives with finely tuned biological properties.[1] More particularly, compounds with a quaternary anomeric position show strong promises in glycobiology,[2] but classical approaches to their preparation, where *quaternarisation* is performed *before glycosylation*, almost exclusively deliver α-ketopyranosides because of major drawbacks.[3,4] We recently reported a new approach toward carbohydrates with quaternary anomeric position involving a shift in the retrosynthetic paradigm.[5] Ketopyranosides in both α and β series could indeed be obtained by *reversing the order of the key C–O and C–C bond-forming reactions.*[6] This quaternarization by insertion of a Rh(II)-carbene into the anomeric C–H bond was recently optimized to obtain lactones 7 and 8 in α-*manno* and β-*gluco* series on preparative scale.[7] Accordingly, the C–H functionalization process first requires preparation of diazo sugars 5 and 6 from the starting methyl 4,6-*O*-benzylidene-3-*O-tert*-butyldimethylsilyl glycosides 1 and 2. In this two steps procedure, acylation and diazo transfer are performed independently. First, bromoacetylation of 1 and 2 gives 2-*O*-bromoacetates 3 and 4, respectively. It is worth mentioning that use of pyridine is necessary to achieve full conversion,[8] and bromoacetyl bromide must be used to avoid a mixture of bromo- and chloroacetate at position 2, when bromoacetyl chloride is used instead. Diazo transfer is next achieved with *N,N'*-ditosylhydrazine and 1,8-diazabicyclo[5.4.0]undec-7-ene in tetrahydrofuran (THF), to give carbene precursors 5 and 6 in 59 and 63% yield, respectively, over two steps after chromatography. Rh(II)-catalyzed decomposition of 5 and 6 is next performed in refluxing 1,2-dichloroethane to give lactones 7 and 8 by 1,5-C–H insertion of the highly reactive transient metallo-carbene.

EXPERIMENTAL

GENERAL METHODS

Reactions requiring anhydrous conditions were conducted under argon. Compounds were pre-dried under high vacuum (<1 mbar) at room temperature before use. Dichloromethane and THF were dried over a MBRAUN MB SPS-800 solvent purification system. Bromoacetyl bromide, anhydrous pyridine, 1,8-diazabicyclo[5.4.0] unde-7-ene, and dirhodium(II)tetraacetate were used as received. Anhydrous 1,2-dichloroethane, stored in sealed 1-L bottles, was purchased from Sigma-Aldrich (ref. 284505) and dried additionally over molecular sieves (4 Å, 8–12 mesh) before use. Crystalline *N,N'*-ditosylhydrazine was prepared following Fukuyama's procedure.[9] Deloxan® was purchased from Strem chemicals. Reactions were monitored by thin-layer chromatography on silica gel 60 F254 pre-coated aluminum plates (0.25 mm). Visualization was performed under UV light and by phosphomolybdic acid oxidation. Crude products were adsorbed on silica (1 g/g) and chromatographed (silica, 40–63 μm, VWR). ^{1}H NMR spectra were recorded at 300 MHz, and ^{13}C NMR spectra at 75 MHz, on a Bruker Avance-III spectrometer. Abbreviations used for peak multiplicities are s: singlet, d: doublet, t: triplet, and m: multiplet. Coupling constants (*J*) are in Hz and chemical shifts are given in ppm. Chemical shifts are reported relative to $CHCl_3$ (residual solvent signals). Carbon multiplicities were assigned by distortionless enhancement by polarization transfer experiments. The ^{1}H and ^{13}C signals were assigned by COSY and HSQC experiments. Accurate mass measurements (HRMS) were performed with a Q-TOF analyzer. Elemental analyses were performed on a Thermo EAGER 300 Flash 2000 apparatus. Infrared spectra (IR) were recorded by application of foams on a Single Reflection Attenuated Total Reflectance Accessories, and data are reported in cm^{-1}. Optical rotations were determined with a water-jacketed 10-cm cell. Specific rotations are reported in 10^{-1} deg cm^2 g^{-1} and concentrations in g per 100 mL. Melting points are uncorrected. Solutions in organic solvents were dried with anhydrous $MgSO_4$ and concentrated at reduced pressure at <40°C.

DRYING OF 1,2-DICHLOROETHANE

Commercial, dry 1,2-dichloroethane was additionally dried as follows: In a 500-mL round-bottom flask, 4 Å molecular sieves (10 g) were heated with a heat-gun (600°C) under high vacuum (<1 mbar) for 2 h. The system was depressurized with argon and allowed to cool to room temperature. Dry 1,2-dichloroethane (350 mL) was added *via* a cannula, and the mixture was stirred for 2 h.

ADDITION OF DIAZO SUGARS AND REACTION VESSEL (FIGURE 11.1):

To prevent dimerization of the Rh(II)-carbene, reverse addition of diazo sugars **5** or **6** to a highly diluted solution of $Rh_2(OAc)_4$ is performed with a KDS 100 SE syringe infusion pump at a rate not exceeding 15 μmol/h, via a carefully positioned addition needle. A one-piece glassware, composed of a 500-mL round-bottom flask that is sealed to a condenser, can also be used as reaction vessel.

METHYL 4,6-*O*-BENZYLIDENE-2-*O*-DIAZOACETYL-3-*O*-*TERT*-BUTYLDIMETHYLSILYL-α-D-MANNOPYRANOSIDE (5)

In a 250-mL round-bottom flask, bromoacetyl bromide (0.585 mL, 6.695 mmol) was added dropwise over a period of 5 min to a solution of methyl 4,6-*O*-benzylidene-3-*O*-*tert*-butyldimethylsilyl-α-D-mannopyranoside **1** (1.77 g, 4.47 mmol) and pyridine

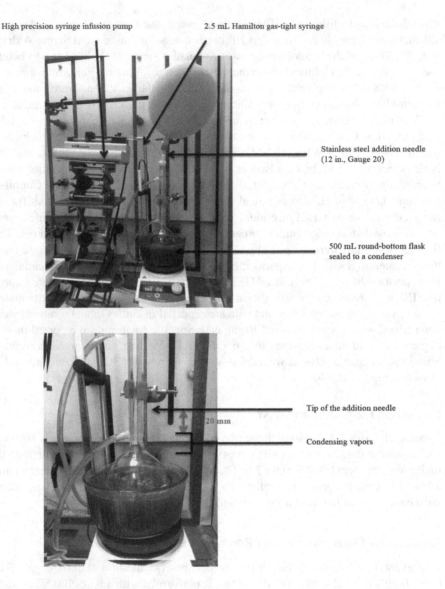

High precision syringe infusion pump   2.5 mL Hamilton gas-tight syringe

Stainless steel addition needle
(12 in., Gauge 20)

500 mL round-bottom flask
sealed to a condenser

Tip of the addition needle

20 mm

Condensing vapors

FIGURE 11.1   Reaction apparatus assembly.

(1.083 mL, 13.39 mmol) in CH_2Cl_2 (30 mL) at −30°C. The mixture was stirred at
−30°C until TLC (5:1 cyclohexane–EtOAc) showed complete conversion of the starting material. Methanol (0.5 mL) was then added dropwise and, after 30 min, aqueous
hydrochloric acid (1 M, 15 mL) and CH_2Cl_2 (15 mL) were added. The aqueous phase
was extracted with CH_2Cl_2 (2 × 15 mL), and the combined organic layers were dried
and concentrated under reduced pressure. The crude product was chromatographed
(40 x 80 mm column, 5:1 cyclohexane–EtOAc) to give methyl 4,6-O-benzylidene-2-
O-bromoacetyl-3-O-tert-butyldimethylsilyl-α-D-manno-pyranoside **3**.

In a 250-mL round-bottom flask, 1,8-diazabicyclo[5.4.0]unde-7-ene (3.34 mL, 22.32 mmol) was added dropwise over a period of 10 min to a solution of 3 and N,N'-ditosylhydrazine (3.04 g, 8.93 mmol) in THF (35 mL) at 0°C. The bright yellow reaction mixture was stirred at 0°C until TLC (5:1 cyclohexane–EtOAc) showed complete conversion of the starting material (2–3 h). A saturated solution of NaHCO$_3$ (30 mL) and CH$_2$Cl$_2$ (30 mL) were then added, and the phases were separated after 30 min of vigorous stirring at 0°C.[*] The aqueous layer was extracted with ethyl acetate (3 × 20 mL), and the combined organic layers were concentrated under reduced pressure. The crude product was chromatographed (35 × 125 mm column, 5:1 cyclohexane–EtOAc) to give methyl 4,6-O-benzylidene-2-O-diazoacetyl-3-O-tert-butyldimethylsilyl-α-D-mannopyranoside 5 (1.29 g, 2.8 mmol, 63% over two steps) as yellow syrup; $[\alpha]_D^{20}$ –3 (c 1.0, CHCl$_3$); R_f 0.52 (3:1 cyclohexane–EtOAc). ^{1}H NMR (CDCl$_3$): δ 7.60–7.30 (m, 5H, H$_{arom.}$), 5.60 (s, 1H, H-7), 5.26 (dd, 1H, $J_{2,3}$ 3.7, $J_{1,2}$ 1.1 Hz, H-2), 4.89 (br s, 1H, H-8), 4.71 (d, 1H, $J_{1,2}$ 1.1 Hz, H-1), 4.29 (d, 1H, J_5,6eq 5.7 Hz, H-6$_{eq}$), 4.22 (dd, 1H, $J_{3,4}$ 9.0 Hz, $J_{2,3}$ 3.7 Hz, H-3), 3.92–3.70 (m, 3H, H-4, H-5, H-6$_{ax}$), 3.41 (s, 3H, OMe), 0.88 [s, 9H, C(CH$_3$)$_3$], 0.10 (s, 3H, CH$_3$Si), 0.07 (s, 3H, CH$_3$Si); ^{13}C NMR (CDCl$_3$): δ 166.2 (C=O), 137.5 (C$_{quat. arom}$), 128.9 (CH$_{arom}$), 128.1 (CH$_{arom}$), 126.1 (CH$_{arom}$), 101.8 (C-7), 100.0 (C-1), 79.4 (C-4), 72.9 (C-2), 68.8 (C-6), 68.1 (C-3), 63.6 (C-5), 55.2 (OMe), 46.5 (C-8), 25.6 [C(CH$_3$)$_3$], 18.2 [SiC(CH$_3$)$_3$], –4.7 (SiCH$_3$), –5.1 (SiCH$_3$); IR (film): 2930; 2113; 1698; 1471; 1384; 1282; 1249; 1216; 1173; 1133; 1079; 1033; 1008; 978; 914; 888; 863; 838; 779; 758; 734; 698; 668; HRMS: m/z calcd for [M + Na]$^+$ C$_{22}$H$_{32}$N$_2$O$_7$SiNa: 487.18765; found, 487.1872.

METHYL 4,6-O-BENZYLIDENE-2-O-DIAZOACETYL-3-O-TERT-BUTYLDIMETHYLSILYL-β-D-GLUCOPYRANOSIDE (6)

In a 500-mL round-bottom flask, pyridine (2.68 mL, 33.82 mmol) was added in one portion to a solution of methyl 4,6-O-benzylidene-3-O-tert-butyldimethylsilyl-β-D-glucopyranoside 2 (4.47 g, 11.27 mmol) in CH$_2$Cl$_2$ (100 mL) at –30°C, followed by dropwise addition of bromoacetyl bromide (1.477 mL, 16.908 mmol) over a period of 15 min. The mixture was stirred at –30°C until TLC (5:1 cyclohexane–EtOAc) showed complete conversion of the starting material (overnight). Methanol (1 mL) was then added dropwise and, after 30 minutes, aqueous hydrochloric acid (1 M, 50 mL), and CH$_2$Cl$_2$ (50 mL) added. The aqueous phase was extracted with CH$_2$Cl$_2$ (2 × 50 mL), and the combined organic layers were concentrated under reduced pressure. The crude product was chromatographed (60 × 80 mm column, 5:1 cyclohexane–EtOAc) to give methyl 4,6-O-benzylidene-2-O-bromoacetyl-3-O-tert-butyldimethylsilyl-β-D-glucopyranoside 4.

In a 500-mL round-bottom flask, 1,8-diazabicyclo[5.4.0]unde-7-ene (8.43 mL, 56.35 mmol) was added dropwise over a period of 10 min to a solution of 4 and N,N'-ditosylhydrazine (7.675 g, 22.54 mmol) in THF (90 mL) at 0°C. The bright yellow mixture was stirred at 0°C until TLC (5:1 cyclohexane–EtOAc) showed complete conversion of the starting material (2–3 h). Saturated NaHCO$_3$ (60 mL) and CH$_2$Cl$_2$

---

[*] If required, water might be added to obtain both layers clear.

(60 mL) were added, and the phases were separated after 30 min of vigorous stirring at 0°C. The aqueous layer was extracted with EtOAc (3 × 50 mL), and the combined organic layers were concentrated under reduced pressure. Chromatography (48 × 125 mm column, 6:1 cyclohexane–EtOAc) gave methyl 4,6-O-benzylidene-2-O-diazoacetyl-3-O-$tert$-butyldimethylsilyl-β-D-glucopyranoside **6** (3.1 g, 6.7 mmol, 59% over two steps) as yellow syrup; $[\alpha]_D^{20}$ –37 (c 1.0, CHCl$_3$); R_f 0.46 (3:1 cyclohexane–EtOAc). ^{1}H NMR (CDCl$_3$): δ 7.60–7.31 (m, 5H, H$_{arom}$), 5.54 (s, 1H, H-7), 5.02 (t, 1H, $J_{1,2}$ $J_{2,3}$ 8.8 Hz, H-2), 4.81 (br s, 1H, H-8), 4.47–4.30 (m, 2H, H-1, H-6$_{eq}$), 3.89 (t, 1H, $J_{2,3}$ = $J_{3,4}$ 8.8 Hz, H-3), 3.81 (t, 1H, $J_{5,6ax}$ = $J_{6ax,6eq}$ 10.2 Hz, H-6$_{ax}$), 3.57 (m, 1H, H-4), 3.53 (s, 3H, OMe), 3.44 (dt, 1H, $J_{5,6ax}$ = $J_{4,5}$ 10.2 Hz, $J_{5,6eq}$ 4.8 Hz, H-5), 0.85 [s, 9H, C(CH$_3$)$_3$], 0.05 (s, 3H, SiCH$_3$), 0.01 (s, 3H, SiCH$_3$); ^{13}C NMR (CDCl$_3$): δ 165.3 (C=O), 137.1 (C$_{quat.\ arom}$), 129.1 (CH$_{arom}$), 128.2 (CH$_{arom}$), 126.3 (CH$_{arom}$), 102.5 (C-1), 101.8 (C-7), 81.4 (C-4), 74.5 (C-2), 72.8 (C-3), 68.7 (C-6), 66.4 (C-5), 57.1 (OMe), 46.5 (C-8), 25.6 [C(CH$_3$)$_3$], 18.1 [SiC(CH$_3$)$_3$], –4.2 (SiCH$_3$), –5.1 (SiCH$_3$); IR (film): 2929; 2112; 1709; 1471; 1380; 1237; 1134; 1098; 1029; 1009; 838; 777; 731; 698; 668; HRMS: m/z calcd for [M + Na]$^+$ C$_{22}$H$_{32}$N$_2$O$_7$SiNa: 487.18765; found: 487.1874.

(2R,4AR,5AS,8AS,9AR)-9-[($TERT$-BUTYLDIMETHYLSILYL)OXY]-5A-METHOXY-2-PHENYLHEXAHYDROFURO[2′,3′:5,6] PYRANO[3,2-D][1,3]DIOXIN-7(8AH)-ONE (7)

A solution of diazo sugar **5** (700 mg, 1.605 mmol) in anhydrous 1,2-dichloroethane (2 mL)† was transferred into a gas-tight Hamilton syringe connected to a 30-cm stainless steel needle. Dirhodium(II)tetraacetate (7 mg, 0.016 mmol) was suspended in anhydrous 1,2-dichloroethane (320 mL)† in the reaction vessel§ (Figure 11.1). The flask was heated at 110°C and, when a continuous and stable refluxing regime was established, the needle used for addition of the substrate was placed ~20 mm above the condensing vapors (Figure 11.1). Dropwise addition of the substrate was then ensured with the aid of a high precision infusion syringe pump at a rate of 12 μmol/h. When the addition was complete, the syringe was rinsed with anhydrous 1,2-dichloroethane (2 mL, rate of addition 0.2 mL/h).§ After cooling to room temperature, the mixture was transferred into a 1-L round-bottom flask and concentrated under vacuum to ~20 mL. Deloxan® (10 mg) was added, the reaction mixture was stirred for 2 h at room temperature and filtered. Concentration of the filtrate (10 mbar, 30°C) and drying under high vacuum (<1 mbar) overnight gave lactone **7** (590 mg, 84%) as white foam;[10] $[\alpha]_D^{20}$ –75 (c 1.0, CHCl$_3$); R_f 0.41 (3:1 cyclohexane–EtOAc). ^{1}H NMR (CDCl$_3$): δ 7.56–7.33 (m, 5H, H$_{arom}$), 5.57 (s, 1H, H-7), 4.47 (d, 1H, $J_{2,3}$ 4.0 Hz, H-2), 4.33–4.24 (m, 1H, H-6), 4.19 (dd, 1H, $J_{3,4}$ 9.5 Hz, $J_{2,3}$ 4.0 Hz, H-3), 3.93–3.69 (m, 3H, H-4, H-5, H-6), 3.38 (s, 3H, OMe), 2.82 (d, 1H, $J_{8,8'}$ 16.3 Hz, H-8), 2.70 (d, 1H, $J_{8,8'}$ 16.3 Hz, H-8′), 0.90 [s, 9H, C(CH$_3$)$_3$], 0.12 (s, 3H, SiCH$_3$), 0.07 (s, 3H, SiCH$_3$); ^{13}C NMR (CDCl$_3$): δ 172.0 (C=O), 137.2 (C$_{quat.\ arom}$), 129.0 (CH$_{arom}$), 128.2 (CH$_{arom}$), 126.1 (CH$_{arom}$), 103.7 (C-1), 101.7 (C-7),

---

† See General Methods section for more details.

82.4 (C-2), 77.6 (C-4), 68.5 (C-3), 68.2 (C-6), 64.7 (C-5), 51.8 (OMe), 40.6 (C-8), 25.7 [(CH$_3$)$_3$C], 18.3 [(CH$_3$)$_3$CSi], −4.4 (SiCH$_3$), −5.1 (SiCH$_3$); IR (film): 2930; 1795; 1457; 1386; 1250; 1213; 1174; 1118; 1099; 1027; 838; 779; 697; 669; HRMS: *m/z* calcd for [M + Na]$^+$ C$_{22}$H$_{32}$O$_7$SiNa: 459.1815; found: 459.1811. Anal. calcd for C$_{22}$H$_{32}$O$_7$Si: C, 60.53; H, 7.39; found: C, 60.57; H, 7.36.

(2*R*,4A*R*,5A*R*,8A*R*,9*S*,9A*R*)-9-[(*TERT*-BUTYLDIMETHYLSILYL)OXY]-5A-METHOXY-2-PHENYLHEXAHYDROFURO[2′,3′:5,6] PYRANO[3,2-*D*][1,3]DIOXIN-7(8A*H*)-ONE (8)

A solution of diazo sugar **6** (700 mg, 1.605 mmol) in anhydrous 1,2-dichloroethane (2 mL) was transferred into a gas-tight Hamilton syringe connected to a 30-cm stainless steel needle (as discussed previously and Figure 11.1). Dirhodium(II)tetraacetate (7 mg, 0.016 mmol) was suspended in rigorously anhydrous 1,2-dichloroethane (320 mL) in the reaction vessel§ (Figure 11.1). The flask was heated at 110°C, and when a continuous and stable refluxing regime was established, the needle used for addition of the substrate was placed ~20 mm above the condensing vapors (Figure 11.1). Dropwise addition of the substrate was then ensured with the aid of a high precision infusion syringe pump at a rate of 12 μmol/h. When the addition was complete, the syringe was rinsed with additional anhydrous 1,2-dichloroethane (2 mL, rate of addition 0.2 mL/h).§ After cooling to room temperature, the reaction mixture was transferred into a 1-L round-bottom flask, and concentrated under vacuum to ~20 mL. Deloxan® (10 mg) was then added, the reaction mixture was stirred for 2 h at room temperature and filtered. Concentration of the filtrate (10 mbar, 30°C) and drying under high vacuum (<1 mbar) overnight gave lactone **8** (620 mg, 89%) as a white foam[10]; [α]$_D^{20}$ −8 (*c* 1.0, CHCl$_3$); *R$_f$* 0.49 (3:1 cyclohexane–EtOAc). ^{1}H NMR (CDCl$_3$): δ 7.57–7.32 (m, 5*H*, H$_{arom.}$), 5.58 (s, 1*H*, H-7), 4.47–4.29 (m, 2*H*, H-2, H-6), 3.89 (dd, 1*H*, *J*$_{3,4}$ 9.6 Hz, *J*$_{2,3}$ 5.8 Hz, H-3), 3.83–3.71 (m, 2*H*, H-4, H-6), 3.69–3.55 (m, 1*H*, H-5), 3.41 (s, 3*H*, OMe), 2.86 (d, 1*H*, *J*$_{8,8'}$ 17.2 Hz, H-8), 2.68 (d, 1*H*, *J*$_{8,8'}$ 17.2 Hz, H-8′), 0.90 [s, 9*H*, C(CH$_3$)$_3$], 0.13 (s, 3*H*, SiCH$_3$), 0.06 (s, 3*H*, SiCH$_3$); ^{13}C NMR (CDCl$_3$): δ 171.8 (C=O), 136.9 (C$_{quat.\ arom}$), 129.2 (CH$_{arom}$), 128.2 (CH$_{arom}$), 126.1 (CH$_{arom}$), 103.8 (C-1), 101.8 (C-7), 85.4 (C-2), 78.6 (C-4), 74.1 (C-3), 68.8 (C-6), 66.5 (C-5), 50.5 (OMe), 36.5 (C-8), 25.7 [(CH$_3$)$_3$C], 18.2 [(CH$_3$)$_3$CSi], -4.5 (SiCH$_3$), -4.9 (SiCH$_3$); IR (film): 2930; 1800; 1462; 1386; 1282; 1254; 1095; 1063; 1004; 964; 867; 837; 778; 752; 698; 669; HRMS: *m/z* calcd for [M + Na]$^+$ C$_{22}$H$_{32}$O$_7$SiNa: 459.1815; found: 459.1812; Anal. calcd for C$_{22}$H$_{32}$O$_7$Si: C, 60.53; H, 7.39; found: C, 60.55; H, 7.41.

ACKNOWLEDGMENTS

LABEX SynOrg (ANR-11-LABX-0029) and ANR (JCJC-2013-QuatGlcNAc) are gratefully acknowledged for their financial support. Thomas Lecourt also thanks CNRS for a research fellowship.

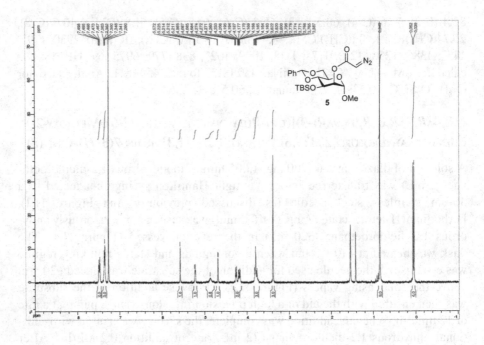

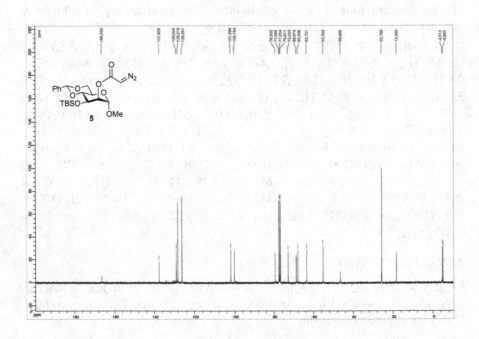

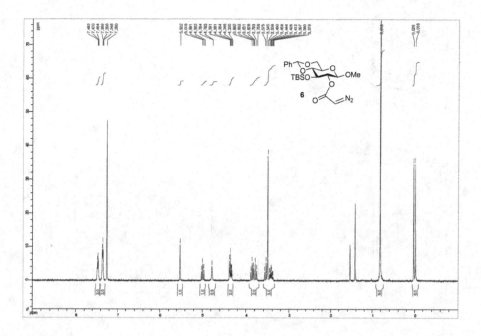

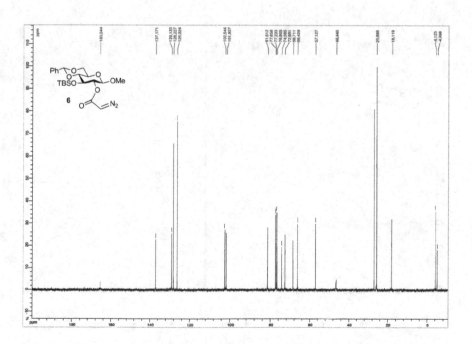

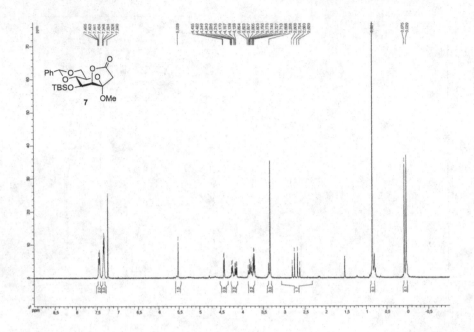

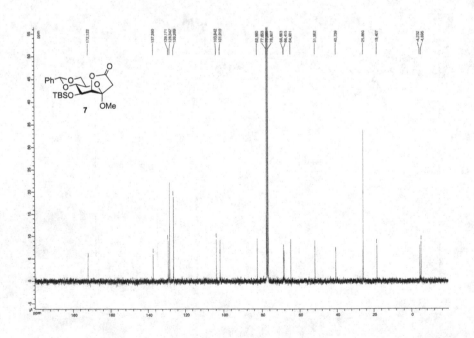

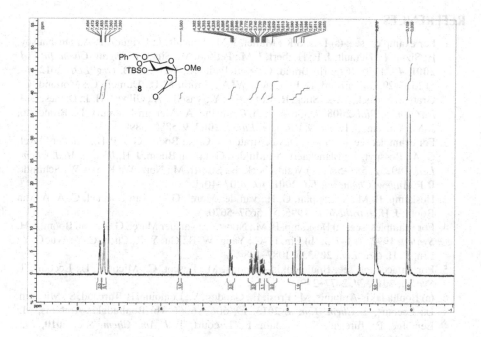

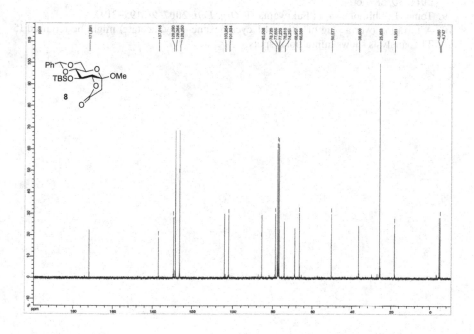

REFERENCES

1. For examples, see: (a) Das, S.K.; Mallet, J.-M.; Esnault, J.; Driguez, P.-A.; Duchausoy, P.; Sizun, P.; Hérault, J.-P.; Herbert, J.-M.; Petitou, M.; Sinaÿ, P. *Angew. Chem. Int. Ed.* **2001**, *40*, 1670–1673; (b) Brand, C.; Kettelhoit, K.; Werz, D. B. *Org. Lett.* **2012**, *14*, 5126–5129; (c) Tailford, L. E.; Offen, W. A.; Smith, N. L.; Dumon, C.; Morland, C.; Gratien, J.; Heck, M.-P.; Stick, R. V.; Blériot, Y.; Vasella, A.; Gilbert, H. J.; Davies, G. J. *Nat. Chem. Biol.* **2008**, *4*, 306–312; (d) Caravano, A.; Mengin-Lecreulx, D.; Brondello, J.-M.; Vincent, S. P.; Sinaÿ, P. *Chem. Eur. J.* **2003**, *9*, 5888–5898.
2. For example, see: (a) Noort, D; van Straten N. C. R.; Boons, G. J. P. H.; van der Marel, G. A.; Bossuyt, X.; Blanckaert, N.; Mulder, G. J.; van Boom, J. H. *Bioorg. Med. Chem. Lett.* **1992**, *2*, 583–588; (b) Waldscheck, B.; Streiff, M.; Notz, W.; Kinzy, W.; Schmidt, R.R. *Angew. Chem. Int. Ed.* **2001**, *40*, 4007–4011.
3. Heskamp, B. M.; Veeneman, G. H.; van der Marel, G. A.; van Boeckel, C. A. A; van Boom, J. H. *Tetrahedron*, **1995**, *51*, 5657–5670.
4. For examples, see: (a) Heskamp, B. M.; Noort, D.; van der Marel, G. A.; van Boom, J. H. *Synlett*, **1992**, 713–715; (b) Lin, H.-C.; Yang, W.-B.; Gu, Y.-F.; Chen, C.-Y.; Wu, C.-Y.; Lin, C.-H. *Org. Lett.* **2003**, *5*, 1087–1089.
5. For an account, see: Boultadakis-Arapinis, M.; Lescot, C.; Micouin, L.; Lecourt, T. *Synlett*, **2013**, *24*, 2477–2491.
6. (a) Boultadakis-Arapinis, M.; Prost, E.; Gandon, V.; Lemoine, P.; Turcaud, S.; Micouin, L.; Lecourt, T. *Chem. Eur. J.* **2013**, *19*, 6062–6066; (b) Boultadakis-Arapinis, M.; Lemoine, P.; Turcaud, S.; Micouin, L.; Lecourt, T. *J. Am. Chem. Soc.* **2010**, *132*, 15477–15479.
7. Mébarki, K.; Gavel, M.; Heis, F.; Joosten, A. Y. P.; Lecourt, T. *J. Org. Chem.* **2017**, *82*, 9030–9037.
8. Boultadakis-Arapinis, M.; Lescot, C.; Micouin, L.; Lecourt, T. *J. Carbohydr. Chem.* **2011**, *30*, 587–604.
9. Toma, T.; Shimokawa, J.; Fukuyama, T. *Org. Lett.* **2007**, *9*, 3195–3197.
10. A filtration over a plug of silica (4:1 cyclohexane–ethyl acetate) might be required if TLC analysis shows minor polar spots.

Part II

Synthetic Intermediates

12 Glucose and Glucuronate 2,2,2-Trichloroethyl Sulfates: Precursors for Multiply Sulfated Oligosaccharides

Kenya Matsushita
Graduate School of Engineering, Tottori University,
Tottori, Japan

*Jun-ichi Tamura**
Graduate School of Engineering, Tottori University,
Tottori, Japan
Faculty of Agriculture, Tottori University,
Tottori, Japan

Serena Traboni[a]
Dipartimento di Scienze Chimiche,
Università degli Studi di Napoli Federico II,
Napoli, Italy

CONTENTS

* Corresponding author: jtamura@tottori-u.ac.jp
[a] Checker: serena.traboni@unina.it

Sulfated oligo- and polysaccharides, *e.g.*, fucoidan, heparan sulfate, and chondroitin sulfate, are common constituents of agar and animal tissues. These substances, where the sulfate replaces original hydroxyl groups at specific positions, exhibit numerous biological activities.

Pharmaceutical applications of various synthetic regiospecifically sulfated oligosaccharides are yet to be established. Protected sulfate groups, which may be introduced at the desired positions in the early stage of glycan syntheses, may serve at the later stages as protecting groups.

The use of variously protected sulfate groups ($-SO_3R$) such as phenyl,[1,2] neopentyl,[3] isobutyl,[3] and 2,2,2-trifluoroethyl[4–6] groups has been described. A previous study[7] demonstrated the possibility to obtain protected sulfates by the oxidation of their corresponding alkyl sulfites. The 2,2,2-trichloroethyl sulfate ($-SO_3CH_2CCl_3$, $-SO_3TCE$)[8–10] groups reported by Taylor's group is promising. The TCE group may be removed under mild conditions, such as hydrogenolysis[9] or reduction with Zn.[8] Protected sulfation using 2,2,2-trichloroethylsulfuryl-1-methylimidazolium triflate[8] and 2,2,2-trichloroethylsulfuryl-1,2-dimethylimidazolium triflate,[10] as well as the stability of protected sulfate groups under various conditions have also been reported. We found unexpected reactions with SO_3TCE during glycan synthesis, which demonstrates the possibility of synthesizing sulfated oligosaccharides.[11,12] We also clarified regioselectivities in the introduction of protected sulfates to various di- and triols. In addition, we demonstrated that protected sulfates tolerate many reaction conditions, such as a weak base ($H_2NNH_2 \cdot AcOH$), reduction ($PhBCl_2$ with Et_3SiH), and oxidation [TEMPO (2,2,6,6-tetramethylpiperidine 1-oxyl)] with NaClO or BAIB (bisacetoxyiodobenzene)] but acted as potential leaving groups under more basic conditions.[11,12]

We herein describe the synthesis of methyl glucuronate with protected sulfates via TEMPO oxidation of the corresponding diol and subsequent esterification. The uronate with protected sulfates acts as a pivotal precursor for the synthesis of multiply sulfated oligosaccharides. The known bis-2,2,2-trichloroethyl sulfate (**1**)[10] was converted to diol (**2**)[10] by the acid hydrolysis of benzylidene acetal. We purified the product by recrystallization from $CHCl_3$ and obtained **2** as needles melting sharply at 140.5 – 141.0°C and showing $[\alpha]_D$ −19, which produced correct combustion analysis data. These physical constants are in sharp contrast with those found for the much lower melting, positively rotating "white solid," for which the proof of purity by combustion analysis was not provided.[10] We cannot explain the vast difference in the physical constant found for these two materials prepared independently, when our NMR data and those found previously are virtually identical.

The diol (2) was oxidized by TEMPO and BAIB to give the corresponding uronic acid, which was esterified with TMSCHN$_2$ to give methyl uronate (3)[13] in 82% yield over 2 steps. Compound 3 was further characterized as the corresponding 4-O-acetyl derivative. No side reactions were observed during the oxidation and acetylation.

EXPERIMENTAL

GENERAL METHODS

Optical rotations were measured at 22±3°C with the HORIBA automatic polarimeter SEPA-500. Melting points were measured with Büchi B-545 and were uncorrected. ^{1}H and ^{13}C NMR assignments were confirmed by two-dimensional HH COSY and HSQC experiments using Bruker AVANCE II 600 MHz spectrometers. ^{1}H and ^{13}C NMR chemical shifts were referenced to the signals of Me$_4$Si at 0 ppm. ESI-high resolution mass spectra were recorded using Exactive-Orbitrap (Thermo Fisher SCIENTIFIC) in the positive ion mode. Analytical thin-layer chromatography was performed using Silica gel 60 F$_{254}$ glass plates. Compounds were visualized by UV light (254 nm) and charring with a solution of Ce(NH$_4$)$_4$(SO$_4$)$_4$•2H$_2$O (10 g) and (NH$_4$)$_6$Mo$_7$O$_{24}$•4H$_2$O (15 g) in 6% H$_2$SO$_4$ (1 L). Silica gel chromatography was performed on columns of Silica gel 60 (Merck) and Wakogel C-300 (Wako Chemical). Sephadex LH-20 was purchased from GE Healthcare. TMSCHN$_2$ solution 2.0 M in Et$_2$O was purchased from Sigma-Aldrich. Solutions in organic solvents were washed consecutively with aq. NaHCO$_3$ and brine, and dried over anhyd. MgSO$_4$.

4-METHOXYPHENYL 2,3-BIS-O-(2,2,2-

TRICHLOROETHYL)SULFONYL-β-D-GLUCOPYRANOSIDE (2)

Camphorsulfonic acid (176 mg, 0.758 mmol) was added to a suspension of 4-methoxyphenyl 4,6-O-benzylidene-2,3-bis-O-(2,2,2-trichloroethyl)sulfonyl-β-D-glucopyranoside (1, 1.00 g, 1.25 mmol) in MeOH (10 mL) and CH$_2$Cl$_2$ (10 mL) and stirred at room temperature. The starting material gradually dissolved (~48 h) and after total of 11 days, when TLC (1:1 toluene–EtOAc) showed that the reaction was complete, the reaction was quenched by the addition of Et$_3$N (158 μL, 1.14 mmol), the crude mixture was concentrated under reduced pressure, and the residue was chromatographed (Silica gel 60, 100:0 to 96:4 v/v CH$_2$Cl$_2$–MeOH) to give pure compound 2 as needles (727 mg, 90%), mp 140.5–141.0°C (from CHCl$_3$); [α]$_D$ −19 (c 1.08, CHCl$_3$); R$_f$ 0.38, 1:1 toluene–EtOAc). Lit.[10] Yield, 80%, white solid, mp

62–64°C, $[\alpha]_D$ +64.3 (c 1.0, CHCl$_3$); ^{1}H NMR (CDCl$_3$) δ 7.02 (m, 2H, Ph), 6.85 (m, 2H, Ph), 5.08 (d, 1H, $J_{1,2}$ 7.5 Hz, H-1), 4.94, 4.92 (ABq, 2H, J 11.0 Hz, CH$_2$CCl$_3$), 4.91 (dd, 1 H, $J_{2,3}$ = $J_{3,4}$ 9.0 Hz, H-3), 4.86 (dd, 1H, $J_{2,3}$ 9.4 Hz, H-2), 4.84, 4.81 (ABq, 2H, J 11.1 Hz, CH$_2$CCl$_3$), 4.13 (ddd, 1H, $J_{3,4}$ = $J_{4,5}$ 9.1 Hz, $J_{4,OH}$ 4.0 Hz, H-4), 3.99 (ddd, 1H, $J_{5,6a}$ 3.8 Hz, $J_{6a,6b}$ 12.2 Hz, $J_{6a,OH}$ 5.4 Hz, H-6a), 3.93 (ddd, 1H, $J_{5,6b}$ 3.8 Hz, $J_{6b,OH}$ 7.7 Hz, H-6b), 3.78 (s, 3H, PhOMe), 3.56 (ddd, 1H, H-5), 3.29 (d, 1H, OH-4), 1.94 (dd, 1H, OH-6); ^{13}C NMR (CDCl$_3$) δ 156.3 (qPh), 149.9 (qPh), 118.6 (tPh), 114.8 (tPh), 98.9 (C-1), 92.5 (CCl$_3$), 92.4 (CCl$_3$), 86.1 (C-3), 80.6, 80.6, 80.4 (C-2, 2CH$_2$CCl$_3$), 74.9 (C-5), 69.2 (C-4), 61.6 (C-6), 55.7 (PhOMe); HRMS (ESI): [M + K]$^+$ calcd for C$_{17}$H$_{21}$Cl$_6$O$_{14}$S$_2$K, 744.8111; found, 744.8095. Anal. calcd for C$_{17}$H$_{21}$Cl$_6$O$_{14}$S$_2$: C, 28.75; H, 2.98. Found: C, 28.81; H, 2.87.

METHYL {4-METHOXYPHENYL 2,3-BIS-O-(2,2,2-TRICHLOROETHYL)SULFONYL-β-D-GLUCOPYRANOSYL}URONATE (3)[13]

TEMPO (378 mg, 2.42 mmol) and BAIB (9.76 g, 30.3 mmol) were added to a two-phase, stirred solution of 2 (8.57 g, 12.1 mmol) in CH$_2$Cl$_2$ (180 mL) and H$_2$O (90 mL), and the stirring was continued for 1.5 h. A fresh portion of BAIB (1.95 g, 6.05 mmol) was added and the mixture was stirred until TLC (1:1 toluene–EtOAc, and 1:5 toluene–EtOAc with 5% AcOH) analysis showed that the conversion was complete (1.5 h). Na$_2$S$_2$O$_3$ solution (1 M, 380 mL) was added, and the mixture was extracted with EtOAc. The organic phase was washed with brine, dried, and concentrated. To a cooled (0°C) solution of the residue in toluene (570 mL) and MeOH (190 mL) was added a 2.0-M solution of TMSCHN$_2$ in Et$_2$O (18 mL) and the mixture was stirred for 30 min. After concentration, the residue was chromatographed (Silica gel 60, 1:0–1:2 toluene-EtOAc) to give 3 (7.82 g, 82% over 2 steps), mp 161.0–161.5°C (from toluene); $[\alpha]_D$ –4.8 (c 0.50, CHCl$_3$); (R_f 0.50, 3:1 toluene-EtOAc); ^{1}H NMR (CDCl$_3$) δ 7.08 (m, 2H, Ph), 6.85 (m, 2H, Ph), 5.08 (m, 1H, H-1), 4.94, 4.91 (ABq, 2H, J 10.9 Hz, CH$_2$CCl$_3$), 4.90 (m, 2H, H-2,3), 4.87, 4.84 (ABq, 2H, J 11.1 Hz, CH$_2$CCl$_3$), 4.27 (m, 1H, H-4), 4.03 (d, 1H, $J_{4,5}$ 9.7 Hz, H-5), 3.86 (s, 3H, COOMe), 3.79 (s, 3H, PhOMe), 3.64 (d, 1H, $J_{4,OH}$ 2.8 Hz, OH-4); ^{13}C NMR (CDCl$_3$) δ 168.5 (C-6), 119.0 (Ph), 114.8 (Ph), 99.5 (C-1), 83.9 (C-3), 80.5, 80.5, 80.4 (C-2, 2CH$_2$CCl$_3$), 73.1 (C-5), 69.7 (C-4), 55.7 (PhOMe), 53.5 (COOMe); HRMS (ESI): [M + K]$^+$ calcd for C$_{18}$H$_{20}$Cl$_6$O$_{14}$S$_2$K, 772.8063; found, 772.8044. Anal. calcd for C$_{18}$H$_{20}$Cl$_6$O$_{14}$S$_2$: C, 29.33; H, 2.73. Found: C, 29.45; H, 2.78.

METHYL {4-METHOXYPHENYL 4-O-ACETYL-2,3-BIS-O-(2,2,2-TRICHLOROETHYL)SULFONYL-β-D-GLUCOPYRANOSYL}URONATE (4)

A mixture of acetic anhydride (1 mL) and 3 (82.5 mg, 112 μmol) in pyridine (1 mL) was stirred overnight. Volatiles were removed under diminished pressure and the residue was chromatographed (Wakogel C-300, 50:1–30:1 toluene-EtOAc) to give 4 (70.8 mg, 81%), mp 122.4 – 124.6°C (from toluene-n-hexane); $[\alpha]_D$ –13 (c 0.88, CHCl$_3$);

(R_f 0.68, 3:1 toluene-EtOAc). ^{1}H NMR (CDCl$_3$) δ 7.04 (m, 2H, Ph), 6.84 (m, 2H, Ph), 5.48 (dd, 1H, $J_{3,4}$ = $J_{4,5}$ 9.4 Hz, H-4), 5.13 (d, 1H, $J_{1,2}$ 7.1 Hz, H-1), 5.07 (dd, 1H, $J_{2,3}$ =$J_{3,4}$ 9.2 Hz, H-3), 4.95 (dd, 1H, H-2), 4.90, 4.83 (ABq, 2H, J 11.1 Hz, CH$_2$CCl$_3$), 4.84, 4.81 (ABq, 2H, J 11.0 Hz, CH$_2$CCl$_3$), 4.14 (d, 1H, $J_{4,5}$ 9.6 Hz, H-5), 3.78 (s, 3H, PhOMe), 3.74 (s, 3H, COOMe), 2.15 (s, 3H, Ac); ^{13}C NMR (CDCl$_3$) δ 169. 2 (CO), 166.0 (CO), 156.5 (Ph), 149.7 (Ph), 119.1 (Ph), 114.8 (Ph), 99.1 (C-1), 92.4 (2CH$_2$CCl$_3$), 81.7 (C-3), 80.8, 80.5 (CH$_2$CCl$_3$), 80.2 (C-2), 72.1 (C-5), 68.5 (C-4), 55.7 (PhOMe), 53.3 (COOMe), 20.5 (COCH$_3$); HRMS (ESI): [M + Na]$^+$ calcd for C$_{20}$H$_{22}$O$_{15}$S$_2$Cl$_6$Na, 798.8429; found, 798.8411. Anal. calcd for C$_{20}$H$_{22}$O$_{15}$S$_2$Cl$_6$: C, 30.83; H, 2.85. Found: C, 30.88; H, 2.81.

ACKNOWLEDGMENTS

This study was supported by Otsuka Chemical Co., Ltd., Otsuka Pharmaceutical Co., Ltd., and by a Grant-in-Aid for Scientific Research in Innovative Areas (23110003) from MEXT, Japan. The authors thank Mrs. Mayumi Ikenari, Mrs. Mizuki Yokono, and Mrs. Miyuki Tanmatsu (Research Center for Bioscience and Technology, Division of Instrumental Analysis, Tottori University) for performing ESI-HRMS and the elemental analysis.

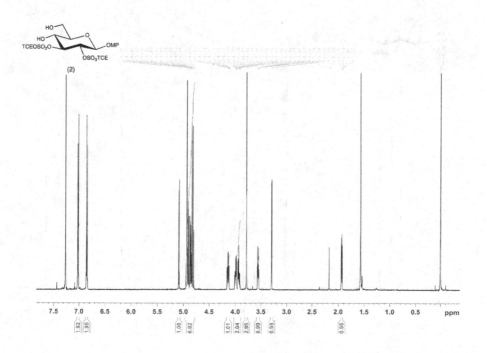

(2)

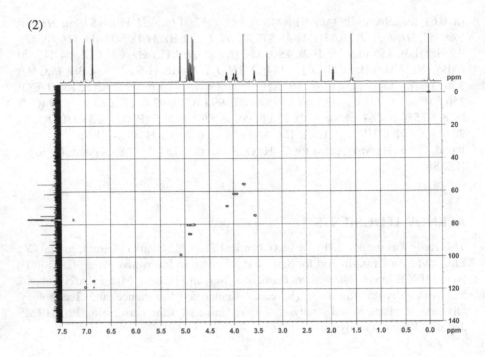

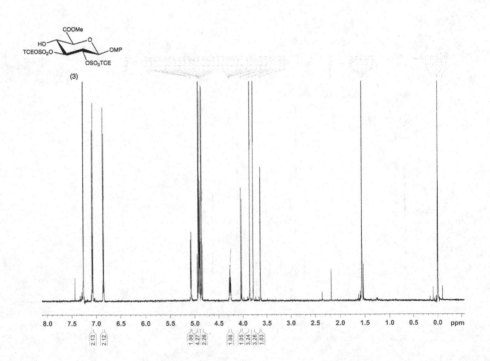

(3)

(3)

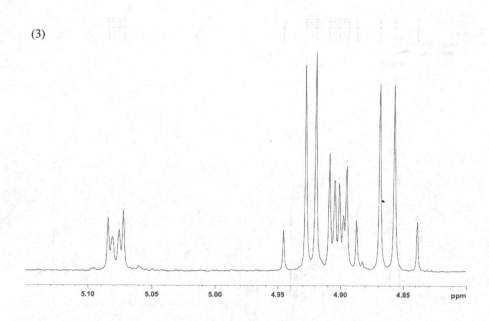

(3)

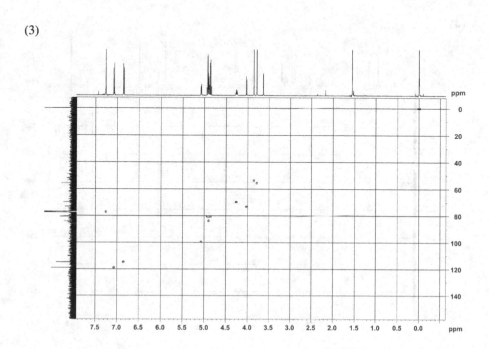

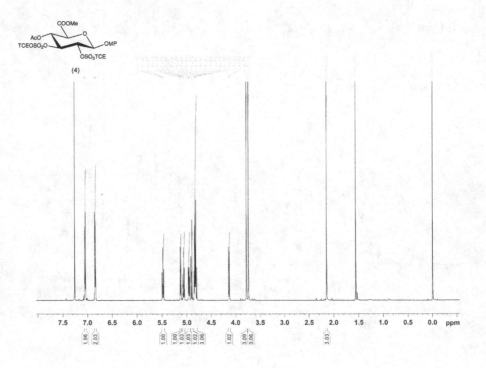

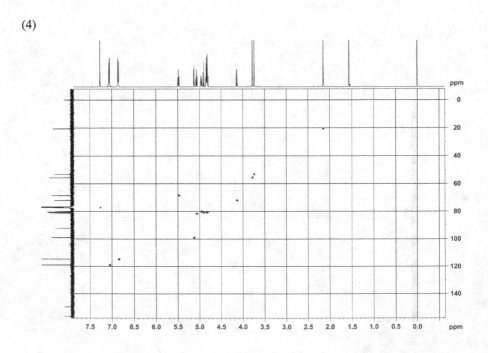

REFERENCES

1. Penney, C. L.; Perlin, A. S. *Carbohydr. Res.* **1981**, *93*, 241–246.
2. Abdel-Malik, M. M.; Perlin, A. S. *Carbohydr. Res.* **1989**, *190*, 39–52.
3. Simpson, L. S.; Widlanski, T. S. *J. Am. Chem. Soc.* **2006**, *128*, 1605–1610.
4. Karst, N. A.; Islam, T. F.; Avic F. Y.; Linhardt, R. J. *Tetrahedron Lett.* **2004**, *45*, 6433–6437.
5. Liu, Y.; Lien, I-F. F.; Ruttgaizer, S.; Dove, P.; Taylor, S. D. *Org. Lett.* **2004**, *6*, 209–212.
6. Proudd, A. D.; Prodger, J. C.; Flitsch, S. L. *Tetrahedron Lett.* **1997**, *38*, 7243–7246.
7. Huibers, M.; Manuzi, Á.; Rutjes, F. P. J. T.; van Delft, F. L. *J. Org. Chem.* **2006**, *71*, 7473–7476.
8. Ingram, L. J.; Taylor, S. D. *Angew. Chem. Int. Ed.*, **2006**, *45*, 3503–3506.
9. Ingram, L. J.; Desoky, A. Y.; Ali, A. M.; Taylor, S. D. *J. Org. Chem.* **2009**, *74*, 6479–6485.
10. Desoky, A. Y.; Taylor, S. D. *J. Org. Chem.* **2009**, *74*, 9406–9412.
11. Matsushita, K.; Sato, Y.; Funamoto, S.; Tamura, J. *Carbohydr. Res.* **2014**, *396*, 14–24.
12. Matsushita, K.; Nakata, T.; Tamura, J. *Carbohydr. Res.* **2015**, *406*, 76–85.
13. Tamura, J.; Sorajo, K.; Akagi, H.; Sato, Y.; Goto, F. *Jpn. Kokai Tokkyo Koho*, **2014**, JP 2014047155A.

13 Synthesis of Allyl α-(1→2)-Linked α-Mannobioside from a Common 1,2-Orthoacetate Precursor

Nino Trattnig, and Paul Kosma*
Department of Organic Chemistry, University of Natural Resources and Life Sciences,
Vienna, Austria

Aisling Ní Cheallaigh[a]
School of Chemistry, University College Dublin,
Belfield, Dublin 4, Ireland

CONTENTS

Oligomannosides are common constituents of plant, invertebrate, and vertebrate *N*-glycoproteins such as in high mannose-, hybrid-, paucimannosidic-, and complex-type glycans but also occur as antigenic determinants in the viral glycan shield of HIV 1.[1-3] Furthermore, α-mannosides are frequently found components of bacterial and fungal polysaccharides.[4-5] Glycosylation reactions leading to α-mannosides are well developed, but there is still a need to improve the assembly of oligomannosides with respect to scalability, ease of purification, and simplification of protecting group patterns.[6]

---

* Corresponding author: paul.kosma@boku.ac.at
[a] Checker: aisling.nicheallaigh@ucd.ie; present contact: aisling.nicheallaigh@manchester.ac.uk

Trichloroacetimidate donor **3** has previously been prepared in four steps from benzoylated 1,2-*O*-ethylidene-β-D-mannopyranose in 68% yield[7] but can also be directly obtained in a two-step sequence from the related 1,2-orthoacetate **1** in near theoretical yield.[8] For the preparation of donor **3**, orthoester **1** is first treated with aqueous AcOH to give the known 2-*O*-acetyl derivative **2**.[7,9] Any volatiles can be removed by simple evaporation and the intermediate lactol **2** can be directly treated with CCl₃CN and solid K₂CO₃. After complete conversion into the imidate donor **3**, the solid base is removed by filtration and the filtrate is taken to dryness. The crude donor **3** is of sufficient purity for a subsequent glycosylation reaction. The glycosyl acceptor **4** can also be generated from orthoester **1** exploiting a simultaneous introduction of the aglycon and cleavage of the 2-*O*-acetyl protecting group. Upon activation with TMSOTf direct formation of α-(1→2)-linked disaccharides from peracetylated 1,2-orthoester derivatives in modest to fair yields has been reported.[10-12] Treatment of the orthoester **1** with allyl alcohol and stoichiometric amounts of BF₃·Et₂O, however, afforded the partially deprotected allyl glycoside **4** in 57% yield as a 6.3:1 α/β anomeric mixture. Previously, compound **4** had been synthesized from a perbenzoylated ethylidene acetal in 21% yield over four steps[12] or in two steps and 87% yield from a less readily available trichloroacetimidate precursor.[13] Coupling of donor **3** with acceptor **4** promoted by TMSOTf gave selectively the known α-(1→2)-connected mannobioside **5**, which was isolated by column chromatography in 76% yield.[14-16] Compound **5** allows for a further elongation at position 2 upon selective de-*O*-acetylation as well as for a conversion into a disaccharide glycosyl donor after selective cleavage of the anomeric allyl group.

EXPERIMENTAL

GENERAL METHODS

All purchased chemicals were used without further purification unless stated otherwise. BF₃·Et₂O was used as a solution in diethyl ether (≥46% as per the manufacturer). CH₂Cl₂ was dried over activated 4 Å. Concentration of organic solutions was performed under reduced pressure <40°C. Optical rotations were measured with a PerkinElmer 243 B or a Anton Paar MCP100 polarimeter. $[\alpha]_D^{20}$ values are given in units of 10^{-1} deg cm^2 g^{-1}. All reactions were monitored by thin layer chromatography on Merck pre-coated plates (5×10 cm^2, layer thickness: 0.25 mm, silica gel 60F254). Spots were detected by dipping into anisaldehyde-H_2SO_4 followed by charring on a hot plate. For column chromatography, silica gel (0.040–0.063 mm) was used. NMR spectra were recorded on a Bruker Avance III 600 instrument (600.2 MHz for ^1H, 150.9 MHz for ^{13}C) using standard Bruker NMR software. ^1H spectra were referenced to $\delta = 0$ using the TMS signal and ^{13}C spectra were referenced to 77.00 for solutions in $CDCl_3$. Assignments were based on gCOSY, gHSQC, and gHMBC experiments. ESI-MS data were obtained on a Waters Micromass Q-TOF Ultima Global instrument.

ALLYL 2-O-ACETYL-3,4,6-TRI-O-BENZOYL-α-D-MANNOPYRANOSYL-(1→2)-3,4,6-TRI-O-BENZOYL-α-D-MANNOPYRANOSIDE (5)

Compound $\mathbf{1}^{17}$ (0.7 g; 1.28 mmol) was dissolved in acetone (3 mL), and 60% aq AcOH (12 mL) was added. The resulting suspension was stirred at room temperature until the mixture turned into a clear solution (~6 h). The solvent was removed *in vacuo* and the residue was co-evaporated with toluene three times and further dried for 16 h under vacuum (0.5 mbar) to give **2** (0.683 g) as a colorless amorphous glass. The crude 2-O-acetyl-3,4,6-tri-O-benzoyl-D-mannopyranose (**2**), thus obtained, was used in the next steps without further purification. $R_f = 0.11$ (3:1 hexane–EtOAc). NMR data was in accordance to literature data[7]; ^1H NMR (600 MHz, $CDCl_3$): δ 8.07–8.04 (m, 2*H*, Ar), 7.96–7.94 (m, 2*H*, Ar), 7.92–7.89 (2*H*, Ar), 7.59–7.47 (m, 3*H*, Ar), 7.42–7.33 (m, 6*H*, Ar), 5.96 (t, 1*H*, $J_{3,4} = J_{4,5}$ 10.0 Hz, H-4) 5.87 (dd, 1*H*, $J_{3,4}$ 10.0, $J_{3,2}$ 3.5 Hz, H-3), 5.50 (dd, 1*H*, $J_{2,3}$ 3.5 Hz, $J_{2,1}$ 1.9 Hz, H-2), 5.37 (bd, 1*H*, $J_{1,2}$ 1.9 Hz, H-1), 4.66 (dd, 1*H*, $J_{6a,6b}$ 12.2 Hz, $J_{6a,5}$ 2.9 Hz, H-6a), 4.45 (dd, 1*H*, $J_{6b,6a}$ 12.2 Hz, $J_{6b,5}$ 4.6 Hz, H-6b), 4.65 (ddd, 1*H*, $J_{5,4}$ 10 Hz, $J_{5,6b}$ 4.6 Hz, $J_{5,6a}$ 2.9 Hz, H-5), 2.11 (s, 3*H*, CH_3CO); ^{13}C NMR (75 MHz, $CDCl_3$): δ 170.2 (CH_3CO), 166.6, (ArCO), 165.7 (2 C, ArCO), 133.6, 133.4, 133.3 130.0, 129.9, 129.8, 129.3, 129.3, 129.12, 128.6, 128.5, 128.5 (18 C, C-Ar), 92.3 (C-1), 70.6 (C-2), 69.8 (C-3), 69.0 (C-5), 67.2 (C-4), 63.3 (C-6), and 20.9 (CH_3CO).

K_2CO_3 (0.531 g; 3.84 mmol) was added to a solution of **2** (0.683 g; 1.28 mmol) in dry CH_2Cl_2 (6 mL) and the suspension was stirred under argon at room temperature for 10 min. CCl_3CN (0.513 mL; 5.12 mmol) was added dropwise, and the stirring was continued for 5 h. After filtration through a bed of silica (10:1 CH_2Cl_2–MeOH), the solvent was removed *in vacuo* to afford 2-O-acetyl-3,4,6-tri-O-benzoyl-D-mannopyranosyl 2,2,2-trichloroacetimidate (**3**) (0.826 g; 95% from **1**) as a colorless amorphous solid. $R_f = 0.28$ (20:1 toluene–EtOAc). NMR data was in accordance to literature[7,16]; ^1H NMR (600 MHz, $CDCl_3$): δ 8.82 (s, 1*H*, NH), 8.04–8.02 (m, 2*H*, Ar), 7.97–7.94 (m, 2*H*, Ar), 7.91–7.88 (m, 2*H*, Ar), 7.56–7.48 (m, 3*H*, Ar), 7.42–7.34 (m, 6*H*, Ar), 6.42 (d, 1*H*, $J_{1,2}$ 1.9 Hz, H-1), 6.02 (dd, 1*H*, $J_{4,3} = J_{4,5}$ 10.3 Hz, H-4), 5.85 (dd, 1*H*, $J_{3,4}$ 10.3 Hz, $J_{3,2}$ 3.4 Hz, H-3), 5.71 (dd, 1*H*, $J_{2,3}$ 3.4 Hz, $J_{2,1}$ 1.9 Hz, H-2), 4.64 (dd, 1*H*, $J_{6a,6b}$ 12.0 Hz, $J_{6a,5}$ 3.2 Hz, H-6a), 4.57 (ddd, 1*H*, $J_{5,4}$ 10.3 Hz, $J_{5,6b}$ 4.9 Hz,

$J_{5,6a}$ 3.2 Hz, H-5), 4.48 (dd, 1H, $J_{6b,6a}$ 12.0 Hz, $J_{6b,5}$ 4.9 Hz, H-6b), 2.14 (s, 3H, CH_3CO); ^{13}C NMR (150 MHz, CDCl$_3$): δ 169.6 (CH$_3$CO), 166.2 (ArCO), 165.6 (2 C, ArCO), 160.0 (C=NH), 133.7, 133.6, 133.3, 130.0, 129.9, 129.8, 129.1, 128.9, 128.6, 128.5 (18 C, C-Ar), 94.7 (C-1), 71.7 (C-5), 69.8 (C-3), 68.5 (C-2), 66.5 (C-4), 62.9 (C-6), and 20.7 (CH_3CO); ESI-TOF-MS: m/z calcd for C$_{29}$H$_{25}$O$_9$ [M–O(C=N)CCl$_3$]$^+$: 517.1493; found 517.1517.

Allyl alcohol (0.43 mL; 6.40 mmol) was added to a suspension of 1^{17} (0.7 g; 1.28 mmol) and powdered freshly activated MS 4 Å (1 g) in dry CH$_2$Cl$_2$ (7 mL) and the mixture was stirred under argon at room temperature for 30 min. After cooling (0°C), BF$_3$·Et$_2$O (0.162 mL; 1.28 mmol) was added dropwise, the suspension was allowed to warm to room temperature, and the stirring was continued for 16 h. The suspension was filtered through Celite®, the filtrate was concentrated, and the crude product was chromatographed (10:1 toluene–EtOAc) to give allyl 3,4,6-tri-O-benzoyl-D-mannopyranoside (4) (0.388 g; α/β = 6.3:1; 57%) as colorless foam. R_f = 0.10 (20:1 toluene–EtOAc). NMR data was in accordance to literature data^{13}; ^{1}H NMR (600 MHz, CDCl$_3$): δ 8.04–8.01 (m, 2H, Ar), 7.98–7.96 (m, 2H, Ar), 7.95–7.93 (2H, Ar), 7.55–7.47 (m, 3H, Ar), 7.41–7.33 (m, 6H, Ar), 6.00–5.93 (m, 1H, CH=CH$_2$), 5.94 (t, 1H, $J_{4,3}$ = $J_{4,5}$ 10.0 Hz, H-4), 5.71 (dd, 1H, $J_{3,4}$ 10.0 Hz, $J_{3,2}$ 3.4 Hz, H-3), 5.34 (dq, 1H, $^3J_{trans}$ 17.2 Hz, ^4J = ^2J 1.5 Hz, CH=CH_2), 5.25 (dq, 1H, $^3J_{cis}$ 10.5 Hz, ^4J = ^2J 1.5 Hz, CH=CH_2), 5.04 (d, 1H, $J_{1,2}$ 1.7 Hz, H-1), 4.59 (dd, 1H, $J_{6a,6b}$ 12.1 Hz, $J_{6a,5}$ 3.0 Hz, H-6a), 4.49 (dd, 1H, $J_{6b,6a}$ 12.1 Hz, $J_{6b,5}$ 5.5 Hz, H-6b), 4.39 (ddd, 1H, $J_{5,4}$ 10.0 Hz, $J_{5,6b}$ 5.5 Hz, $J_{5,6a}$ 3.0 Hz, H-5), 4.34 (dd, 1H, $J_{2,3}$ 3.4 Hz, $J_{2,1}$ 1.7 Hz, H-2), 4.31 (ddt, 1H, ^2J 12.8 Hz, ^3J 5.0 Hz, ^4J 1.5 Hz, OCH_2CH=CH$_2$), 4.12 (ddt, 1H, ^2J 12.8 Hz, ^3J 6.3 Hz, ^4J 1.5 Hz OCH_2CH=CH$_2$); ^{13}C NMR (150 MHz CDCl$_3$): δ 166.3, 165.8, 165.7 (3 C, COAr), 133.5, 133.5, 133.4, 133.2, 130.0, 130.0, 129.9, 129.4, 129.3, 128.6, 128.5, 128.5 (19 C, C-Ar, CH=CH$_2$) 118.9 (CH=CH_2), 98.9 (C-1), 72.8 (C-3), 69.7 (C-2), 69 (C-5), 68.8 (OCH_2CH=CH$_2$), 67.2 (C-4), 63.7 (C-6); ESI-TOF-MS: m/z calcd for C$_{30}$H$_{29}$O$_9$ [M + H]$^+$: 533.1806; found 533.1822.

Dry CH$_2$Cl$_2$ (3 mL) was added to a suspension of 3 (0.382 g; 0.56 mmol), 4 (0.250 g; 0.47 mmol; α/β = 6.3:1), and molecular sieves 4 Å (0.3 g) the suspension was stirred under argon at room temperature for 10 min. TMSOTf (8 µL; 0.05 mmol) was added dropwise, and the stirring was continued for 2 h at room temperature. The reaction was quenched by the addition of NEt$_3$ (50 µL), the mixture was filtered over a bed of Celite® and the solvent was removed in vacuo. The crude product was chromatographed (20:1 toluene–EtOAc) to give the title disaccharide 5 (0.372 g; 76%) as colorless amorphous solid. R_f = 0.26 (20:1 toluene–EtOAc), $[\alpha]_D^{20}$ +3.6 (c 1.1, CHCl$_3$); $[\alpha]_D^{20}$ +3.0 (c 1.3, CHCl$_3$)^{11}; NMR data were in accordance with literature data^{12,16}; ^{1}H NMR (600 MHz, CDCl$_3$): δ 8.10–7.92 (m, 12H, Ar), 7.53–7.32 (m, 18H, Ar), 5.99 (t, 1H, $J_{3,4}$ = $J_{4,5}$ 9.5 Hz, H-4^{II}), 5.92–5.83 (m, 4H, H-3^I, H-3^{II}, H-4^I, CH=CH$_2$), 5.71 (dd, 1H, $J_{2,3}$ 3.1 Hz, $J_{2,1}$ 2.0 Hz, H-2^{II}), 5.27 (dq, 1H, $^3J_{trans}$ 17.2 Hz, ^4J = ^2J 1.4 Hz, CH=CH_2), 5.20 (dq, 1H, $^3J_{cis}$ 10.5 Hz, ^4J = ^2J 1.4 Hz, CH=CH_2), 5.17 (d, 1H, $J_{1,2}$ 1.9 Hz, H-1^I), 5.12 (d, 1H, $J_{1,2}$ 2.0 Hz, H-1^{II}), 4.63 (dd, 1H, $J_{6a,6b}$ 11.7 Hz, $J_{5,6a}$ 2.9 Hz, H-6^{II}a), 4.60–4.52 (m, 3H, H-6^{II}b,

H-6^{I}a, H-5^I), 4.48 (dd, 1H, $J_{6b,6a}$ 11.7 Hz, $J_{6b,5}$ 5.5 Hz, H-6^{I}b), 4.40–4.37 (m, 1H, H-5^{II}), 4.38 (dd, 1H, $J_{2,3}$ 3.4 Hz, $J_{2,1}$ 1.9 Hz, H-2^I), 4.18 (ddt, 1H, ^2J 12.8 Hz, ^3J 6.0 Hz, ^4J 1.4 Hz, OCH_2CH=CH$_2$), 3.92 (ddt, 1H, ^2J 12.8 Hz, ^3J 6.3 Hz, ^4J 1.4 Hz OCH_2CH=CH$_2$), 2.04 (s, 3H, CH_3CO); ^{13}C NMR (150 MHz, CDCl$_3$): δ 169.4 (CH$_3$$C$O), 166.4, 166.2, 165.8, 165.7, 165.4, 165.1 (ArCO), 133.6, 133.5, 133.4, 133.3, 133.2, 133.1, 130.1, 130.0, 129.9, 129.8, 129.5, 129.2, 129.0, 128.7, 128.6, 128.5 (37 C, C-Ar, CH=CH$_2$), 118.2 (CH=CH$_2$), 99.7 (C-1^{II}), 98.0 (C-1^I), 77.0 (C-2^I), 71.0 (C-3^{II}), 69.9 (C-5^I), 69.7 (2 C, C-2^{II}, C-3^I), 69.1 (C-5^{II}), 68.8 (OCH$_2$–CH=CH$_2$), 67.7 (C-4^{II}), 67.3 (C-4^I), 63.8 (C-6^{II}), 63.5 (C-6^I); ESI-TOF-MS: m/z calcd for C$_{59}$H$_{56}$NO$_{18}$ [M + NH$_4$]$^+$: 1066.3492; found 1066.3496. Anal. calcd for C$_{59}$H$_{52}$O$_{18}$: C, 67.55; H, 5.00; found: C, 67.42, H, 5.02.

ACKNOWLEDGMENTS

Financial support of this work by the Austrian Science Fund FWF (grant P26919-N28) is gratefully acknowledged.

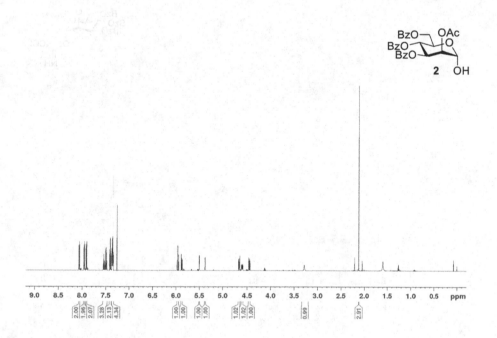

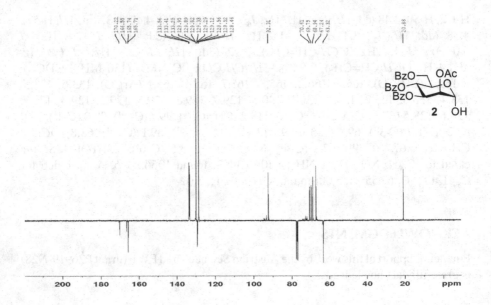

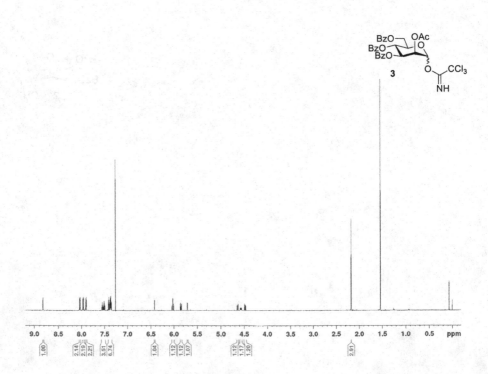

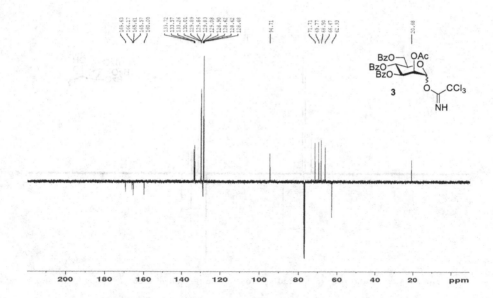

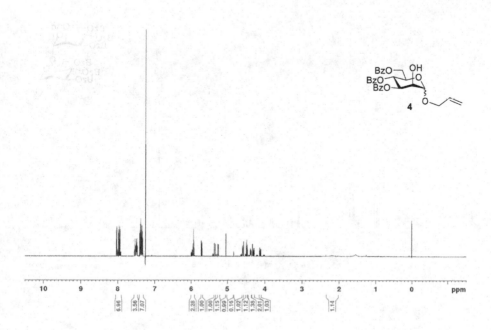

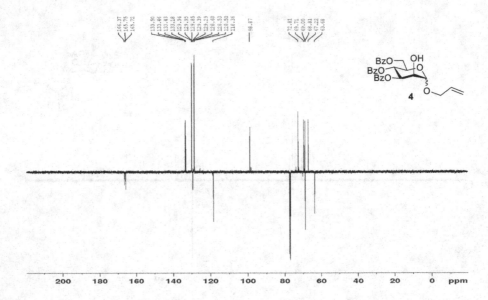

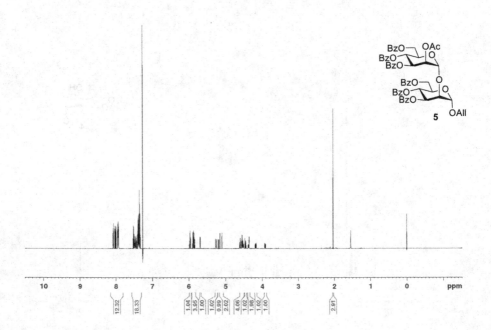

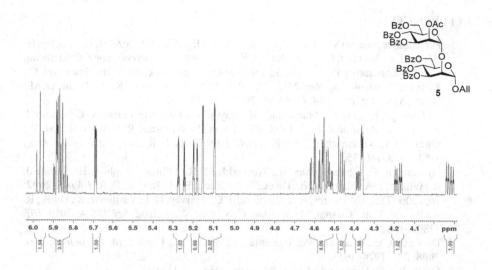

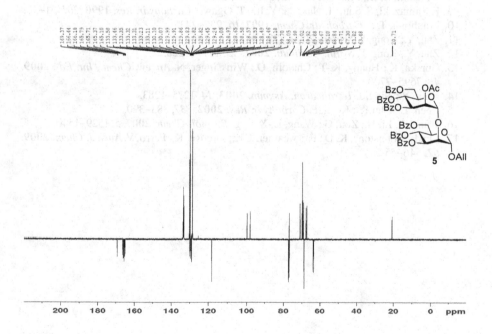

REFERENCES

1. For examples see: (a) Varki, A.; Cummings, R. D.; Esko, J. D.; Freeze, H. H.; Stanley, P.; Bertozzi, C. R.; Hart, G. W.; Etzler, M. E. In *Essentials of Glycobiology*, Cold Spring Harbour Laboratory Press, Cold Spring Harbour, New York, **2009**; (b) Unverzagt, C.; Kajihara, Y. *Chem. Soc. Rev.* **2013**, *42*, 4408–4420; (c) Moremen, K. W.; Tiemeyer, M.; Nairn, A. V. *Nat. Rev. Mol. Cell Biol.* **2012**, *13*, 448–462.
2. (a) Lerouge, P.; Cabanes-Macheteau, M.; Rayon, C.; Fischette-Laine, A.-C.; Gomord, V.; Faye, L. *Plant Mol. Biol.* **1998**, *38*, 31–48; (b) Williams, P. J.; Wormald, M. R.; Dwek, R. A.; Rademacher, T. W.; Parker, G. F.; Roberts, D. R., *Biochim. Biophys. Acta*, **1991**, *1075*, 146–153.
3. (a) Scanlan, C. N.; Pantophlet, R.; Wormald, M. R.; Ollmann Saphire, E.; Stanfield, R.; Wilson, I. A.; Katinger, H.; Dwek, R. A.; Rudd, P. M.; Burton, D. R. *J. Virol.* **2002**, *76*, 7306–7321; (b) Doores, K. J.; Bonomelli, C.; Harvey, D. J.; Vasiljevic, S.; Dwek, R. A.; Burton, D. R.; Crispin, M.; Scanlan, C. N. *Proc. Natl. Acad. Sci. U.S.A.* **2010**, *107*, 13800–13805.
4. De Castro, C.; Molinaro, A.; Lanzetta, R.; Silipo, A.; Parrilli, M. *Carbohydr. Res.* **2008**, *343*, 1924–1933.
5. Gorin, P. *Adv. Carbohydr. Chem. Biochem.* **1981**, *38*, 13–104.
6. For examples see: (a) T. Ogawa, T. Nukada, *Carbohydr. Res.* **1985**, *136*, 135–152.; (b) X. Geng, V. Y. Dudkin, M. Mandal, S. J. Danishefsky, *Angew. Chem. Int. Ed.*, **2004**, *43*, 2562–2565; (c) D. Ratner, O. Plante, P. H. Seeberger, *Eur. J. Org. Chem.* **2002**, 826–833; (d) O. Kanie, Y. Ito, T. Ogawa, *J. Am. Chem. Soc.* **1994**, *116*, 12073–12074; (e) J. Zhang, Z. Ma, F. Kong, *Carbohydr. Res.* **2003**, *338*, 1711–1718.
7. Heng, L.; Ning, J.; Kong, F. *J. Carbohydr. Chem.* **2001**, *20*, 285–296.
8. Johnston, B. D.; Pinto, B. M. *J. Org. Chem.* **2000**, *65*, 4607–4617.
9. F. Yamazaki, S. Sato, T. Nukada, Y. Ito, T. Ogawa, *Carbohydr. Res.* **1990**, *201*, 31–50.
10. Lindhorst, T. *J. Carbohydr. Chem.* **1997**, *16*, 237–243.
11. Zhu, Y.; Kong, F. *Synlett*, **2000**, 1783–1787.
12. Zhu, Y.; Kong, F. *Synth. Commun.* **2002**, *32*, 1219–1226.
13. Gorska, K.; Huang, K.-T.; Chaloin, O.; Winssinger, N. *Angew. Chem., Int. Ed.*, **2009**, *48*, 7695–7700.
14. Xing, Y.; Ning, J. *Tetrahedron: Asymm.* **2003**, *14*, 1275–1283.
15. Chen, L.; Zhu, Y.; Kong, F. *Carbohydr. Res.* **2002**, *337*, 383–390.
16. Wang, J.; Li, H.; Zou, G.; Wang, L.-X. *Org. Biomol. Chem.* **2007**, *5*, 1529–1540.
17. Liu, L.; Johnstone, K. D.; Fairweather, J. K.; Dredge, K.; Ferro, V. *Aust. J. Chem.* **2009**, *62*, 546–552.

14 Synthesis of 2^I-O-Propargylcyclo-Maltoheptaose and Its Peracetylated and Hepta(6-O-Silylated) Derivatives

Juan M. Casas-Solvas*, and Antonio Vargas-Berenguel*
Department of Chemistry and Physics, University of Almería, Almería, Spain

Milo Malanga[a]
Cyclolab Ltd,
Budapest, Hungary

CONTENTS

Cyclomaltoheptaose or β-cyclodextrin (β-CD) is a naturally occurring cyclooligosaccharide composed of seven α-(1–4)-linked D-glucopyranose units. In aqueous solutions, it can form inclusion complexes with many hydrophobic organic molecules of suitable size and geometry. In last decades, β-CD has become a valuable building block for a variety of nanodevices which operate through the entrapment of organic

---

* Corresponding authors: jmcasas@ual.es; avargas@ual.es
[a] Checker: malaga@cyclolab.hu

109

Reagents and conditions: (a) Propargyl bromide, LiH, LiI, DMSO, 55 °C, 15-30 h, (33-37%). (b) Ac₂O, Py, 4-DMAP, rt, 12 h, (81%). (c) TBDMSCl, Py, rt, 24-48 h, (75%). Glucose moieties are identified by Roman numerals for NMR assignment purposes.

molecules in water, such as molecular sensors, switches, and delivery systems.[1–6] Many β-CD applications require the covalent linkage of one or more functional appendages on its structure to provide biological, photochemical, catalytic, or redox abilities. In other cases, attaching the macrocycle to bigger structures such as dendrimers, polymers, surfaces, or nanoparticles is desired. The three types of hydroxyl groups available on the two different sides of β-CD—seven primary hydroxyl groups at the narrower rim (OH-6) and fourteen secondary ones at the wider opening (OH-2 and OH-3)—offer convenient modification alternatives to achieve such aims. However, differences in the reactivity of these OH groups are sometimes subtle, and obtaining homogeneous regioisomers or face-selective derivatives can be challenging, especially when monofunctionalization is needed.[7,8] Alkylation or substitution of just one of the twenty-one OH groups is pivotal in the construction of sensors,[4] switchers,[5] and dimers,[9] among others. Supramolecular properties of the resulting conjugate can be quite contrasting, depending on which rim is modified,[9–11] given that guest molecules preferably penetrate the cavity through the wider opening. However, examples of regioselective modification on the secondary face are considerably rarer than those involving OH-6 groups, due to the marked difference in reactivity between the primary and secondary hydroxyl groups.[7,8]

Herein, we present the synthesis of 2¹-O-propargylcyclomaltoheptaose 2 as a convenient building block for the construction of β-CD derivatives functionalized on the secondary face.[12] The terminal alkyne group offers the possibility to attach a variety of appendages.[12–28] Compound 2 can be prepared directly from native β-CD 1 without using protecting groups, by treatment with propargyl bromide and a judiciously measured amount of lithium hydride in dry DMSO in the presence of catalytic lithium iodide. These conditions lead mainly to regioselective alkylation at C-2 of the macrocycle, despite multiple hydroxyl groups are present.[29] It is known that the concentration of the base and the solvent affect the regioselectivity of the partial alkylation of β-CD.[30] Because species other than the 2-*O*-monopropargylated derivative are also formed, isolation of the desired product from the reaction mixture remains an intricate problem. In our case, thin layer chromatography (TLC) developed with 10:5:2 $CH_3CN–H_2O$–30% aq. NH_3 showed four spots with R_f values of 0.75, 0.66, 0.50, and 0.31, the last two corresponding to 2¹-*O*-propargylcyclomaltoheptaose **2** and the unchanged starting β-CD **1**, respectively (Figure 14.1). This mixture was chromatographed, giving **2** in 37% yield. In addition, 41% of starting β-CD was recovered in purity sufficient to be reused, increasing the yield of **2** to 62% based on the β-CD consumed. Due to the loss of the macrocycle C_7 symmetry upon the monosubstitution, ^{1}H and ^{13}C NMR spectra of the product are complex. Location of the propargyl group in compound **2** was confirmed by NMR. The signal for the anomeric proton of the substituted moiety (referred to as H-1¹), which appeared at lower field (δ 4.98 ppm) than those for H-1[II–VII] (δ 4.84–4.82 ppm), allowed the identification of protons and carbons of this glucose unit by selective 1D TOCSY and 2D gHMQC experiments (not shown), respectively. The signal for C-2¹ (δ 79.1 ppm) showed a strong paramagnetic shift of 6.4 ppm when compared with those for C-2[II–VII] of the nonsubstituted moieties (δ 73.3–71.7 ppm), while shifts for C-3¹ and C-6¹ remained unchanged, indicating that alkylation had taken place at position C-2.

Compound **2** is soluble in water, DMF and DMSO, and can be directly used in reactions compatible with the presence of hydroxyl groups such as Sonogashira-type oxidative coupling and Cu^I-catalyzed azide–alkyne Huisgen [3+2] cycloaddition.[12–28] Nevertheless, compound **2** can be fully protected as eicosakis(2[II–VII],3[I–VII],6[I–VII]-*O*-acetyl) derivative **3**, if needed. When protection of only the primary rim of the macrocycle is required, compound **2** can be converted to heptakis(6-*O*-*tert*-butyldimethylsilyl) derivative **4** by treatment with TBDMSCl in pyridine. However, as in the case of derivatization of the native β-CD,[31] this transformation usually leads to a mixture of over- and undersilylated by-products, along with formation of the desired compound **4**, as revealed by TLC (15:2:1 EtOAc–96% aq. EtOH–H_2O).* Additional portions of TBDMSCl must be added until spots corresponding to compounds more polar than **4** (R_f < 0.6, Figure 14.2) are no longer present.† The oversilylated species (R_f = 0.87 and 0.80) can be easily removed by filtration through a short pad of silica (5 cm) using 40:40:20:4 CH_2Cl_2–CH_3CN–96% aq. EtOH–30% aq. NH_3 as eluent. This mixture retains compound **4** at the top of the pad, if traces of pyridine and water had previously been removed (concentration with toluene, 3×). Subsequent elution of the pad with 40:40:20:4 CH_2Cl_2–CH_3CN–96% aq. EtOH–H_2O affords heptakis(6-*O*-*tert*-butyldimethylsilyl) derivative **4**.

EXPERIMENTAL

GENERAL METHODS

TLC was performed on Merck silica gel 60 F_{254}-coated aluminum sheets*. Spots were visualized by charring with ethanolic sulfuric acid (5% v/v). Solutions in organic solvents were dried with $MgSO_4$ and concentrated at <40°C. Flash column

---

* Clear separation of the spots requires thorough drying of the TLC plates before development.
† Depending on quality of the silylating reagent, more than one extra portion might be required.

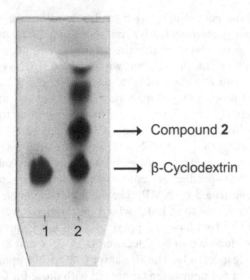

FIGURE 14.1 Monitoring of mono-2-O-propargylation of β-cyclodextrin by TLC, 10:5:2 CH_3CN–H_2O–30% v/v aqueous NH_3. Lane 1: starting material, and lane 2: reaction mixture after 5 h of reaction time.

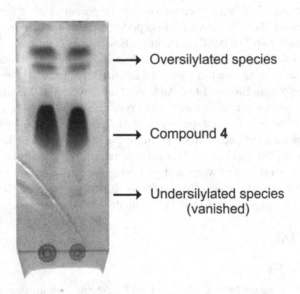

FIGURE 14.2 Monitoring in duplicate of silylation of 2^I-O-propargylcyclomaltoheptaose by TLC, 15:2:1 EtOAc–96% v/v EtOH-H_2O. Reaction time: 6 h after addition of an extra portion of TBDMSCl.

chromatography was performed on Merck[‡] silica gel (230–400 mesh, ASTM). Optical rotation was recorded with a Jasco P-1030 polarimeter at room temperature. The $[\alpha]_D$ values are given in 10^{-1} deg cm^{-1} g^{-1}. Infrared spectra were recorded with a Bruker Alpha FTIR equipped with a Bruker universal ATR sampling accessory. ^{1}H and ^{13}C NMR spectra were recorded with a Bruker Avance 500 Ultrashield spectrometer equipped with an inverse TBI ^{1}H/^{31}P/BB probe, and a Bruker Avance III HD 600 MHz spectrometer equipped with a QCI ^{1}H/^{13}C/^{15}N/^{31}P proton-optimized quadrupole inverse cryoprobe with ^{1}H and ^{13}C cryochannels. Standard Bruker software was used for acquisition and processing data. Chemical shifts (δ) are given in ppm and referenced to the signals from DMSO-d$_6$ (δ 2.50 and 39.52 ppm for ^{1}H and ^{13}C, respectively) and CDCl$_3$ (δ 7.26 and 77.16 ppm for ^{1}H and ^{13}C, respectively). Coupling constants (J) are given in Hertz (Hz). Signal multiplicities are described by the following abbreviations: broad (br), singlet (s), doublet (d), doublet of doublets (dd), triplet (t), and multiplet (m). 1D NMR TOCSY and 2D NMR gCOSY, gHMQC, and gHMBC experiments were used for unequivocal assignment. MALDI-TOF mass spectra were recorded on an Applied Biosystems Voyager DR-RP spectrometer with α-cyano-4-hydroxycinnamic acid as the matrix. Elemental analyses were performed on an Elementar Vario Micro CHNS. β-CD (1) was purchased from Cyclodextrin Research and Development Laboratory Ltd (Cyclolab, Budapest, Hungary) and dried at 50°C in vacuum in the presence of P$_2$O$_5$ until constant weight. Lithium hydride (Fluka, powder, 97%), anhydrous lithium iodide (Fluka, ≥98%), and propargyl bromide (Aldrich, solution in toluene, 80% w/w) were purchased from commercial sources. Ac$_2$O was obtained from PanReac Applichem, and TBDMSCl was purchased from Acros. All reagents were used as received. DMSO was dried by distillation from CaH$_2$ under high vacuum using a fractionation column[32] and stored over 4 Å molecular sieves. Pyridine was dried by refluxing over KOH for 3 h and fractionally distilled.[30] Dry solvents were kept under inert atmosphere in the dark.

2¹-O-PROPARGYLCYCLOMALTOHEPTAOSE (2)

Lithium hydride[§] (53 mg, 6.608 mmol) was added to the solution[**] of β-CD 1 (5 g, 4.405 mmol) in dry DMSO (75 mL). The resulting suspension was stirred under N$_2$ at 65°C until it became clear (ca. 10 min) and then cooled down to 30°C. Propargyl bromide (solution in toluene, 80% w/w, 491 µL, 4.405 mmol) and a catalytic amount of lithium iodide (~ 5 mg) were then added and the mixture was stirred at 55°C in the absence of light for 5 h. TLC (10:5:2 CH$_3$CN–H$_2$O–30% aq. NH$_3$) showed four spots with R$_f$ values of 0.75, 0.66, 0.50, and 0.31 (Figure 14.1), the last two corresponding to monopropargylated and nonpropargylated β-CD, respectively. The solution was poured into acetone (800 mL) and the precipitate was

---

[‡] Our experience suggests that this silica gives the best results in columns that use aqueous mixtures as eluents.
[§] This compound was weighted under N$_2$ using an inverted funnel.
[**] It may take up to 30 min for a clear solution to form. Efficient stirring is required.

filtered[††] through paper filter with the aid of a Büchner funnel and washed thoroughly with acetone (3 × 50 mL). The resulting solid was transferred into a round-bottom flask[‡‡] and dissolved in a minimum volume of water. Silica gel (10 g) was added and the solvent was removed under vacuum until a powdered residue was obtained. This crude mixture was applied on top of a column of silica (25 × 6 cm), and chromatography (10:5:2 $CH_3CN-H_2O-30\%$ aq. NH_3 as eluent, 1.8 L) gave, after freeze-drying, 2^I-O-propargylcyclomaltoheptaose **2** (1.912 g, 1.628 mmol, 37%) as white solid; this compound was stored in high vacuum in the presence of P_2O_5 to avoid hydration; the material decomposes at 239–245°C; $[\alpha]_D^{25}$ +126 (c 0.25, H_2O); R_f = 0.50 (10:5:2 $CH_3CN-H_2O-30\%$ aq. NH_3); IR (KBr): 3397, 2923, 2117, 1646, 1156, 1081, 1029 cm^{-1}; ^1H NMR (500 MHz, DMSO-d_6) δ 5.98 (br d, 1H, J 5.0 Hz, OH), 5.88 (br s, 1H, OH), 5.79–5.69 (m, 10H, OH), 4.98 (d, 1H, ^3J 3.6 Hz, H-1^I), 4.84–4.82 (br s, 6H, H-1^{II-VII}), 4.54 (t, 1H, J 5.6 Hz, OH), 4.50–4.45 (m, 8H, OH, CHC≡), 4.38 (dd, 1H, ^2J 15.8 Hz, ^4J 2.4 Hz, CHC≡), 3.78 (t, 1H, ^3J 9.8 Hz, H-3^I), 3.64–3.53 (m, 27H; H-3^{II-VII},5^{I-VII},6a^{I-VII},6b^{I-VII}), 3.51 (t, 1H, ^4J 2.4 Hz, ≡CH), 3.43–3.40 (m, 2H, H-2^I,4^I), 3.36–3.29 ppm (m, 12H, H-2^{II-VII},4^{II-VII}); ^{13}C NMR (125 MHz, DMSO-d_6) δ 102.0–101.7 (C-1^{II-VII}), 100.1 (C-1^I), 82.2–81.4 (C-4^{I-VII}), 79.9 (C≡), 79.1 (C-2^I), 77.8 (≡CH), 73.3–71.7 (C-2^{II-VII},3^{II-VII},5^{I-VII}), 72.6 (C-3^I), 60.1–59.7 (C-6^{I-VII}), 58.8 ppm (CH_2C≡); MALDI-TOF: m/z mass [M + Na]$^+$ calcd for $C_{45}H_{72}O_{35}Na$, 1195.4; found: 1195.6.

Later eluted was unchanged β-CD (2.03 g, 1.79 mmol, 41%).

EICOSAKIS(2^{II-VII},3^{I-VII},6^{I-VII}-O-ACETYL)-2^I-O-PROPARGYLCYCLOMALTOHEPTAOSE (3)

A solution of 2^I-O-propargylcyclomaltoheptaose **2** (100 mg, 0.085 mmol) and a catalytic amount of 4-DMAP (~ 5 mg) in a mixture of Ac_2O (1 mL) and dry pyridine (2 mL) was stirred under N_2 at room temperature for 24 h, when TLC (EtOAc) showed presence of only one product (R_f = 0.45). The solution was poured into a mixture of 5% aqueous HCl (30 mL) and ice, stirred until the ice melted and extracted with CH_2Cl_2 (2 × 40 mL). Combined organic phases were washed with H_2O (2 × 40 mL), dried and concentrated. Traces of pyridine were removed by co-evaporation with toluene (2 × 30 mL). Resulting material was mixed with silica gel (300 mg), suspended in CH_2Cl_2, and the solvent was removed under vacuum. Chromatography (15 × 2 cm, 20:1 EtOAc–hexane 250 mL → EtOAc 50 mL) gave a material that was remarkably electrostatic and quickly adsorbed water. Thus, it was suspended in methanol (20 mL) and rotary evaporated to dryness at 60°C twice to yield eicosakis(2^{II-VII}, 3^{I-VII},6^{I-VII}-O-acetyl)-2^I-O-propargylcyclomaltoheptaose **3** (160 mg, 0.079 mmol, 93%) as a glassy, not electrostatic and water free white solid; $[\alpha]_D^{25}$ +109 (c 1.0, CH_2Cl_2), lit.[31] +109 (c 1.0, $CHCl_3$); R_f = 0.45 (EtOAc); IR (ATR): 2960, 1738, 1369,

---

[††] Filtration must be fast and stirring of the remaining suspension in acetone must continue until the suspension is completely filtered in order to prevent the formation of a sticky layer on the bottom of the beaker.

[‡‡] A yellowish, creamy solid that can be managed with a spatula is initially obtained. If it turns into a sticky syrup-like material after a few minutes in air, it helps to add a small amount of water on the Büchner funnel to dissolve the material and pass it directly through the filter into the flask.

1213, 1027, 734 cm^{-1}; ^{1}H NMR (600 MHz, CDCl$_3$) δ 5.34–5.25 (m, 6H, H-3$^{\text{II–VII}}$), 5.21 (app t, 1H, $^3J_{\text{app}}$ 9.4 Hz, H-3^{I}), 5.09–5.00 (6 × d, 6H, ^3J 3.8–4.2 Hz, H-1$^{\text{II–VII}}$), 4.99 (d, 1H, ^3J 3.4 Hz, H-1^{I}), 4.81–4.73 (m, 6H, H-2$^{\text{II–VII}}$), 4.58–4.48 (m, 7H, H-6a$^{\text{I–VII}}$), 4.30–4.18 (m, 7H, H-6b$^{\text{I–VII}}$), 4.15 (app dd, 2H, J_{app} 4.6 Hz, J_{app} 2.4 Hz, CH$_2$C≡), 4.13–4.08 (m, 6H, H-5$^{\text{I,III–VII}}$), 4.02 (ddd, 1H, ^3J 9.8 Hz, ^3J 4.0 Hz, ^3J 1.6 Hz, H-5^{II}), 3.72–3.67 (m, 6H, H-4$^{\text{II–VII}}$), 3.62 (app t, 1H, $^3J_{\text{app}}$ 9.4 Hz, H-4^{I}), 3.52 (dd, 1H, ^3J 10.0 Hz, ^3J 3.4 Hz, H-2^{I}), 2.44 (t, 1H, ^4J 2.3 Hz, ≡CH), 2.10–2.09 (m, 22H, CH$_3$CO-6$^{\text{I–VII}}$), 2.07–2.06 (m, 22H, CH$_3$CO-2$^{\text{II–VII}}$,3^{I}), 2.03–2.00 (m, 18H, CH$_3$CO-3$^{\text{II–VII}}$); ^{13}C NMR (150 MHz, CDCl$_3$) δ 170.9–170.7 (CO-2$^{\text{II–VII}}$), 170.5–170.2 (CO-6$^{\text{I–VII}}$), 169.4–169.3 (CO-3$^{\text{II–VII}}$), 169.1 (CO-3^{I}), 98.6 (C-1^{I}), 97.1–96.5 (C-1$^{\text{II–VII}}$), 79.5 (C≡), 78.5 (C-4^{II}), 77.8 (C-4^{I}), 77.2 (overlapped with solvent, C-2^{I}), 76.9–76.2 (C-4$^{\text{III–VII}}$), 75.3 (≡CH), 72.4 (C-3^{I}), 71.1–70.3 (C-2$^{\text{II–VII}}$,3$^{\text{III–VII}}$), 70.0 (C-5^{II}), 69.7–69.5 (C-3^{II},5$^{\text{I,III–VII}}$), 62.8 (C-6^{I}), 62.5–62.4 (C-6$^{\text{II–VII}}$), 58.4 (CH$_2$C≡), 21.0–20.7 (CH$_3$CO-2$^{\text{II–VII}}$,3$^{\text{I–VII}}$,6$^{\text{I–VII}}$); MALDI-TOF: m/z mass [M + Na]$^+$ calcd for C$_{85}$H$_{112}$O$_{55}$Na, 2035.6; found: 2035.5. Anal. calcd for C$_{85}$H$_{112}$O$_{55}$: C, 50.70; H, 5.60. Found: C, 50.73; H, 5.69.

Heptakis(6$^{\text{I-VII}}$-O-tert-butyldimethylsilyl)-2^{I}-O-propargylcyclomaltoheptaose (4)

2^{I}-O-Propargylcyclomaltoheptaose 2 (100 mg, 0.085 mmol) was suspended in dry pyridine (2 mL) under N$_2$ and stirred at room temperature until a clear solution formed (30 min). Tert-butyldimethylsilyl chloride (108 mg, 0.714 mmol) was then added in one portion, and the resulting suspension was stirred at room temperature for 6 h. TLC (15:2:1 EtOAc–96% aq. EtOH–H$_2$O) showed formation of the title product (R$_f$ = 0.65), along with materials both more and less polar corresponding to under- and oversilylated species, respectively. Portions of TBDMSCl (27 mg, 0.179 mmol;†)were added every 6 h until the former spots completely disappeared (Figure 14.2). The solution was then poured into a mixture of 5% aqueous HCl (30 mL) and ice, stirred until the ice melted, and extracted with CH$_2$Cl$_2$ (2 × 40 mL). Combined organic phases were washed with H$_2$O (2 × 40 mL), dried and concentrated. Traces of pyridine were removed by co-evaporation with toluene (2 × 30 mL). The resulting material was mixed with silica gel (300 mg), suspended in CH$_2$Cl$_2$, and the solvent was removed under vacuum. Chromatography (5 × 2 cm) was performed using first 40:40:20:4 CH$_2$Cl$_2$–CH$_3$CN–96% aq. EtOH–30% aq. NH$_3$ as eluent until compounds with TLC (15:2:1 EtOAc–96% aq. EtOH–H$_2$O) spots at R$_f$ = 0.87 and 0.80 had been eluted (200 mL). Subsequent elution with 40:40:20:4 CH$_2$Cl$_2$–CH$_3$CN–96% aq. EtOH-H$_2$O (150 mL) as solvent yielded heptakis(6$^{\text{I-VII}}$-O-tert-butyldimethylsilyl)-2^{I}-O-propargylcyclomaltoheptaose 4 as a white, amorphous powder which was dried at 100°C under high vacuum for 6 h (126 mg, 0.064 mmol, 75%); the material decomposes at 232–236°C; [α]$_D^{25}$ +96 (c 1.0, CH$_2$Cl$_2$); R$_f$ = 0.65 (15:2:1 EtOAc–96% v/v EtOH–H$_2$O); IR (ATR): 3313, 2953, 2930, 2887, 2857, 1253, 1155, 1083, 1038, 833, 777, 735 cm^{-1}, lit.^{34} (KBr) 3420, 3325, 1473, 1254, 1086, 1040, 835 cm^{-1}; ^{1}H NMR (500 MHz, CDCl$_3$) δ 5.34 (br s, OH), 5.05 (d, 1H, $^3J_{1,2}$ 3.2 Hz, H-1^{I}), 4.89–4.88 (m, 6H, H-1$^{\text{II–VII}}$), 4.50 (dd, 1H, ^2J 16.7 Hz, ^4J 2.3 Hz, CHC≡), 4.41 (dd, 1H, ^2J 16.7 Hz, ^4J 2.3 Hz, CHC≡), 4.11–3.82

(m, 14H, H-3$^{\text{I-VII}}$,6a$^{\text{I-VII}}$), 3.74–3.49 (m, 28H, H-2$^{\text{I-VII}}$,4$^{\text{I-VII}}$,5$^{\text{I-VII}}$,6b$^{\text{I-VII}}$), 2.40 (t, 1H, ^4J = 2.3 Hz, ≡CH), 0.88–0.86 [m, 63H, SiC(CH$_3$)$_3$], 0.04–0.02 (m, 42H, SiCH$_3$); ^{13}C NMR (125 MHz, CDCl$_3$) δ 103.1–102.0 (C-1$^{\text{II-VII}}$), 101.3 (C-1^{I}), 82.1–81.7 (C-4$^{\text{III-VII}}$), 80.5–80.4 (C-4$^{\text{I,II}}$), 79.6 (C≡), 75.2 (≡CH), 74.0–72.5 (C2$^{\text{I-VII}}$,3$^{\text{I-VII}}$,5$^{\text{I-VII}}$), 62.3–61.6 (C-6$^{\text{I-VII}}$), 59.9 (CH$_2$C≡), 26.0 [SiC(CH$_3$)$_3$], 18.5–18.3 [SiC(CH$_3$)$_3$], −4.9–(−5.1) (SiCH$_3$); MALDI-TOF: m/z mass [M + Na]$^+$ calcd for C$_{87}$H$_{170}$O$_{35}$Si$_7$Na, 1995.0; found: 1995.0. Anal. calcd for C$_{87}$H$_{170}$O$_{35}$Si$_7$: C, 52.97; H, 8.69. Found: C, 52.93; H, 8.71.

ACKNOWLEDGMENTS

The authors acknowledge the Spanish Ministry of Economy and Competitiveness-ERD Fund (Grant CTQ2013-48380-R and CTQ2017-90050-R), and the Marie Curie ITN program (CYCLON Hit 608407) for financial support.

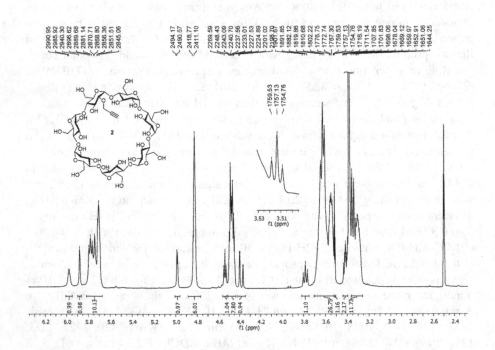

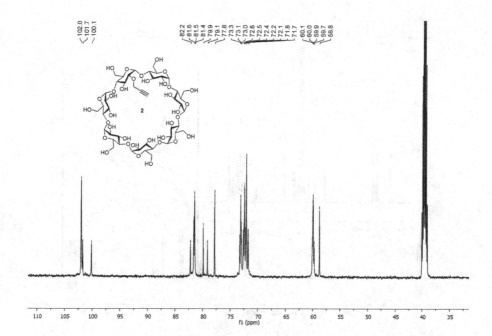

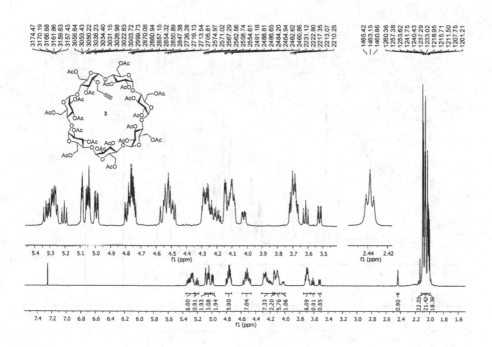

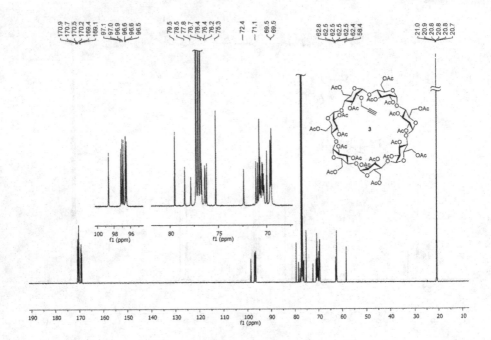

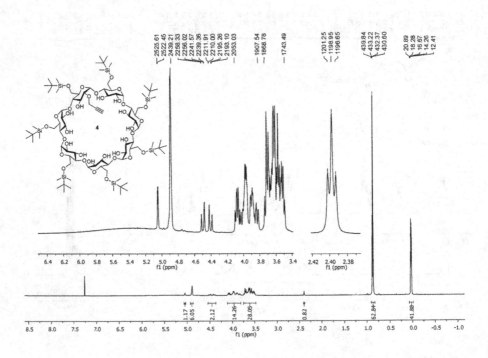

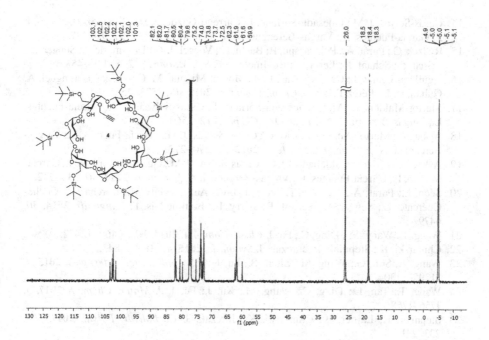

REFERENCES

1. Sliwa, W.; Girek, T. *Cyclodextrins: Properties and Applications*; Wiley-VCH Verlag GmbH & Co. KGaA: Weinheim, 2017.
2. Dodziuk, H. Ed.; *Cyclodextrins and Their Complexes: Chemistry, Analytical Methods, Applications*; Wiley-VCH Verlag GmbH & Co. KGaA: Weinheim, 2006.
3. Cutrone, G.; Casas-Solvas, J. M.; Vargas-Berenguel, A. *Int. J. Pharm.* **2017**, *531*, 621–639.
4. Ogoshi, T.; Harada, A. *Sensors* **2008**, *8*, 4961–4982.
5. Tian, H.; Wang, Q.-C. Cyclodextrin-based switches. In *Molecular Switches*, 2nd ed.; Feringa, B. L., Browne, W. R., Eds.; Wiley-VCH Verlag GmbH & Co. KGaA: Weinheim, 2011.
6. Schmidt, B. V. K. J.; Barner-Kowollik, C. *Angew. Chem. Int. Ed.*, **2017**, *56*, 8350–8369.
7. Guieu, S.; Sollogoub, M. Advances in cyclodextrin chemistry. In *Modern Synthetic Methods in Carbohydrate Chemistry: From Monosaccharides to Complex Glycoconjugates*; Werz, E. B., Vidal, S., Eds.; Wiley-VCH Verlag GmbH & Co. KGaA: Weinheim, 2013.
8. Řezanka, M. *Eur. J. Org. Chem.* **2016**, 5322–5334.
9. Liu, T.; Chen, Y. *Acc. Chem. Res.* **2006**, *39*, 681–691.
10. Park, J. W.; Lee, S. Y.; Song, H. J.; Park, K. K. *J. Org. Chem.* **2005**, *70*, 9505–9513.
11. Liu, Y.; Li, X.-Y.; Guo, D.-S.; Chi, H. *Supramol. Chem.* **2008**, *20*, 609–617.
12. Casas-Solvas, J. M.; Vargas-Berenguel, A. *Tetrahedron Lett.* **2008**, *49*, 6778–6780.
13. Casas-Solvas, J. M.; Ortiz-Salmerón, E.; Fernandez, I.; García-Fuentes, L.; Santoyo-González, F.; Vargas-Berenguel, A. *Chem. Eur. J.* **2009**, *15*, 8146–8162.

14. Casas-Solvas, J. M.; Quesada-Soriano, I.; Carreño-Gázquez, D.; Giménez-Martínez, J. J.; García-Fuentes, L.; Vargas-Berenguel, A. *Langmuir*, **2011**, *27*, 9729–9737.
15. Rydzek, G.; Parat, A.; Polavarapu, P.; Baehr, C.; Voegel, J.-C.; Hemmerlé, J.; Senger, B.; Frisch, B.; Schaaf, P.; Jierry, L.; Boulmedais, F. *Soft Matter* **2012**, *8*, 446–453.
16. Aguilera-Sigalat, J.; Casas-Solvas, J. M.; Morant-Miñana, M. C.; Vargas-Berenguel, A. Galian, R. E.; Pérez-Prieto, J. *Chem. Commun.* **2012**, *48*, 2573–2575.
17. Martos-Maldonado, M. C.; Quesada-Soriano, I.; Casas-Solvas, J. M.; García-Fuentes, L.; Vargas-Berenguel, A. *Eur. J. Org. Chem.* **2012**, 2560–2571.
18. Aykaç, A.; Martos-Maldonado, M. C.; Casas-Solvas, J. M.; García-Fuentes, L.; Vargas-Berenguel, A. *J. Drug Del. Sci. Tech.* **2012**, *22*, 270–272.
19. Aykaç, A.; Martos-Maldonado, M. C.; Casas-Solvas, J. M.; Quesada-Soriano, I.; García-Maroto, F.; García-Fuentes, L.; Vargas-Berenguel, A. *Langmuir*, **2014**, *30*, 234–242.
20. Séon, L.; Parat, A.; Gaudère, F.; Voegel, J.-C.; Auzély-Velty, R.; Lorchat, P.; Coche-Guérente, L.; Senger, B.; Schaaf, P.; Jierry, L.; Boulmedais, F. *Langmuir*, **2014**, *30*, 6479–6488.
21. Wang, T.; Wang, M.; Ding, C.; Fu, J. *Chem. Commun.* **2014**, *50*, 12469–12472.
22. Chmurski, K.; Stepniak, P.; Jurczak, J. *Synthesis*, **2015**, *47*, 1838–1843.
23. Wang, T.; Sun, G.; Wang, M.; Zhou, B.; Fu, J. *ACS Appl. Mater. Interfaces*, **2015**, *7*, 21295–21304.
24. Wang, T.; Tan, L.; Ding, C.; Wang, M.; Xu, J.; Fu, J. *J. Mater. Chem. A* **2017**, *5*, 1756–1768.
25. Stepniak, P.; Lainer, B.; Chmurski, K.; Jurczak, J. *Carbohydr. Polym.* **2017**, *164*, 233–241.
26. Aykaç, A.; Noiray, M.; Malanga, M.; Agostoni, V.; Casas-Solvas, J. M.; Fenyvesi, É.; Gref, R.; Vargas-Berenguel, A. *Biochim. Biophys. Acta, Gen. Subj.* **2017**, *1861*, 1606-1616.
27. Wang, T.; Wang, C.; Zhou, S.; Xu, J. Jiang, W.; Tan, L; Fu, J. *Chem. Mater.* **2017**, *29*, 8325-8337.
28. Gallego-Yerga, L.; Benito, J. M.; Blanco-Fernández, L.; Martínez-Negro, M.; Vélaz, I.; Aicart, E.; Junquera, E.; Ortiz Mellet, C.; Tros de Ilarduya, C.; García Fernández, J. M. *Chem. Eur. J.* **2018**, *24*, 3825-3835.
29. Hanessian, S.; Benalil, A.; Laferrière, C. *J. Org. Chem.* **1995**, *60*, 4786–4797.
30. Jindrich, J.; Pitha, J.; Lindberg, B.; Seffers, P.; Harata, K. *Carbohydr. Res.* **1995**, *266*, 75–80.
31. Casas-Solvas, J. M.; Vargas-Berenguel, A.; Malanga, M. Synthesis of heptakis(6-*O-tert*-butyldimethylsilyl)cyclomaltoheptaose. In *Carbohydrate Chemistry: Proven Synthetic Methods*, *Volume 4*; Vogel, C., Murphy, P., Eds; Kováč, P., Series Ed; CRC Press, Taylor & Francis Group, Boca Raton, 2017. Chapter 26, pp. 209–216.
32. Perrin, D. D.; Armarego, W. F. L. *Purification of Laboratory Chemicals*, 3rd ed., Pergamon Press, Oxford, 1989.
33. Zhou, Y.; Marinescu, L.; Pedersen, C. M.; Bols, M. *Eur. J. Org. Chem.* **2012**, 6383–6389.
34. Trotta, F.; Martina, K.; Robaldo, B.; Barge, A.; Cravotto, G. *J. Incl. Phenom. Macrocyclic Chem.* **2007**, *57*, 3–7.

15 Synthesis of 1,3,4,6-Tetra-O-Acetyl-2-Azido-2-Deoxy-α,β-D-Galactopyranose

Enrique Mann, and Jose Luis Chiara*
Instituto de Química Orgánica General, IQOG-CSIC
Madrid, Spain

Stella Verkhnyatskaya[a]
Stratingh Institute of Chemistry, University of Groningen,
Groningen, The Netherlands

CONTENTS

The title compound is an important synthetic intermediate for the synthesis of D-galactosamine-containing oligosaccharides. It may be directly used as glycosyl donor[1] or can be readily converted into more efficient donors, including glycosyl halides (fluoride,[2] chloride,[3] bromide,[3] or iodide[4]), glycosyl trichloroacetimidate,[5] and thioglycosides,[6] among others. The presence of the nonparticipating azido group facilitates a stereoselective α-glycosylation reaction, while allowing to subsequently unmask the amino function by reduction. Earlier syntheses of this family of compounds involved either azidonitration[7] or addition of halogeno azides[8] to costly D-galactal. Still, a simpler and more cost-effective procedure employs the diazo-transfer reaction from activated sulfonyl azides to readily available D-galactosamine. However, despite successive improvements,[9] the original method using trifluoromethanesulfonyl azide[10] suffered from a serious drawback due to the instability and very hazardous nature of this reagent, which needs to be prepared in situ and should always be used in solution.[11] Several safer and

---

* Corresponding author: jl.chiara@csic.es
[a] Checker, under supervision of Dr. M. T. C. Walvoort: m.t.c.walvoort@rug.nl

i) $CF_3(CF_2)_3SO_2N_3$,
cat. $CuSO_4$, $NaHCO_3$
ii) Ac_2O, pyridine
⟶
63-70%

1   **2**

i) $CF_3(CF_2)_3SO_2N_3$, cat. $CuSO_4 \cdot 5H_2O$, $NaHCO_3$,
H_2O/MeOH/Et_2O, rt, 6h. ii) Ac_2O, pyridine, 0 °C to rt, 4 h.

more cost-effective diazotransfer reagents have been described more recently, including imidazole-1-sulfonyl azide salts,[12] 2-azido-1,3-dimethylimidazolinium hexafluorophosphate,[13] and nonafluorobutanesulfonyl azide.[14] The first have gained popularity due to their low cost, but, in spite of recent developments regarding the nature of the anion to improve stability,[15] concerns still remain about their shelf life. They are moisture-sensitive, which may lead to inadvertent formation of the very hazardous and explosive hydrazoic acid, as reported[16] for the originally described hydrochloride salt. Although 2-azido-1,3-dimethylimidazolinium hexafluorophosphate seems to be a safer alternative, it has a narrower reactivity scope than the other reagents, requiring the presence of a strong organic base for the diazotization of highly nucleophilic, sterically unhindered primary amines.[13] In comparison, nonafluorobutanesulfonyl azide has nearly the same reactivity as trifluoromethanesulfonyl azide and displays the best stability and safety profile of all diazotransfer reagents described to date.[17]

Nonafluorobutanesulfonyl azide is a colorless liquid with a characteristic pungent odor, similar to that of trifluoromethanesulfonyl azide, which boils at 100°C without decomposition (it decomposes at 152°C without detonation); it is insensitive to impact and to electrostatic discharge and shows longer shelf life.[17] Thus, we have safely stored >40-g batches of the pure reagent for >3 years at 4°C without any detectable decomposition. This compound is readily prepared by reaction of sodium azide with nonafluorobutanesulfonyl fluoride,[18] a fairly cheap, bench stable, and easy to handle, commercially available reagent. The Cu^{2+}-catalyzed diazotransfer reaction of nonafluorobutanesulfonyl azide with 2-amino-2-deoxy-D-galactopyranose hydrochloride affords under mildly basic homogeneous conditions (in a H_2O–MeOH–Et_2O solvent mixture) the corresponding 2-azido-2-deoxy-pyranose in good overall yield. To facilitate purification, the crude product is acetylated and chromatographed. This methodology has also been successfully applied to the preparation of the 2-azido-2-deoxy-D-*gluco*-derivative.[14b]

EXPERIMENTAL

GENERAL METHODS

Infrared (FT-IR) spectra were measured with a Perkin Elmer Spectrum One spectrophotometer and are reported in cm^{-1}. ^{1}H, ^{19}F, and ^{13}C NMR spectra were recorded with a Bruker Avance III-400 (400, 376, and 100 MHz, respectively) spectrometer.

Chemical shifts are expressed in parts per million (δ scale) downfield from tetramethylsilane and are referenced to residual peaks of the deuterated NMR solvent used or to internal tetramethylsilane. Data are presented as follows: chemical shift, multiplicity (s = singlet, d = doublet, t = triplet, m = multiplet and/or multiple resonances, b = broad), coupling constants in Hertz (Hz). Assignments are based on gCOSY, gHSQC, and gHMBC correlation experiments. Thin layer chromatography (TLC) was performed with Merck Silica Gel 60 F254 plates. Spots were visualized by treatment with a solution of ammonium molybdate (50 g) and cerium(IV) sulfate (1 g) in 5% aqueous H_2SO_4 (1 L) followed by charring on a hot plate. Alternatively, for the detection of azides, the chromatograms were first dipped in a 1% (w/v) solution of Ph_3P in EtOAc, dried at rt, then dipped in a 1 or 5% (w/v) solution of ninhydrin in 95% aqueous EtOH and finally charred on a hot plate. Solutions in organic solvents were dried with $MgSO_4$ or Na_2SO_4 and concentrated at <40°C. Column chromatography was performed with Merck silica gel, grade 60, 230–400 mesh. High-resolution mass spectra (HRMS) were recorded on an Agilent 6520 Q-TOF instrument with an ESI source. Elemental analyses were determined in a Heraus CHNO analyzer. Solvents were of HPLC grade and were used as provided. All reactions were carried out under argon with magnetic stirring and in oven-dried glassware.

NONAFLUOROBUTANESULFONYL AZIDE

Nonafluorobutanesulfonyl fluoride (20 mL, 111 mmol) was added with stirring to a suspension of NaN_3 (8.0 g, 123 mmol) in MeOH (220 mL). After stirring overnight (15 h), the mixture was poured onto ice water (800 mL). If a stable emulsion was formed, it was broken by quick filtration through a layer of Na_2SO_4 to give two separate layers. The colorless oily layer of nonafluorobutanesulfonyl azide was separated and dried [alternatively, the reagent can be isolated by diluting the reaction mixture with water (600 mL), followed by CH_2Cl_2 (2 × 130 mL) extraction (the last aqueous layer was filtered over $MgSO_4$ to promote layer separation). The combined organic layers were dried and concentrated at reduced pressure (100 mbar, 45°C)]. Yield, 22.0–22.5 g (60–62%). The crude reagent (bp. 100–102°C, $d = 1.74$ g/mL; $d = 1.69$ g/mL when using the alternative solvent-extraction procedure) can be kept at 4 °C for >3 years without decomposition and is of high purity (as checked by ^{19}F NMR) to be used without further purification for the diazo-transfer reaction. ^{19}F NMR ($CDCl_3$): δ −81.8 (3F), −110.5 (2F), −122.0 (2F), −127.0 (2F).

1,3,4,6-TETRA-*O*-ACETYL-2-AZIDO-2-DEOXY-α,β-D-GALACTOPYRANOSE (2)

To a stirred solution of D-galactosamine hydrochloride (1.0 g, 4.6 mmol) in water (12 mL), in a 1-L round-bottom flask was added in sequence MeOH (17 mL), $NaHCO_3$ (1.53 g, 18.4 mmol), a solution of nonafluorobutanesulfonyl azide (2.34 g, 6.9 mmol) in Et_2O (18 mL) and $CuSO_4 \cdot 5H_2O$ (210 mg, 0.09 mmol). The reaction mixture turned blue and the consumption of starting material was followed by TLC (3:2:2 EtOAc–MeOH–25% NH_4OH, $R_f = 0.33$). After stirring the reaction mixture at room temperature for 6 h, TLC control indicated complete conversion of starting material into one major spot (5:1 EtOAc–MeOH, R_f = 0.58). The mixture was concentrated at reduced pressure (<30 mbar, 30°C waterbath; higher bath temperatures cause significant decomposition of the diazo-transfer product). The oily residue was suspended in dry pyridine (50 mL) and Ac_2O (6.3 mL, 67.5 mmol) was added at 0°C with stirring. After 4 h, TLC control indicated the conversion of starting material into one major spot (2:1 pentane–EtOAc, R_f = 0.53). The mixture was diluted with CH_2Cl_2 (150 mL) and washed with aqueous 1-M HCl (2 × 150 mL). The combined aqueous layers were extracted with CH_2Cl_2 (3 × 100 mL), the organic layers were washed with saturated aqueous $NaHCO_3$ solution (200 mL) and brine (100 mL), dried and concentrated at reduced pressure. Residual pyridine was removed as azeotrope with toluene at reduced pressure. The residue was purified by flash column chromatography (hexane → 7:3 hexane–EtOAc), to give 1.039–1.093 g (60–63%, α/β = 1:3 anomeric mixture) of **2** as a viscous oil. FT-IR (film): ν (cm^{-1}) 2118 (s), 1754 (vs), 1373 (m), 1220 (vs), 1087 (m), 1044 (m). ^{1}H NMR ($CDCl_3$): δ 6.31 (d, 0.25H, $J_{1,2}$ 3.6 Hz, H-1, α-anomer), 5.54 (d, 0.75H, $J_{1,2}$ 8.4 Hz, H-1, β-anomer), 5.47 (d, 0.25H, $J_{4,3}$ 2.4 Hz, H-4), 5.36 (d, 0.75H, $J_{4,3}$ 2.8 Hz, H-4), 5.31 (dd, 0.25H, $J_{3,4}$ 3.5 Hz, $J_{3,2}$ 11.2 Hz, H-3, α-anomer), 4.89 (dd, 0.75H, $J_{3,4}$ 3.2 Hz, $J_{3,2}$ 10.9 Hz, H-3, β-anomer), 4.27 (t, 0.25H, $J_{5,6}$ 6.2 Hz, H-5, α-anomer), 4.16–4.06 (m, 2H, H-6, H-6'), 4.00 (t, 0.75H, $J_{5,6}$ 6.8 Hz, H-5, β-anomer), 3.93 (dd, 0.25H, $J_{2,1}$ 3.5 Hz, $J_{2,3}$ 11.0 Hz, H-2, α-anomer), 3.82 (dd, 0.75H, $J_{2,1}$ 8.7 Hz, $J_{2,3}$ 11.0 Hz, H-2, β-anomer), 2.19 (s, 2.25 H, AcO, β-anomer), 2.16 (s, 0.75H, AcO, α-anomer), 2.15 (s, 3H, AcO, α+β anomers), 2.06 (s, 0.75H, AcO, α-anomer), 2.05 (s, 2.25H, AcO, β-anomer), 2.03 (s, 3H, AcO, α+β anomers); ^{13}C NMR (100 MHz, $CDCl_3$): δ 170.43 (α-anomer, CO), 170.42 (β-anomer, CO), 170.06 (α-anomer, CO), 170.03 (β-anomer, CO), 169.9 (α-anomer, CO), 169.7 (β-anomer, CO), 168.8 (α-anomer, CO), 168.7 (β-anomer, CO), 93.0 (β-anomer, C-1), 90.5 (α-anomer, C-1), 71.9 (β-anomer, C-5), 71.4 (β-anomer, C-3), 68.9 (α-anomer, C-3), 68.8 (α-anomer, C-5), 67.0 (α-anomer, C-4), 66.3 (β-anomer, C-4), 61.2 (α-anomer, C-6), 61.1 (β-anomer, C-6), 59.8 (β-anomer, C-2), 57.0 (α-anomer, C-2), 21.0 (CH_3CO), 21.0 (CH_3CO), 20.75 (CH_3CO), 20.7 (CH_3CO), 20.68 (CH_3CO); HRMS (ESI): m/z: $[M+NH_4]^+$ calcd for $C_{14}H_{23}N_4O_9$, 391.1459; found, 391.1470. Anal. calcd for $C_{14}H_{19}N_3O_9$: C, 45.04; H, 5.13; N, 11.26. Found: C, 44.82; H, 5.135; N, 10.86.

ACKNOWLEDGMENTS

This research was supported by the Spanish Ministerio de Economía y Competitividad (project MAT2017-83856-C3-1-P).

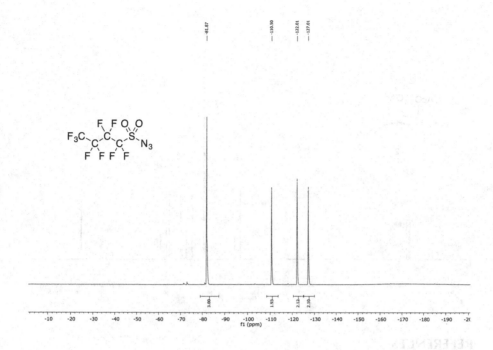

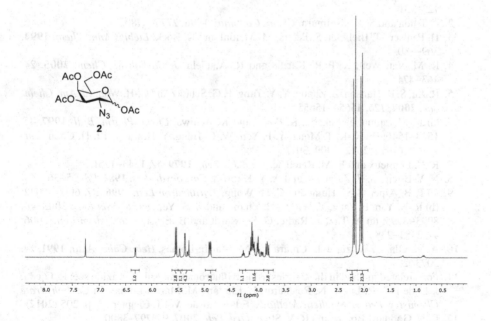

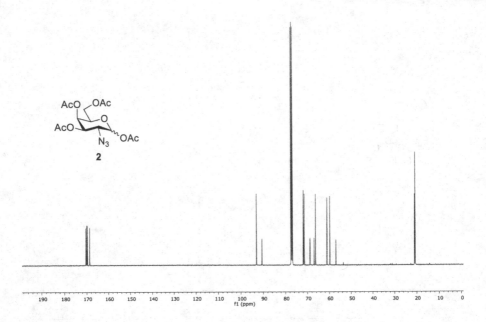

REFERENCES

1. J. M. Wojnar, C. W. Evans, A. L. Devries and M. A. Brimble, *Aust. J. Chem.* **2011**, *64*, 723–731.
2. T. Tsuda and S.-I. Nishimura, *Chem. Commun.* **1996**, 2779–2780.
3. H. Paulsen, T. Bielfeldt, S. Peters, M. Meldal and K. Bock, *Liebigs Ann. Chem.* **1994**, 369–380.
4. R. M. Van Well, K. P. R. Kartha and R. A. Field, *J. Carbohydr. Chem.* **2005**, *24*, 463–474.
5. R. Xu, S. R. Hanson, Z. Zhang, Y.-Y. Yang, P. G. Schultz and C.-H. Wong, *J. Am. Chem. Soc.* **2004**, *126*, 15654–15655.
6. (a) K. Miyajima, T. Nekado, K. Ikeda and K. Achiwa, *Chem. Pharm. Bull.* **1997**, *45*, 1544–1546; (b) K.-K. T. Mong, Y.-F. Yen, W.-C. Hung, Y.-H. Lai and J.-H. Chen, *Eur. J. Org. Chem.* **2012**, 3009–3017.
7. R. U. Lemieux and R. M. Ratcliffe, *Can. J. Chem.* **1979**, *57*, 1244–1251.
8. N. V. Bovin, S. E. Zurabyan and A. Y. Khorlin, *Carbohydr. Res.* **1981**, *98*, 25–36.
9. (a) P. B. Alper, S.-C. Hung and C.-H. Wong, *Tetrahedron Lett.* **1996**, *37*, 6029–6032; (b) R.-B. Yan, F. Yang, Y. Wu, L.-H. Zhang and X.-S. Ye, *Tetrahedron Lett.* **2005**, *46*, 8993–8995; (c) A. Titz, Z. Radic, O. Schwardt and B. Ernst, *Tetrahedron Lett.* **2006**, *47*, 2383–2385.
10. A. Vasella, C. Witzig, J. L. Chiara and M. Martin-Lomas, *Helv. Chim. Acta*, **1991**, *74*, 2073–2077.
11. For a related method in these series using trifluoromethanesulfonyl azide see: R. Ojeda, J. L. de Paz, R. Lucas, N. Reichardt, L. Liu, and M. Martín-Lomas. In *Carbohydrate Chemistry. Proven Synthetic Methods*, Ed. P. Kovác, Vol 1, Chapter 22, p. 205 (**2012**).
12. E. D. Goddard-Borger and R. V. Stick, *Org. Lett.* **2007**, *9*, 3797–3800.
13. M. Kitamura, S. Kato, M. Yano, N. Tashiro, Y. Shiratake, M. Sando and T. Okauchi, *Org. Biomol. Chem.* **2014**, *12*, 4397–4406.

14. (a) S. Yekta, V. Prisyazhnyuk and H.-U. Reissig, *Synlett*, **2007**, 2069–2072; (b) J. R. Suarez, B. Trastoy, M. E. Perez-Ojeda, R. Marin-Barrios and J. L. Chiara, *Adv. Synth. Catal.* **2010**, *352*, 2515–2520; (c) B. Trastoy, M. E. Perez-Ojeda, R. Sastre and J. L. Chiara, *Chem. - Eur. J.* **2010**, 16, 3833–3841.
15. (a) N. Fischer, E. D. Goddard-Borger, R. Greiner, T. M. Klapoetke, B. W. Skelton and J. Stierstorfer, *J. Org. Chem.* **2012**, *77*, 1760–1764; (b) G. T. Potter, G. C. Jayson, G. J. Miller and J. M. Gardiner, *J. Org. Chem.* **2016**, *81*, 3443–3446.
16. E. D. Goddard-Borger and R. V. Stick, *Org. Lett.* **2011**, 13, 2514.
17. J. R. Suarez, D. Collado-Sanz, D. J. Cardenas and J. L. Chiara, *J. Org. Chem.* **2015**, *80*, 1098–1106.
18. S.-Z. Zhu, *J. Chem. Soc., Perkin Trans. 1.* **1994**, 2077–2081.

16 Regioselective Palladium Catalyzed Oxidation at C-3 of Methyl Glucoside

Nittert Marinus, Marthe T. C. Walvoort,
Martin D. Witte, and Adriaan J. Minnaard*
Stratingh Institute for Chemistry, University of Groningen,
Groningen, The Netherlands

J. Hessel M. van Dijk[a]
Leiden Institute of Chemistry, Leiden University,
Leiden, The Netherlands

CONTENTS

Functional group transformations of carbohydrates often rely on the use of protecting groups, and the direct modification of unprotected carbohydrates remains a formidable challenge.[1] We have developed a method in which unprotected glycopyranosides can be oxidized regioselectively by a cationic palladium neocuproine catalyst using *p*-benzoquinone as the oxidant. Under these conditions the C-3 OH is oxidized to the corresponding ketone.[2,3] Oxygen can be used as well, although it negatively impacts the reaction rate.[4]

The transformation has been successfully applied to various unprotected and partially protected saccharides.[5–8] Most reactions were performed on *gluco*-configured sugars, with methyl α-D-glucopyranoside (**1**) as the benchmark example. The C-3 selective oxidations of methyl 2-acetamido-2-deoxy-α-D-glucopyranoside and phenyl 1-thio-β-D-glucopyranoside demonstrate the good functional group compatibility with amides and thioglycosides, albeit the latter required a higher catalyst loading (6.5 mol% instead of 2.5 mol%).[3] The reducing sugar α-D-glucose was also successfully oxidized at the C-3 position.[5] Interestingly, the oxidation of

---

* Corresponding author: a.j.minnaard@rug.nl
[a] Checker: j.h.m.van.dijk@lic.leidenuniv.nl

the corresponding β-anomer was not regioselective, and oxidation also occurred on the C-1 position resulting in the corresponding gluconolactone. The oxidation protocol was also successfully applied to oligosaccharides with exceptionally high regioselectivity, wherein only the C-3 OH of the terminal residue at the nonreducing end is oxidized.[3,6] Saccharides in other than the glucose configuration, namely, methyl α-D-mannopyranoside and methyl β-D-galactopyranoside, could also selectively be oxidized at the C-3 position by this method. However, in this case, the corresponding ketoses are prone to side reactions, such as over-oxidation and rearrangements.[7]

The new keto functionality is a versatile handle for further ligation and modification without the use of protecting groups. For example, it allowed the synthesis of D-allose from α-D-glucose in two steps through an oxidation and reduction cycle.[5]

This oxidation protocol suffers from a challenging purification, since a high boiling solvent, DMSO or water, has to be removed and the product is highly polar.[3] Herein we describe the C-3 regioselective oxidation of methyl α-D-glucopyranoside (1) with improved reaction conditions that are scalable and offer straightforward purification. In comparison with previously reported, lower catalyst loading, near equimolar amounts of p-benzoquinone and methanol as the solvent are used. The product, methyl α-D-*ribo*-hex-3-ulopyranoside (2), is purified by column chromatography with ethyl acetate–pentane as the eluent, thereby avoiding chlorinated solvents.

EXPERIMENTAL

GENERAL METHODS

Methanol for the reaction was of reagent grade (Macron Fine Chemicals). Methyl α-D-glucopyranoside was supplied by Sigma. [(2,9-dimethyl-1,10-phenanthroline)Pd(μ-OAc)]$_2$(OTf)$_2$ was prepared according to a procedure from Waymouth and coworkers.[9] p-Benzoquinone (Sigma-Aldrich) was recrystallized from ethanol before use.[10*] TLC was performed on Merck silica gel 60, 0.25-mm plates, and visualization was done by staining with potassium permanganate stain (a mixture of 3 g KMnO$_4$, 10 g K$_2$CO$_3$, and 300 mL water) or with anisaldehyde stain (a mixture of 300 mL AcOH, 6 mL H$_2$SO$_4$, and 3 mL anisaldehyde) and heating. Flash column chromatography was performed on silica (silica 60M, Macherey-Nagel). Celite (Celite® 545)

---

* 15 g p-benzoquinone was recrystallized from 45 mL hot ethanol and was obtained as dark yellow crystals (mp 112-114°C, lit., 115°C). It was stored at –20°C and protected from light.

was purchased from Merck. NMR spectra were recorded on a Varian AMX400 at 25°C at the following frequencies: 400 MHz (^{1}H) and 100.6 MHz (^{13}C). Chemical shift values are reported in ppm with the solvent resonance as the internal standard (MeOD-$d4$: δ 3.31 for ^{1}H, δ 49.00 for ^{13}C). High resolution mass spectra (HRMS) were recorded on a Thermo Scientific LTQ Orbitrap XL. Optical rotations were measured on a Schmidt+Haensch polarimeter (Polartronic MH8) with a 10-cm cell (c given in g/100 mL) at ambient temperature (~20°C). Melting points were recorded on a Stuart SMP 11 apparatus and infrared spectra were recorded on a PerkinElmer FT-IR spectrometer.

METHYL α-D-*RIBO*-HEX-3-ULOPYRANOSIDE (2)

Methyl α-D-glucopyranoside (**1**) (5.00 g, 25.7 mmol, 1 equiv.) and *p*-benzoquinone (2.92 g, 27.0 mmol, 1.05 equiv.) were placed in a round-bottom flask equipped with a magnetic stirring bar. MeOH (103 mL, 0.25 M) was added and, after stirring for 15 min, [(2,9-dimethyl-1,10-phenanthroline)Pd(μ-OAc)]$_2$(OTf)$_2$ (135 mg, 129 μmol, 0.5 mol%) was added to the orange solution.[†] No efforts were made to exclude water or oxygen from the reaction. The reaction mixture became darker over time and, after 1 h, the starting material was consumed (TLC, 15% MeOH in DCM) and one less polar product was formed.[‡] Twenty grams of Celite® was added to the black reaction mixture and the slurry was concentrated to dryness at 40°C. The resulting green solid Celite-product mixture was pulverized and placed on top of a silica column made of 200-g silica (column volume: ~420 mL, bed volume: 470 mL). Hydroquinone was eluted with 1-L 10% pentane in EtOAc[§] and subsequent elution with 2-L 3% MeOH in EtOAc provided the keto saccharide (**2**, 4.56 g, 92%) as a white solid, mp 120–121°C (EtOAc–MeOH, after drying at 45°C/1 mbar); [α]$_D$ +148 (c 2.0, H$_2$O); compound **2** is dimorphous: lit.[11] mp 91–92°C, [α]$_D^{25}$ +155 (c 2.6, H2O); lit.[12] mp 92.5–93°C, [α]$_{578}^{21}$ +149 (c 1.0, H$_2$O); ^{1}H NMR (CD$_3$OD): δ 5.05 (d, 1H, $J_{1,2}$ 4.2 Hz, H-1), 4.40 (dd, 1H, $J_{2,1}$ 4.3 Hz, $J_{2,4}$ 1.5 Hz, H-2), 4.23 (dd, 1H, $J_{4,5}$ 9.7 Hz, $J_{4,2}$ 1.5 Hz, H-4), 3.88 (dd, 1H, $J_{6a,6b}$ 12.0 Hz, $J_{6a,5}$ 2.2 Hz, H-6a), 3.80 (dd, 1H, $J_{6b,6a}$ 12.1 Hz, $J_{6b,5}$ 4.6 Hz, H-6b), 3.65 (ddd, 1H, $J_{5,4}$ 9.7 Hz, $J_{5,6b}$ 4.5 Hz, $J_{5,6a}$ 2.2 Hz, H-5), 3.40 (s, 3H, OCH$_3$); ^{13}C NMR (CD$_3$OD): δ 207.0 (C-3), 103.8 (C-1), 76.72 (C-5), 76.05 (C-2), 73.34 (C-4), 62.47 (C-6), 55.72 (OCH$_3$); IR (powder): v_{max}/cm^{-1} 3451 (w), 3390 (w, br), 3232 (w, br), 2938 (w), 1738 (s); lit.[13] IR: v_{max}/cm^{-1} 1736; HRMS (ESI): m/z [M + Na]$^+$ calcd for C$_7$H$_{12}$O$_6$Na, 215.053; found 215.052. Anal. calcd for C$_7$H$_{12}$O$_6$: C, 43.75; H, 6.29. Found: C, 43.46; H, 6.15.

ACKNOWLEDGMENTS

The authors thank the Netherlands Organization for Scientific Research (NWO) for financial support.

[†] Not all benzoquinone dissolved immediately, but did dissolve completely during the reaction.

[‡] **1** appears as a blue spot (R_f: 0.25) and **2** appears as a brown spot (R_f: 0.5) when the TLC plate is stained with anisaldehyde.

[§] Hydroquinone is best visualized on TLC with the KMnO$_4$ stain.

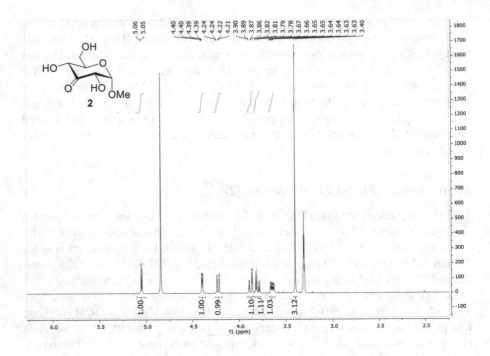

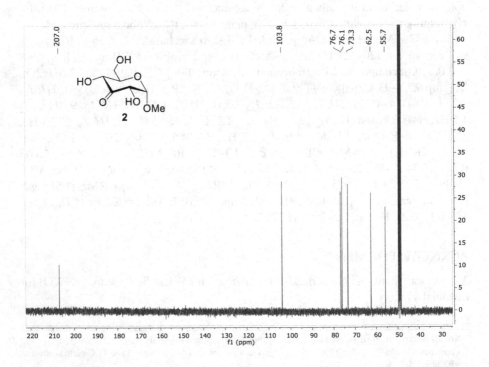

REFERENCES

1. Jäger, M.; Minnaard, A. J. *Chem. Commun.* **2016**, *52*, 656–664.
2. Painter, R. M.; Pearson, D. M.; Waymouth, R. M. *Angew. Chem. Int. Ed.* **2010**, *49*, 9456–9459.
3. Jäger, M.; Hartmann, M.; de Vries, J. G.; Minnaard, A. J. *Angew. Chem. Int. Ed.* **2013**, *52*, 7809–7812.
4. Armenise, N.; Tahiri, N.; Eisink, N. N. H. M.; Denis, M.; Jäger, M.; de Vries, J. G.; Witte, M. D.; Minnaard, A. J. *Chem. Commun.* **2016**, *52*, 2189–2191.
5. Jumde, V. R.; Eisink, N. N. H. M.; Witte, M. D.; Minnaard, A. J. *J. Org. Chem.* **2016**, *81*, 11439–11443.
6. Eisink, N. N. H. M.; Lohse, J.; Witte, M. D.; Minnaard, A. J. *Org. Biomol. Chem.* **2016**, *14*, 4859–4864.
7. Eisink, N. N. H. M.; Witte, M. D.; Minnaard, A. J. *ACS Catal.* **2017**, *7*, 1438–1445.
8. Chung, K.; Waymouth, R. M. *ACS Catal.* 2016, *6*, 4653–4659.
9. Ho, W. C.; Chung, K.; Ingram, A. J.; Waymouth, R. M. *J. Am. Chem. Soc.* **2018**, *140*, 748–757.
10. Minisci, F.; Citterio, A.; Vismara, E.; Fontana, F.; De Bernardinis, S.; Correale, M. *J. Org. Chem.* **1989**, *54*, 728–731.
11. Lindberg, B.; Slessor, K. N. *Acta Chem. Scand.* **1967**, *21*, 910–914.
12. Wirén, E.; Ahrgren, L.; N. De Belder, A. *Carbohydr. Res.* **1976**, *49*, 201–207.
13. Lui, H.-M.; Sato, Y.; Tsuda, Y. *Chem. Pharm. Bull.* **1993**, *41*, 491–501.

17 Synthesis of Dibenzyl 2,3,4,6-Tetra-O-Benzyl-α-D-Mannopyranosyl Phosphate

Sanaz Ahmadipour, and Gavin J. Miller*
Lennard-Jones Laboratory, School of Chemical and
Physical Sciences, Keele University,
Keele, United Kingdom

Bettina Riedl[a]
University of Vienna, Institute of Organic Chemistry,
Vienna, Austria

CONTENTS

Sugar-1-phosphates are essential materials for chemical and chemoenzymatic approaches to native and mimetic sugar-nucleotide targets, and there are a wide range of methodologies available to install the phosphate group at the anomeric position of monosaccharides.[1] Within chemical synthesis approaches, the phosphate is commonly added at a late stage (often in protected form) through the functionalization of a hemiacetal or using an appropriate donor group for reaction with a phosphate aglycone, followed by deprotection. The fully *O*-benzylated form of α-D-mannose-1-phosphate **3** has been synthesized by Kragl[2] using LDA/tetrabenzylpyrophosphate functionalization of the hemiacetal, and both Schmidt[3] and Turner[4] utilized an *O*-trichloroacetimidate donor with dibenzyl phosphate **5** to access **3**. More recently Lowary[5] used a thioethyl donor with NIS/AgOTf and **5** to install an anomeric phosphate, applied across a series of D-Man*p* mimetics. We include here our preparation of

---

* Corresponding author: g.j.miller@keele.ac.uk
[a] Checker: Bettina.riedl@univie.ac.at

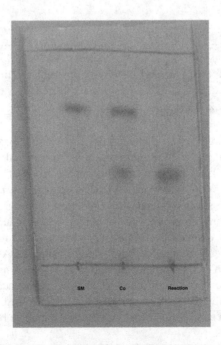

3 using *S*-phenyl thioglycoside donor **2** (readily available from D-mannose **1**[6,7]) and **5** with an NIS/AgOTf activation system. Initially, we found the reaction to give inconsistent yields of **3** (30–50%) and thus considered the quality of commercial **5**, evaluating several different batches with no clear improvements. We thus explored a synthesis of **5** from methyl dichlorophosphate **4** and BnOH in pyridine.[8] Using newly prepared **5** we observed a significant increase in yield for isolated **3** (82–90%) and generally a cleaner conversion (thin layer chromatography [TLC], Figure 17.1) and believe this will be of note for general sugar-1-phosphate syntheses utilizing **5** as the acceptor.

FIGURE 17.1   TLC for conversion of **2** to **3** (1:1 hexane–EtOAc). Left lane is **2**, center lane is co-spot, and right lane is reaction aliquot.

EXPERIMENTAL

GENERAL METHODS

All reagents and solvents which were available commercially were purchased from Acros, Alfa Aesar, Fisher Scientific or Sigma Aldrich. All reactions in nonaqueous solvents were conducted in flame-dried glassware under a nitrogen atmosphere with a magnetic stirring device. Solvents were purified by passing through activated alumina columns and used directly from a Pure Solv-MD solvent purification system and were transferred under nitrogen. Reactions requiring low temperatures used the following cooling baths: −78°C (dry ice/acetone), −30°C (dry ice/acetone), −15°C (NaCl/ice/water), and 0°C (ice/water). Infrared spectra were recorded neat on a Perkin Elmer Spectrum 100 FT-IR spectrometer; selected absorbencies (v_{max}) are reported in cm⁻¹. ¹H NMR spectra were recorded at 400 MHz and ¹³C spectra at 100 MHz, respectively, using a Bruker AVIII400 spectrometer. ¹H NMR signals were assigned with the aid of gDQCOSY. ¹³C NMR signals were assigned with the aid of gHSQCAD. Coupling constants are reported in Hertz. Chemical shifts (δ, in ppm) are standardized against the deuterated solvent peak. NMR data were analyzed using Nucleomatica iNMR software. ¹H NMR splitting patterns were assigned as follows: s (singlet), d (doublet), t (triplet), dd (doublet of doublets), ddd (doublet of doublet of doublets), or m (multiplet and/or multiple resonances). Reactions were followed by TLC using Merck silica gel 60F254 analytical plates (aluminum support) and were developed using standard visualizing agents: short wave UV radiation (245 nm) and 5% sulfuric acid in methanol/Δ. Solutions in organic solvents were dried with Na_2SO_4 or $MgSO_4$ and concentrated at <40°C. Purification *via* flash column chromatography was conducted using silica gel 60 (0.043–0.063 mm). Melting points were recorded using open glass capillaries on a Gallenkamp melting points apparatus and are uncorrected. Optical activities were recorded on automatic polarimeter Rudolph autopol I (concentration in g/100 mL). MS and HRMS (ESI) were obtained on Waters (Xevo, G2-XS Tof) or Waters Micromass LCT spectrometers using a methanol mobile phase. High-resolution (ESI) spectra were obtained on a Xevo, G2-XS Tof mass spectrometer. HRMS was obtained using a lock-mass to adjust the calibrated mass scale. Methyl dichlorophosphate and benzyl alcohol (anhydr. Sureseal) were purchased from Sigma Aldrich and used as received.

DIBENZYL PHOSPHATE (5)

Methyl dichlorophosphate (5.0 mL, 50.0 mmol, 1.0 equiv.) was added dropwise over 15 min under a nitrogen atmosphere to a stirred solution of dry pyridine (50 mL) at 0°C. The mixture was kept at 0°C for 15 min, whereupon a precipitate of *N*-methylpyridinium dichlorophosphate formed. Benzyl alcohol (13.0 mL, 125 mmol, 2.4 equiv.) was added and the mixture stirred overnight at room temperature (18 h). The mixture was poured onto 10% aq. $NaHCO_3$ (200 mL) and the pyridine evaporated under reduced pressure. The solution was diluted with deionized H_2O (200 mL) and then extracted with diethyl ether (2 × 200 mL), to remove benzyl chloride and benzyl alcohol. The aqueous layer was acidified with 2 M HCl (c 70 mL) to pH 1 and extracted with 7:3 chloro-form–n-butanol (2 × 150 mL). The extract was washed with 0.5 M HCl (100 mL) and H_2O (100 mL), dried, filtered, and concentrated under reduced pressure, to give a colorless oil. When stored in the freezer overnight, the oil turned into a pale yellow solid which was triturated with cold hexane (50 mL) to give **5** (6.21 g, 22.3 mmol, 45%) as a white powder; mp 74–76°C (crystallization from solvent was not attempted), Lit[8] 78°C; v_{max}(neat)/cm^{-1}: 3033 (m), 1496 (s), 1455 (s), 1215 (s), 957 (br); ^{1}H NMR (400 MHz, CDCl$_3$) δ 7.34–7.25 (m, 10 H, ArH), 4.99 (d, 4 H, J_{H-P} 7.5 Hz, CH_2Ph); ^{13}C NMR (100 MHz, CDCl$_3$) δ 135.8 (d, J_{C-P} 7.5 Hz, 2C^q, ArC), 128.5 (ArC), 128.4 (ArC), and 127.8 (ArC), 69.1 (d, J_{C-P} 5.4 Hz, 2 × CH_2Ph); ^{31}P NMR (101 MHz, CDCl$_3$) δ −0.02.

DIBENZYL 2,3,4,6-TETRA-*O*-BENZYL-α-D-MANNOPYRANOSYL PHOSPHATE (3)

Thioglycoside donor **2** (0.54 g, 0.85 mmol, 1.0 equiv.) was co-evaporated with toluene (3 × 6 mL) before being dissolved in dry dichloromethane (7.0 mL). Dibenzyl phosphate **5** (0.28 g, 1.1 mmol, 1.2 equiv.) and powdered 4 Å molecular sieves were added, and the mixture was stirred at room temperature for 40 min. After cooling to −30° C, silver trifluoromethanesulfonate[*] (0.10 g, 0.42 mmol, 0.5 equiv.) and *N*-iodosuccinimide (0.28 g, 1.27 mmol, 1.5 equiv.) were added sequentially. The reaction mixture was stirred for 3 h as the temperature was allowed to rise from −30°C to 0°C, and the progress of the reaction was monitored by TLC (1:1 hexane–EtOAc). When the reaction was complete, saturated aqueous $Na_2S_2O_3$ (4.0 mL) was added, the mixture was filtered through Celite®, the filtrate was diluted with dichloromethane (4 mL), and washed with saturated aqueous NaHCO$_3$ (4.0 mL) and saturated aqueous NaCl (4.0 mL). The layers were separated and the organic layer was dried, filtered, and concentrated under reduced pressure. Chromatography (4:1 hexane–EtOAc) furnished **3** (0.56 g, 0.70 mmol, 82%) as a clear oil. R_f 0.38 (1:1 hexane–EtOAc); $[\alpha]_D^{20}$ +23.2 (c 1.0, CHCl$_3$), Lit.[10] +52.1 (c 0.073, CHCl$_3$), Lit.[11] +24.2 (c 1.0, CHCl$_3$); v_{max}(neat)/cm^{-1}: 3030 (m), 2865 (m), 1496 (s), 1453 (s), 1273 (m), 941 (m); ^{1}H NMR (400 MHz, CDCl$_3$) δ 7.39–7.25 (m, 28 H, ArH), 7.21–7.17 (m, 2H, ArH), 5.81 (dd, 1H, J_{H-P} 6.1 Hz, J_{1-2} 1.9 Hz, H-1), 5.03–4.93 (m, 4H, CH_2Ph), 4.85 (d, 1H, J 10.8 Hz, CH_2Ph), 4.69 (s, 1H, CH_2Ph), 4.59 (d, 1H, J 12.0 Hz, CH_2Ph), 4.53–4.40 (m, 4H, CH_2Ph), 4.03 (t, 1H, J 9.7 Hz, H-4), 3.89 (ddd, 1H, J 10.0 Hz, 4.4 Hz, 1.4 Hz, H-5), 3.85 (dd, 1H, J 9.6 Hz, 3.2 Hz, H-3), 3.75–3.72 (m, 2H, H-2, H-6a), 3.59 (dd, 1H, $J_{5,6b}$ 1.7 Hz, $J_{6a,6b}$ 11.0 Hz, H-6b); ^{13}C NMR (100 MHz, CDCl$_3$) δ 138.3, 138.2, 137.8, 136.2, 136.1, 136.0, 129.0, 128.9, 128.8, 128.7, 128.7, 128.3, 128.2, 128.2, 128.2, 128.0, 128.0, 127.9, 96.7 (d, J_{C-P} = 6.1 Hz, C_1), 78.3 (C_3), 75.5 (CH_2Ph), 74.9 (d, J_{C-P} 9.3 Hz, C_2), 74.5 (C_5), 74.3 (C_4), 73.8 (CH_2Ph), 73.1 (CH_2Ph), 72.6 (CH_2Ph), 69.8 (dd, J_{C-P} 5.4, 2.3 Hz, 2 × CH_2Ph), 69.1 (C_6); δ_P (101 MHz CDCl$_3$) −2.61. These data are in agreement with those previously reported[2] and invalidate the optical rotation value given in Ref.[10]. ESI-QTOF MS: *m/z* [M + NH$_4$]$^+$ calcd. for

---

[*] Fresh AgOTf should be used as a significant reduction in the final isolated yield of **3** was observed when using older batches.

$C_{48}H_{53}O_9PN$ 818.3452; found, 818.3475; Anal. calcd For $C_{48}H_{49}O_9P$: C, 71.99; H, 6.17. Found: C, 71.67; H, 6.32.

ACKNOWLEDGMENTS

The EPSRC [EP/P000762/1] are thanked for project grant funding. We also thank the EPSRC UK National Mass Spectrometry Facility (NMSF) at Swansea University.

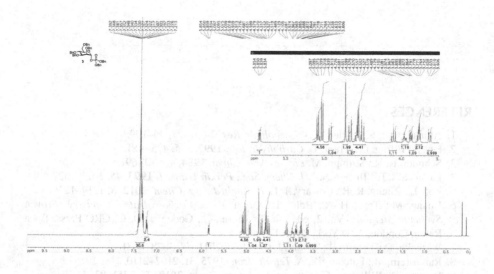

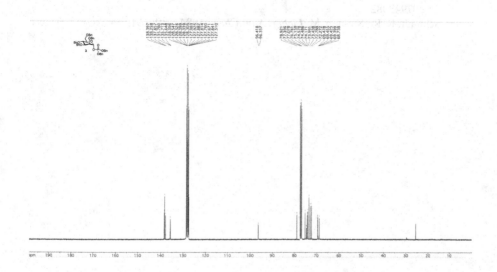

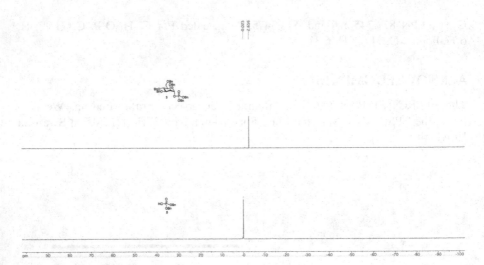

REFERENCES

1. Ahmadipour, S.; Miller, G. J. *Carbohydr. Res.* **2017**, *451*, 95–109.
2. Fey, S.; Elling, L.; Kragl, U. *Carbohydr. Res.* **1997**, *305*, 475–481.
3. Schmidt, R. R.; Stumpp, M. *Liebigs Ann. Chem.* **1984**, *4*, 680–691.
4. Pallanca, J. E.; Turner, N. J. *J. Chem. Soc., Perkin Trans. 1*, **1993**, *45*, 3017–3022.
5. Zou, L.; Zheng, R. B.; Lowary, T. L. *Beilstein J. Org. Chem.* **2012**, *8*, 1219–1226.
6. Huang, M.; Tran, H.-A.; Bohé, L.; Crich, D, in *Carbohydrate Chemistry: Proven Synthetic Methods*, Vol 2, ed. van der Marel, G., Codee, J. D. C, CRC Press, Boca Raton, London, New York, **2014**, 175–181.
7. Waschke, D.; Thimm, J.; Thiem, J. *Org. Lett.* **2011**, *13*, 3628–3631.
8. Rubinstein, M.; Patchornik, A. *Tetrahedron*, **1975**, 31, 2107–2110.
9. Bradley, D.; Williams, G.; Lawton, M. *J. Org. Chem.* **2010**, *75*, 8351–8354.
10. Inuki, S.; Aiba, T.; Kawakami, S.; Akiyama, T.; Inoue, J.; Fujimoto, Y. *Org. Lett.* **2017**, *19*, 3079–3082.
11. Schmidt, R. R.; Stumpp, M. *Liebigs Ann. Chem.* **1984**, 680–691.

18 Synthesis and Characterization of (+)-3-C-Nitromethyl-1,2:5,6-di-O-Isopropylidene-α-D-Allofuranose

Viktors Kumpiņš, Jevgeņija Lugiņina, and Māris Turks*
Faculty of Materials Science and Applied Chemistry,
Riga Technical University,
Riga, Latvia

Nino Trattnig[a]
Department of Chemistry,
University of Natural Resources and Life Sciences,
Vienna, Austria

CONTENTS

Nitro sugars are considered important synthetic intermediates in glycochemistry.[1] Nitro alcohols **2a** and **2b** have been recently used as key starting materials for the synthesis of carbopeptoids and their triazole isosters.[2] Diastereoisomeric mixture **2a,b** can be easily dehydrated by the use of Moffatt protocol and the obtained α,β-unsaturated nitromethylene derivative is susceptible to Michael additions with various nucleophiles.[3] Combination of the latter sequence with isoxazole formation in a 1,3-dipolar cycloaddition reaction opens possibilities for the synthesis of various synthetic glycoconjugates.[4] Title compound **2a** and its isomer **2b** have been used also

---
* Corresponding author: maris.turks@rtu.lv
[a] Checker: nino.trattnig@boku.ac.at

as starting materials in the synthesis of triazolyl carbohydrates.[5] Nevertheless, on many occasions pure isomer **2a** is required. Thus, Liu and coworkers have reported a chiral pool-based multistep synthesis of (3S,4S)-3-((R)-1,2-dihydroxyethyl) pyrrolidine-3,4-diol from compound **2a**.[6] Also a practical access to carbohydrate-based spiro-oxazolidinones starting from nitromethyl furanose **2a** has been developed.[7] Furthermore, nitro alcohol **2a** could be easily converted into (thio)urea organocatalysts which were studied for Friedel–Crafts alkylation of indoles with β-nitrostyrenes.[8]

(+)-C-nitromethyl-1,2:5,6-di-O-isopropylidene-α-D-allofuranose (**2a**) is usually formed as the major isomer in Henry reaction[9b] between ketone **1a** or its hydrate **1b** and nitromethane. This reaction is known for more than 45 years.[9] Various bases have been used to generate nitromethane anion.[9g,9b] Isomerization **2a**↔**2b** under basic reaction conditions has been also observed.[9d] The mechanism of the latter is accepted as a sequence of retro-Henry—Henry addition with adduct **2a** being the kinetic product. Therefore, the choice of the base, solvent, concentration, and reaction temperature showed major influence on reaction diastereoselectivity.[9c] Herein, we describe a more detailed synthetic protocol[7] that provides Henry adduct **2a** with excellent diastereomeric purity and adequate isolated yield. This procedure is easy scalable as selective precipitation is used as the only method for isolation and purification of the title product. The target compound **2a** is easily obtained according to work-up methods A or B. Method A provides product in 54% yield with excellent purity. On the other hand, method B provides higher yield for product **2a** (74%) albeit the isolated substance contains up to 5% of its diastereoisomer **2b**. The product quality obtained by the work-up B is perfectly acceptable for use in many experimental procedures in which the presence of minor amounts of other isomer is not crucial (*e.g.*, subsequent dehydration procedure[3,4]).

EXPERIMENTAL

GENERAL METHODS

The reaction was performed in open beakers without protecting atmosphere using commercially available reagents and solvents. ^1H NMR and ^{13}C NMR spectra were recorded on a *Bruker Avance 300 MHz* spectrometer in $CDCl_3$. Chemical shifts (δ) values are reported in ppm. The residual non-deuterated solvent peaks were used as the internal reference ($CDCl_3$, 7.26 ppm for ^1H NMR and 77.0 ppm for ^{13}C NMR). Coupling constants J are reported in Hz, and coupling patterns are described as s = singlet, d = doublet, and t = triplet. Unambiguous nuclei–signal assignments were made with the aid of Heteronuclear Single Quantum Correlation and Correlation Spectroscopy (COSY) NMR experiments. Pulse calibration (90° pulse) and adjustment of D_1 ($n \times T_1$) for qNMR experiment allowed to reach signal-to-noise-ratio sufficient for the determination of impurities ≤0.5%. IR spectra were recorded in KBr matrix on *FT-IR Perkin Elmer Spectrum BX* spectrometer. Wave numbers are given in reciprocal centimeters (cm^{-1}). Melting points were recorded with a *Fisher Digital Melting Point Analyzer Model 355* apparatus and are uncorrected. Optical rotation was measured at 20°C on *Anton Paar MCP 500* polarimeter (1-dm cell) using a sodium lamp as the light source (589 nm). Elemental analyses were performed using a *Carlo-Erba Instruments EA1108 Elemental Analyzer*.

(+)-3-C-NITROMETHYL-1,2:5,6-DI-O-ISOPROPYLIDENE-α-D-ALLOFURANOSE (2A)

Nitromethane (5.90 mL, 109.9 mmol, 5 equiv.) was added to a cooled (0°C) solution of NaOH (1.75 g, 44.0 mmol, 2 equiv.) in MeOH (20 mL), and the resulting white suspension was stirred for 10 min at 0°C. The latter suspension was slowly added to a precooled solution of ketone hydrate **1a**/ketone **1b** mixture[7,10] (6.0 g, 22.0 mmol, **1a/1b** = 18/82 according to ^{1}H NMR)* in MeOH (10 mL) at −10°C. The reaction mixture was stirred for 3 h while the internal temperature was allowed to rise to −2°C.

Work-up method **A**: the mixture was poured into cold (0 to +5°C) H_2O (45 mL) and solid $(NH_4)_2SO_4$ (15 g) was added. The resulting suspension was vigorously stirred for 20 min at 0 to +5°C. The solid formed was filtered and thoroughly washed with cooled H_2O (4 × 20 mL) to remove residual $(NH_4)_2SO_4$. The solid was dried in oven at 70°C at atmospheric pressure for 24 h to give **2a** (3.79 g, 54%, ^{1}H NMR purity >99.5%).

Work-up method **B**: the reaction mixture was poured into cold (0 to +5°C) H_2O (70 mL), and solid $(NH_4)_2SO_4$ (50 g) was added. The resulting suspension was vigorously stirred for 20 min at 0 to +5°C. The obtained solid was filtered, washed on the filter with cold H_2O (2 × 20 mL). Then the solid was dissolved in DCM (2 × 30 mL) and the organic layer was washed successively with H_2O (2 × 20 mL) and brine (10 mL), dried over Na_2SO_4 and filtered. The filtrate was concentrated and the residue was dried *in vacuo* (1 mbar) for 12 h. Product **2a/2b** (5.19 g, 74%; **2a/2b** = 95:5 according to ^{1}H NMR) was obtained; mp 110°C (Lit.[7] mp 110°C; Lit.[6] mp 110–112°C); $[\alpha]_D^{20}$ +24.4 (*c* 1.0, CHCl$_3$); $[\alpha]_D^{20}$ +24.2 (*c* 1.0, MeOH); [Lit.[7] $[\alpha]_D^{23}$ +25 (*c* 0.53, CHCl$_3$); Lit.[6] $[\alpha]_D^{20}$ +95 (*c* 1.0, MeOH)]. IR (KBr) v, cm^{-1}: 3600–3200 (bs), 2985, 1560, 1375, 1210, 1080, 1015. ^{1}H NMR (300 MHz, CDCl$_3$): δ 5.84 (d, 1H, $^3J_{1,2}$ 3.8 Hz, H-1), 4.96 (d, 1H, AB syst., $^2J_{3'a, 3'b}$ 12.0 Hz, H-3'a), 4.88 (d, 1H, $^3J_{2,1}$ 3.8 Hz, H-2), 4.49 (d, 1H, AB syst., $^2J_{3'b, 3'a}$ 12.0 Hz, H-3'b), 4.13 (dd, 1H, AB syst., $^2J_{6a, 6b}$ 8.2 Hz, $^3J_{6a, 5}$ 5.7 Hz, H-6a), 4.02 (ddd, 1H, $^3J_{5, 4}$ 8.4 Hz, $^3J_{5, 6a}$ 5.7 Hz, $^3J_{5, 6b}$ 4.7 Hz, H-5), 3.95 (dd, 1H, AB syst., $^2J_{6b, 6a}$ 8.2 Hz, $^3J_{6b, 5}$ = 4.7 Hz, H-6b), 3.89 (d, 1H, $^3J_{4, 5}$ 8.4 Hz, H-4), 3.27 (s, 1H, HO-3), 1.60, 1.46, 1.38, 1.35 (4s, 12H, 2 × $(H_3C)_2C$).^{13}C NMR (75.5 MHz, CDCl$_3$): δ 113.4, 110.5 (2 Me$_2$C), 103.7 (C-1), 81.7 (C-4), 79.9 (C-2), 78.6 (C-3), 77.1 (CH$_2$N), 73.0 (C-5), 68.0 (C-6), 26.7, 26.6, 26.5, 25.1 (4 × Me). Anal. calcd for $C_{13}H_{21}NO_8$: C, 48.90; H, 6.63; N, 4.39. Found: C, 48.73; H, 6.55; N, 4.35.

---

* Any ratio of ketone **1a** and its hydrate **1b** can be successfully used. The checker used the starting material with **1a/1b** ratio 98:2 and obtained identical yield and purity of **2a**.

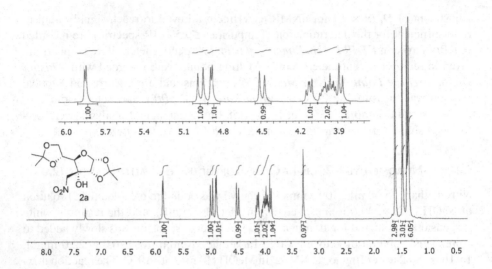

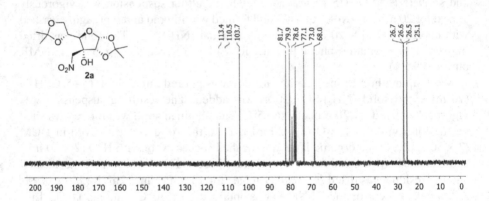

REFERENCES

1. Estevez, A. M.; Wessel, H. P. *Curr. Org. Chem.* **2014**, *18*, 1846–1877.
2. Rjabovs, V.; Ostrovskis, P.; Posevins, D.; Kiseļovs, G.; Kumpiņš, V.; Mishnev, A.; Turks, M. *Eur. J. Org. Chem.* **2015**, 5572–5584.
3. (a) Luginina, J.; Rjabovs, V.; Belyakov, S.; Turks, M. *Carbohydr. Res.* **2012**, *350*, 86–89. (b) Ivanovs, I.; Berziņa, S.; Luginina, J.; Belyakov, S.; Rjabovs, V. *Heterocycl. Commun.* **2016**, *22*, 95–98.
4. Luginina, J.; Rjabovs, V.; Belyakov, S.; Turks, M. *Tetrahedron Lett.* **2013**, *54*, 5328–5331.
5. Grigorjeva, J.; Uzuleņa, J.; Rjabovs, V.; Turks, M. *Chem. Heterocycl. Compd.* **2015**, *51*, 883–890.
6. Zhang, E.; Bai, P. Y.; Sun, W.; Wang, S.; Wang, M. M.; Xu, S. M.; Liu, H. M. *Carbohydr. Res.* **2016**, *434*, 33–36.

7. Turks, M.; Rodins, V.; Rolava, E.; Ostrovskis, P.; Belyakov, S. *Carbohydr. Res.* **2013**, *375*, 5–15.
8. Turks, M.; Rolava, E; Stepanovs, D.; Mishnev, A; Marković, D. *Tetrahedron: Asymmetry*, **2015**, *26*, 952–960.
9. (a) Albrecht, H. P.; Moffatt, J. G. *Tetrahedron Lett.* **1970**, 11, 1063–1066; (b) Rosenthal, A.; Ong, K.-S.; Baker, D. *Carbohydr. Res.* **1970**, 13, 113–125; (c) Sato, K.-I.; Yoshimura, J.; Shin, C.-G. *Bull. Chem. Soc. Jpn.* **1977**, 50, 1191–1194 (d) Sato, K.-I.; Koga, K.; Hashimoto, H.; Yoshimura, *J. Bull. Chem. Soc. Jpn.* **1980**, *53*, 2639–2641; (e) Hart, D. J.; Patterson, S.; Unch, J. P. *Synlett*, **2003**, 1334–1338; (f) Munos, J. W.; Pu, X.; Lio, H.-w. *Bioorg. Med. Chem. Lett.* **2008**, *18*, 3090–3094; (g) Filichev, V. V.; Brandt, M.; Pedersen, E. B. *Carbohydr. Res.* **2001**, *333*, 115–122.
10. (a) Ostrovskis, P.; Mackeviča, J.; Kumpiņš, V.; López, Ó.; Turks, M. An alternative, large scale synthesis of 1,2:5,6-di-*O*-isopropylidene-α-d-ribohex-3-ulofuranose. G. van der Marel, J. Codee (Eds.), *Carbohydrate Chemistry: Proven Synthetic Methods*, Vol. 2, CRC Press, Boca Raton, London, New York (**2014**), pp. 275–281; (b) Both ketone hydrate **1a** (CAS: 10578-85-5) and ketone **1b** (CAS: 2847-00-9) are commercially available.

19 Synthesis and Characterization of Propargyl 2,3,4,6-Tetra-O-Acetyl-β-D-Glucopyranoside

Viktors Kumpiņš, Jevgeņija Luģiņina, and Māris Turks*
Faculty of Materials Science and Applied Chemistry,
Riga Technical University,
Riga, Latvia

Daniela Imperio[a]
Dipartimento di Scienze del Farmaco, Università degli Studi
del Piemonte Orientale "Amedeo Avogadro",
Novara, Italy

CONTENTS

The propargyl function in organic synthesis in general and in carbohydrate chemistry in particular is a well-established dipolarophile component in 1,3-dipolar cycloaddition reaction for the synthesis of different triazole[1] or isoxazole[2] derivatives. Another useful application of propargyl glucopyranoside has been reported in a one-pot three-component reaction of alkyne-aldehyde-aniline for the successful construction of quinoline-based glycoconjugates.[3] Dendritic glycoclusters and other important glycoconjugates could be easily synthesized *via* palladium-catalyzed Sonogashira reactions.[4] Different methods that allow to introduce the propargyl group at the anomeric position using glycosyl halides,[5] trichloroacetimidates,[6] acetates,[7] or partially protected carbohydrates have been developed.[8] However, for a potential scale-up, a cheap starting material and a relatively robust

* Corresponding author: maris.turks@rtu.lv
[a] Checker under supervision of Luigi Panza: luigi.panza@uniupo.it

experimental procedure is required. Herein, we report a more detailed protocol for the synthesis of 2,3,4,6-tetra-O-acetyl-β-D-glucopyranoside **2** than that described,[7] from 1,2,3,4,6-penta-O-acetyl-β-D-glucopyranose **1** and propargyl alcohol in the presence of boron trifluoride etherate.

EXPERIMENTAL

GENERAL METHODS

The reaction was performed using commercial reagents and solvents purified prior to use according to standard procedures.[9] Reactions were monitored by TLC (*E. Merck Kieselgel 60 F$_{254}$*). Solutions in organic solvents were dried with Na_2SO_4 and concentrated at <40°C. Detection was carried out by spraying the plates with a solution containing cerium(IV) sulfate (2.0 g), H_2SO_4 (1.2 mL) in H_2O (18 mL) followed by heating. ^{1}H NMR and ^{13}C NMR spectra were recorded on a *Bruker Avance 300 MHz* spectrometer in CDCl$_3$. Chemical shift (δ) values are reported in ppm. The residual non-deuterated solvent peaks were used as the internal reference (CDCl$_3$, 7.26 ppm for ^{1}H NMR and 77.0 ppm for ^{13}C NMR). Coupling constants *J* are reported in Hz and coupling patterns are described as s = singlet, d = doublet, and t = triplet. Proton signals were assigned with the aid of correlation spectroscopy (COSY) and nuclear Overhauser effect spectroscopy (NOESY) 2D-NMR experiments. Unambiguous carbon–signal assignments were made with the aid of Heteronuclear Single Quantum Correlation (HSQC) NMR experiments. Pulse calibration (90° pulse) and adjustment of D1 = 60 s (D1 ≫ T1) for qNMR experiment allowed to reach signal-to-noise-ratio sufficient for determination of impurities ≤0.5%. IR spectra were recorded in KBr matrix on *FT-IR Perkin Elmer Spectrum BX* spectrometer. Wave numbers are given in reciprocal centimeters (cm^{-1}). Melting points were recorded with a *Fisher Digital Melting Point Analyzer Model 355* apparatus and are uncorrected. Optical rotation was measured at 20°C on *Anton Paar MCP 500* polarimeter (1-dm cell) using a sodium lamp as the light source (589 nm). Elemental analyses were performed using a *Carlo-Erba Instruments EA1108 Elemental Analyzer.*

PROPARGYL 2,3,4,6-TETRA-O-ACETYL-β-D-GLUCOPYRANOSIDE (2)

$BF_3 \cdot OEt_2$ (8.70 mL, 70.44 mmol, 5.5 equiv) was added dropwise with stirring, at 0°C during 5 min, to a mixture of propargyl alcohol (1.30 mL, 21.77 mmol,

1.7 equiv) and 1,2,3,4,6-penta-O-acetyl-β-D-glucopyranose[10] (**1**, 5.00 g, 12.81 mmol, 1.0 equiv) in anhydrous DCM (80 mL) under argon atmosphere. The cooling was removed, and the stirring was continued for 6 h at 20°C (TLC, 1:5 EtOAc–DCM).[*] The mixture was slowly added to ice-cold aqueous $NaHCO_3$ (20 g, ~400 mL). After warming to room temperature, the organic phase was separated and the aqueous phase was extracted with DCM (2 × 100 mL). The organic layers were combined and washed successively with H_2O (100 mL) and brine (2 × 100 mL), dried, filtered, and the filtrate was concentrated. The residue was kept *in vacuo* (1 mbar) for 12–16 h to remove residual propargyl alcohol. The solid formed[†] was crystallized (1:5 DCM–cyclohexane; ~15 mL/g), to give pure (NMR) compound **2** (3.58 g, 72%; mp 116–117°C; $[\alpha]_D^{20}$ −45 (*c* 1.0, $CHCl_3$), (Lit.[11a,b] 113–114°C; Lit.[11c] 114–115°C; Lit.[5a] 116°C; Lit.[11d] 112–113°C [hexane–DCM]; Lit.[11e] 102–104°C; Lit.[11f] 116–117°C [EtOH]; Lit.[11g] 119–121°C; Lit.[11h] 117–118°C); (Lit.[11a] $[\alpha]_D^{25}$ −22.9 [*c* 1.5, $CHCl_3$]; Lit.[11c] $[\alpha]_D^{20}$ −43.2 [*c* 1.0, $CHCl_3$]; Lit.[5a] $[\alpha]_D$ −36.2 [*c* 1.0, $CHCl_3$]; Lit.[11d] $[\alpha]_D^{25}$ −40 [*c* 1.0, $CHCl_3$]; Lit.[11b] $[\alpha]_D^{20}$ −48.4 [*c* 2.0, $CHCl_3$]; Lit.[11f] $[\alpha]_D^{20}$ −43.4 [*c* 0.9, $CHCl_3$]; Lit.[11h] $[\alpha]_D^{22}$ −43.4 [*c* 2.1, $CHCl_3$]); IR (KBr) ν, cm^{-1}: 3475, 2120, 1750, 1370, 1225, 1080, 1070, 1045, 915. ^1H NMR (300 MHz, $CDCl_3$): δ 5.24 (t, 1*H*, $^3J_{3,2} = {^3}J_{3,4}$ 9.6 Hz, H-3), 5.10 (t, 1*H*, $^3J_{4,3} = {^3}J_{4,5}$ 9.6 Hz, H-4), 5.02 (dd, 1*H*, $^3J_{2,3}$ 9.6 Hz, $^3J_{2,1}$ 8.0 Hz, H-2), 4.78 (d, 1*H*, $^3J_{1,2}$ 8.0 Hz, H-1), 4.37 (d, 2*H*, $^4J_{1',3'}$ 2.1 Hz, H-1'), 4.28 (dd, 1*H*, AB syst. $^2J_{6a,6b}$ 12.4 Hz, $^3J_{6a,5}$ 4.5 Hz, H-6a), 4.14 (dd, 1*H*, AB syst. $^2J_{6b,6a}$ 12.4 Hz, $^3J_{6b,5}$ 1.8 Hz, H-6b), 3.73 (ddd, 1*H*, $^3J_{5,4}$ 9.6 Hz, $^3J_{5,6a}$ 4.5 Hz, $^3J_{5,6b}$ 1.8 Hz, H-5), 2.45 (t, 1*H*, $^4J_{3',1'}$ 2.1 Hz, H-3'), 2.09, 2.06, 2.02, 2.01 (4 s, 12*H*, 4 Ac). ^{13}C NMR (75.5 MHz, $CDCl_3$): δ 170.8, 170.4, 169.6, 169.6 (4 CO), 98.2 (C-1), 78.2 (C-2'), 75.6 (C-3'), 72.9 (C-3), 72.0 (C-5), 71.0 (C-2), 68.4 (C-4), 61.9 (C-6), 56.08 (C-1'), 20.9, 20.8, 20.8, 20.7 (4 Ac). Anal. calcd for $C_{17}H_{22}O_{10}$: C 52.85; H 5.74. Found: C 52.76; H 5.70.

---

[*] R_f values of the starting material and the product are very close. Double elution of the TLC plate facilitates monitoring of the reaction progress: R_f (**1**) = 0.7; R_f (**2**) = 0.8.

[†] The main impurity observed in the crude product is 1,2,3,4,6-penta-O-acetyl-α-D-glucopyranose (α-**1**), which is formed during the reaction through anomerization of the starting material.

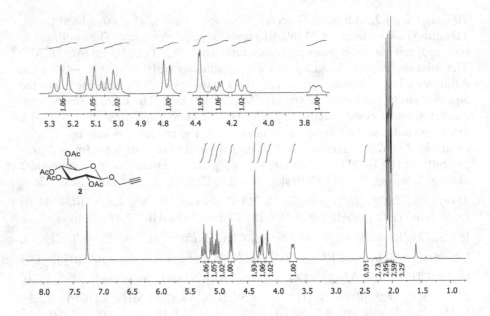

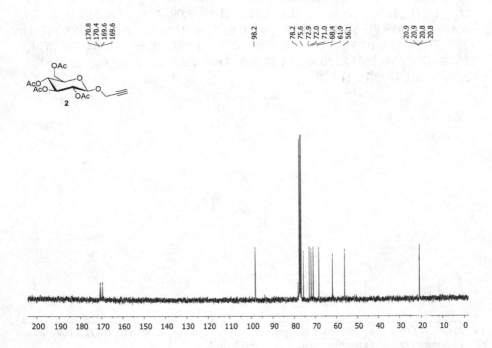

REFERENCES

1. (a) Dondoni, A. *Chem. Asian J.* **2007**, *2*, 700–708. (b) Cheng, J.; Gu, Z.; He, C.; Jin, J.; Wang, L.; Li, G.; Sun, B.; Wang, H.; Bai, J. *Carbohydr. Res.* **2015**, *414*, 72–77. (c) Uruma, Y.; Nonomura, T.; Yen, P. Y. M.; Edatani, M.; Yamamoto, R.; Onuma, K.; Okada, F. *Bioorg. Med. Chem.* **2017**, *25*, 2372–2377. (d) Mackeviča, J.; Ostrovskis, P.; Leffler, H.; Nilsson, U. J.; Rudovica, V.; Viksna, A.; Belyakov, S.; Turks, M. *Arkivoc*, **2014**, *3*, 90–112. (e) Strakova, I.; Kumpiņa, I.; Rjabovs, V.; Lugiņina, J.; Belyakov, S.; Turks, M. *Tetrahedron: Asymmetry* **2011**, *22*, 728–739.

2. (a) Calvo-Flores, F. G.; Isac-García, J.; Hernández-Mateo, F.; Pérez-Balderas, F.; Calvo-Asín, J. A.; Sanchéz-Vaquero, E.; Santoyo-González, F. *Org. Lett.* **2000**, *2*, 2499–2502. (b) Lugiņina, J.; Rjabovs, V.; Belyakov, S.; Turks, M. *Tetrahedron Lett.* **2013**, *54*, 5328–5331.

3. Kumar, K. K.; Das, T. M. *Carbohydr. Res.* **2011**, *346*, 728–732.

4. (a) Sengupta, S.; Sadhukhan, S. K. *Carbohydr. Res.* **2001**, *332*, 215–219. (b) Maggio, B.; Raimondi, M. V.; Raffa, D.; Plescia, F.; Scherrmann, M. C.; Prosa, N.; Lauricella, M.; D'Anneo, A.; Daidone, G. *Eur. J. Med. Chem.* **2016**, *122*, 247–256. (c) Verma, P. R.; Mandal, S.; Gupta, P.; Mukhopadhyay, B. *Tetrahedron Lett.* **2013**, *54*, 4914–4917.

5. (a) Sethi, K. P.; Kartha, K. P. R. *Carbohydr. Res.* **2016**, *434*, 132–135. (b) Tyagi, M.; Khurana, D.; Ravindranathan K. *Carbohydr. Res.* **2013**, *379*, 55–59.

6. (a) Thombal, R. S.; Jadhav, V. H. *J. Carbohydr. Chem.* **2016**, *35*, 57–68. (b) Beckmann, H. S. G.; Wittmann, V. *Org. Lett.* **2007**, *9*, 1–4.

7. (a) Daly, R.; Vaz, G.; Davies, A. M.; Senge, M. O.; Scanlan, E. M. *Chem. Eur. J.* **2012**, *18*, 14671–14679. (b) Rajaganesh, R.; Ravinder, P.; Subramanian, V.; Mohan Das, T. *Carbohydr. Res.* **2011**, *346*, 2327–2336. (c) Giovenzana, G. B.; Luigi, L.; Monti, D.; Palmisano, G.; Panza, L. *Tetrahedron*, **1999**, *55*, 14123–14136. (d) Kaufman, R. J.; Sidhu, R. S. *J. Org. Chem.* **1982**, *47*, 4941–4947. (e) Narayanaperumal, S.; Da Silva, R. C.; Monteiro, J. L.; Correa, A. G.; Paixao, M. W. *J. Braz. Chem. Soc.* **2012**, *23*, 1982–1988.

8. Kumar, K. K.; Kumar, R. M.; Subramanian, V.; Das, T. M. *Carbohydr. Res.* **2010**, *345*, 2297–2304.

9. Armarego, W.L.F.; Chai, L. L. C. *Purification of Laboratory Chemicals.* 5th Edition, Butterworth-Heinemann, Amsterdam, Boston, London, New York, Oxford, Paris, San Diego, San Francisco, Singapore, Sydney, Tokyo, **2003**.

10. (a) Joshi, V. Y.; Sawant, M. R. *Indian J. Chem., Sect B*, **2006**, *45*, 461–465. (b) 1,2,3,4,6-Penta-*O*-acetyl-β-ᴅ-glucopyranose 1 (CAS: 604-69-3) is also commercially available.

11. (a) Thombal, R. S.; Jadhav, V. H. *J. Carbohydr. Chem.* **2016**, *35*, 57–68. (b) Tyagi, M.; Khurana, D.; Ravindranathan K. *Carbohydr. Res.* **2013**, *379*, 55–59. (c) Beaulieu, R.; Attoumbre, J.; Gobert-Deveaux, V.; Grand, E.; Stasik, I.; Kovensky, J.; Giordanengo, P. Novel Solanidine-Derived Compounds. US 2016152659 A1, May 2, 2016. (d) Mandal, P. K. *RSC Adv.* **2014**, *4*, 5803–5814. (e) Rajaganesh, R.; Ravinder, P.; Subramanian, V.; Mohan Das, T. *Carbohydr. Res.* **2011**, *346*, 2327–2336. (f) Hoheisel, T. N.; Frauenrath, H. *Org. Lett.* **2008**, 10, 4525–4528. (g) Maki, T.; Ishida, K. *J. Org. Chem.* **2007**, *72*, 6427–6433. (h) Horisberger, M.; Lewis, B. A.; Smith, F. *Carbohydr. Res.* **1972**, 23, 144–147.

20 3-(2',3',4'-Tri-O-Acetyl-α-L-Fucopyranosyl)-1-Propene

Martina Kašáková, Benedetta Bertolotti, and Jitka Moravcová
University of Chemistry and Technology Prague
(UCT Prague), Czech Republic

Lei Dong, Audric Rousset, and Sébastien Vidal*
Institut de Chimie et Biochimie Moléculaires
et Supramoléculaires, Laboratoire de Chimie
Organique 2—Glycochimie, UMR 5246, CNRS,
Université Claude Bernard Lyon 1,
Villeurbanne, France

Nándor Kánya[a]
Department of Organic Chemistry, University of Debrecen,
Debrecen, Hungary

CONTENTS

C-Glycosyl compounds form a small family of natural products with a broad variety of biological activities. Formally, *C*-glycosyl compounds are analogs of *O*-glycosides with a carbon–carbon bond replacing the enzymatically labile carbon–oxygen anomeric bond.[1] This modification provides also conformational properties differing form the parent *O*-glycosides. *C*-Aryl glycosyl compounds are ubiquitous[2] but alkyl derivatives relatively scarce and are less studied. Glycosylation of 6-deoxy-hexoses such as L-rhamnose or L-fucose, especially the 1,2-*cis*-*O*-glycosylation, is quite challenging

* Corresponding author: sebastien.vidal@univ-lyon1.fr
[a] Checker under supervision of Prof. László Somsák: somsak.laszlo@science.unideb.hu

due to difficult stereocontrol of the anomeric bond. *C*-Allylation of peracetylated fucose **1** has been reported in several studies, but no synthetic procedure was reported.[3]

The present report describes a detailed experimental procedure for the *C*-allylation of peracetylated fucose **1**, which was developed based on studies cited above. Acetylation of L-fucose[4] described here provided a mixture of α-fucopyranose, β-fucopyranose, and α-fucofuranose (85:11:4)* which was directly used for the *C*-allylation. Literature precedence[3] indicated that the use of acetonitrile as solvent in other situations was beneficial for the α-stereoselectivity of *C*-allylation. Nicolaou *et al.*[5] reported that the introduction of a catalytic amount of TMSOTf in combination with BF₃·OEt₂ was highly effective in the syntheses of acetylated *C*-allyl glucopyranose derivatives on large scale (>25 g), and they later used this catalyst in the synthesis of *C*-allyl rhamnose.[6]

The use of TMSOTf was not required in our preparation, and the acetylated *C*-allyl fucoside **2α,β** (82:9:9 mixture of α-fucopyranose, β-fucopyranose, and fucofuranose) was obtained in excellent yield when using allyltrimethylsilane as a reagent under BF₃·OEt₂ catalysis. Because the desired α-anomer could not be obtained pure at this stage, the crude product was deacetylated, and crystallization from isopropanol provided pure **3α** whose acetylation afforded the pure target compound **2α**.

EXPERIMENTAL

GENERAL METHODS

All reagents were of the highest commercially available purity and were used without further purification. Reagent grade acetonitrile was distilled from P_2O_5 or dry acetonitrile 99.9+% (Extra dry, Acroseal®) purchased from Acros Organics was used. Thin-layer chromatography (TLC) was carried out on silica gel-coated aluminum sheets (60 F_{254}, Merck). Spots were revealed by charring with 10% H_2SO_4 in 95% EtOH. Solutions in organic solvents were dried with $MgSO_4$ and concentrated at <40°C. Column chromatography was performed with silica gel 60 (40–63 μm). Optical rotations were measured on a Rudolph Research Analytical Autopol VI system at 589 nm, at 25°C and with a 1-dm cell. NMR spectra were recorded in solvents indicated, at 293 K using a Varian Inova 600 MHz NMR system equipped with

---

* The ratio of three compounds is based on the ¹H NMR data using the methyl (H-6) resonance peaks.

a HCN triple resonance cold probe. Chemical shifts are referenced relative to the residual solvent peaks. The following abbreviations are used for peak multiplicities: s, singlet; d, doublet; t, triplet; q, quadruplet; m, multiplet; p, pseudo; and b, broad. NMR assignments were based on 1D and 2D (COSY, HSQC) experiments.

3-(α-L-FUCOPYRANOSYL)-1-PROPENE (3α)

To a solution of L-fucose (117 mg, 0.713 mmol) in dry pyridine (1.17 mL) was added, at 0°C under argon atmosphere, acetic anhydride (0.475 mL, 4.99 mmol, 7 eq) and N,N-dimethylaminopyridine (9 mg, 0. 087 mmol, 0.1 equiv), and the solution was stirred at rt overnight. The mixture was diluted with EtOAc (2 mL), washed with 1 M HCl (3 × 2 mL) and water (2 mL). The aqueous layers were combined and extracted with EtOAc (5 × 2 mL). The organic layers were combined, dried, filtered, and concentrated under reduced pressure. The crude syrupy 1',2',3',4'-tetra-O-acetyl-L-fucopyranose (**1,** unresolved mixture of isomers, NMR, 236 mg) was sufficiently pure (NMR) for the next step. $R_f = 0.6$ (2:1 PE–EtOAc).

BF$_3$·OEt$_2$ (0.3 mL, 2.5 mmol, 3.5 equiv) followed by allyltrimethylsilane (0.4 mL, 2.45 mmol, 3.45 equiv) were added at 0°C under inert atmosphere (Ar) to a stirred solution of **1** (236 mg, 0.71 mmol) in dry acetonitrile (5 mL). The solution was heated under reflux for 1.5 h, when TLC (3:1 PE–EtOAc) showed disappearance of the starting material. The solution was diluted with CH$_2$Cl$_2$ (5 mL) and the organic layer was washed with 2 M NaOH (2 × 5 mL) and brine (2 × 10 mL), dried (MgSO$_4$), filtered, and concentrated under reduced pressure to afford a syrup (213 mg) containing 3-(2',3',4'-tri-O-acetyl-L-fucopyranosyl)-1-propene (**2**)[†] as the main component; $R_f = 0.6$ (3:1 PE–EtOAc).

Et$_3$N (0.5 mL) was added to a solution of the foregoing crude 3-(2',3',4'-tri-O-acetyl-L-fucopyranosyl)-1-propene (**2α,β,** 0.195 g) in 4:1 MeOH–H$_2$O (2.5 mL), and the mixture was stirred at rt overnight. The solution was concentrated under vacuum and the residue was dissolved in hot 2-propanol (1 mL), and the solution was allowed to cool slowly to room temperature, upon which compound **3α** crystallized.[7] Crystals were filtered off, washed with Et$_2$O (2×5 mL), and dried under high vacuum to give compound (**3α**) (141 mg, 63% from L-fucose); mp 142–144°C, [α]$_D$ –109.6 (c 1.1, MeOH), $R_f = 0.35$ (1:1 PE–EtOAc); ^{1}H NMR (600 MHz, CD$_3$OD): δ 5.83 (ddt, 1H, J 17.1, 10.2, 6.8 Hz, CH=CH$_2$), 5.09 (ddd, 1H, J 17.1, 3.5, 1.6 Hz, =CH$_2$), 5.03 (m, 1H, =CH$_2$), 3.97–3.93 (m, 1H, H-1), 3.89 (dd, 1H, J 9.0, 5.5 Hz, H-2), 3.84 (qd, 1H, J 6.5, 1.9 Hz, H-5), 3.69 (dd, 1H, J 3.5, 1.9 Hz, H-4), 3.66 (dd, 1H, J 9.0, 3.5 Hz, H-3), 2.49–2.43 (m, 1H, CH_2CH=CH$_2$), 2.39–2.33 (m, 1H, CH_2CH=CH$_2$), 1.19 (d, 1H, J 6.5 Hz, H-6). ^{13}C NMR (150 MHz, CD$_3$OD): δ 136.3 (CH=CH$_2$), 116.0 (CH=CH$_2$), 75.3 (C-1), 71.9 (C-4), 71.4 (C-3), 69.1 (C-2), 68.1 (C-5), 30.1 (CH$_2$CH=CH$_2$), 15.8 (C-6). Anal. calcd for C$_{15}$H$_{22}$O$_7$: C, 57.32; H, 7.05. Found: C, 57.47; H, 7.25.

3-(2',3',4'-TRI-O-ACETYL-α-L-FUCOPYRANOSYL)-1-PROPENE (2α)

3-(α-L-Fucopyranosyl)-1-propene **3α** (5.5 g, 29.1 mmol) was dissolved in dry pyridine (60 mL), and acetic anhydride (30 mL, 319 mmol, 11 equiv) was added. The solution was stirred at room temperature until TLC (2:1 PE–EtOAc) showed that the reaction

---

[†] An 82:9:9 mixture of α-fucopyranose, β-fucopyranose, and fucofuranose.

was complete (~2 h). Ice (~30 g) was added slowly, the mixture was diluted with EtOAc (100 mL), and the mixture was washed with 1 M HCl (40 mL). The organic layer was separated and the aqueous layer was further extracted with EtOAc (100 mL). The organic layers were combined and washed with 1 M HCl (40 mL) and brine (100 mL), dried (MgSO$_4$), filtered, and concentrated under reduced pressure to afford pure, syrupy compound **2α** (8.7 g, 95%); [α]$_D$ −92.5 (*c* 1, CHCl$_3$), R_f = 0.8 (2:1 PE–EtOAc), ^{1}H NMR (600 MHz, C$_6$D$_6$): δ 5.69 (dd, 1*H*, *J* 10.2, 5.8 Hz, H-2), 5.64 (ddt, 1*H*, *J* 17.2, 10.2, 6.8 Hz, C*H*=CH$_2$), 5.39 (dd, 1*H*, *J* 10.2, 3.4 Hz, H-3), 5.36 (dd, 1*H*, *J* 3.4, 1.6 Hz, H-4), 5.02–4.95 (m, 2*H*, =CH$_2$), 4.34 (m, 1*H*, H-1), 3.38 (qd, 1*H*, *J* 6.4, 1.6 Hz, H-5), 2.32–2.25 (m, 1*H*, C*H*$_2$CH=CH$_2$), 2.16–2.11 (m, 1*H*, C*H*$_2$CH=CH$_2$), 1.76, 1.65, 1.58 (3 s, 3 × 3*H*, CH$_3$CO), 0.97 (d, 3*H*, *J* 6.4 Hz, H-6). ^{13}C NMR (150 MHz, C$_6$D$_6$): δ 170.2, 169.8, 169.3 (3 s, 3 × C=O), 134.7 (*C*H=CH$_2$), 117.0 (CH=*C*H$_2$), 72.5 (C-1), 71.2 (C-4), 69.0 (C-3), 68.6 (C-2), 65.6 (C-5), 30.6 (*C*H$_2$CH=CH$_2$), 20.4, 20. 2,20.1 (3s, 3 × CO*C*H$_3$), 16.1 (C-6). Anal. calcd for C$_9$H$_{16}$O$_4$: C, 57.43; H, 8.57. Found: C, 57.06; H, 8.57.

ACKNOWLEDGMENTS

The authors thank the Université Claude Bernard Lyon 1 and the CNRS for financial support. Financial support by the GACR 15-17572S project is gratefully acknowledged. L.D. is grateful to the China Scholarship Council for his PhD scholarship. The authors are grateful to Prof. Jean-Marc Lancelin at Institut des Sciences Analytiques (ISA—Université Lyon 1—CNRS) for acquiring the NMR data. The checking in Debrecen was supported by the EU and cofinanced by the European Regional Development Fund under the project GINOP-2.3.2-15-2016-00008.

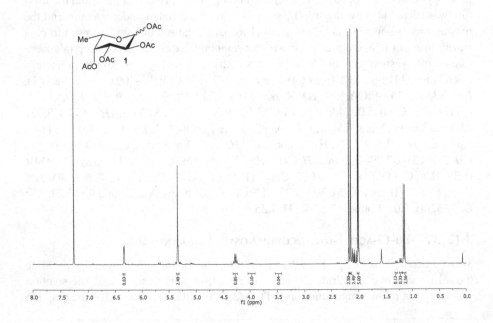

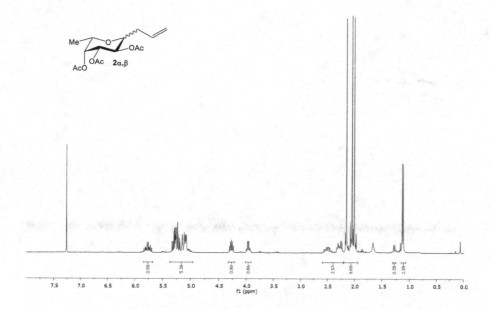

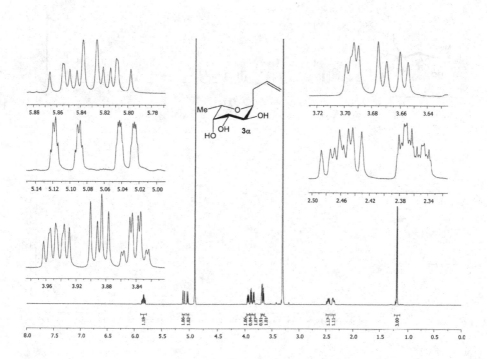

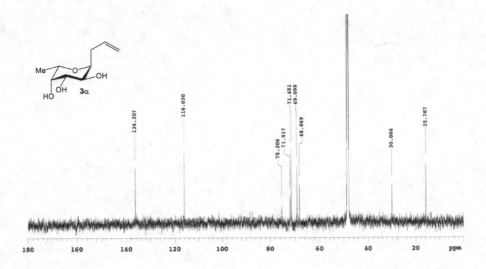

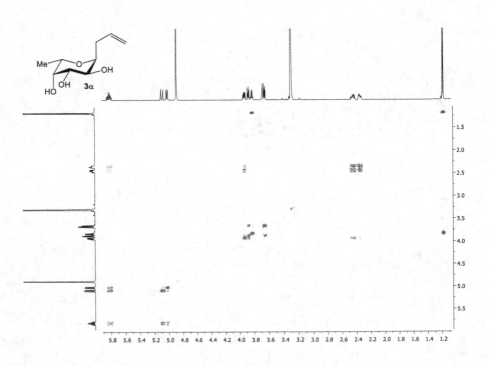

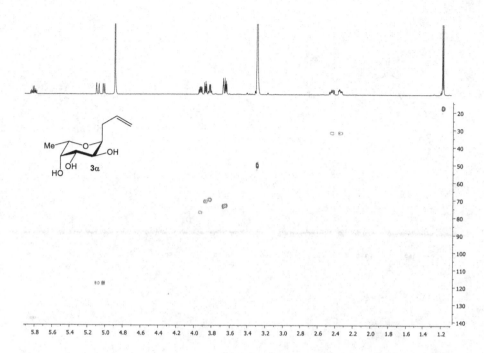

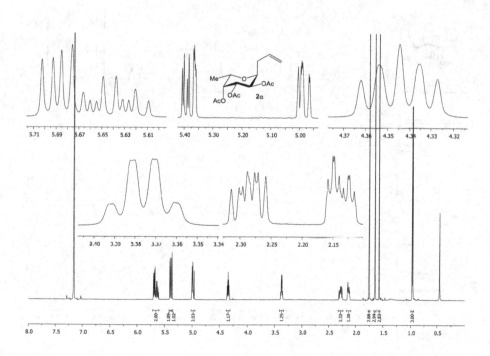

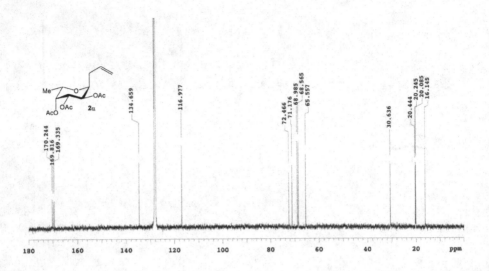

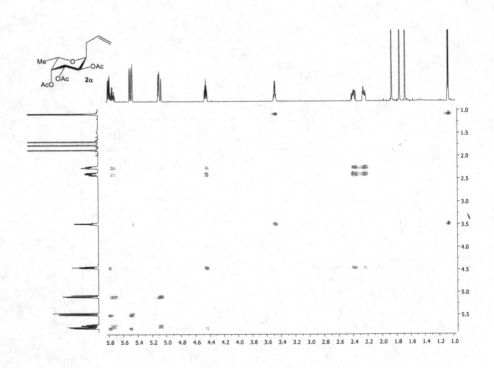

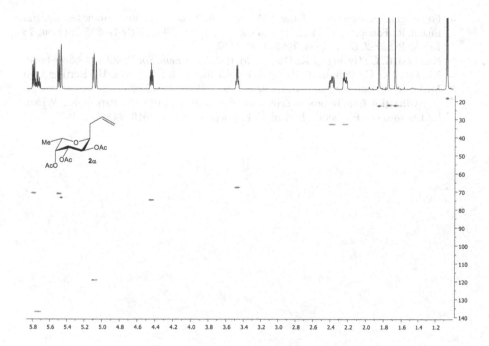

REFERENCES

1. (a) Du, Y.; Linhardt, R. J.; Vlahov, I. R. *Tetrahedron*, **1998**, *54*, 9913–9959. (b) Levy, D. E. Strategies towards C-glycosides. In *The Organic Chemistry of Sugars*, Levy, D. E., Fügedi, P., Eds. CRC Taylor & Francis: Boca Raton, **2006**; pp 269–348.
2. (a) Jaramillo, C.; Knapp, S. *Synthesis*, **1994**, 1–20. (b) Suzuki, K.; Matsumoto, T., Synthesis of glycosylarenes. In *Preparative Carbohydrate Chemistry*, Hanessian, S., Ed. Marcel Dekker Inc.: New York, **1997**; pp 527–542. (c) Parker, K. A. Aryl *C*-Glycosides by the reverse polarity approach. In *Glycomimetics: Modern Synthetic Methodologies*, Roy, R., Ed. American Chemical Society: Washington, **2005**; Vol. 896, pp 93–105. (d) Lee, D. Y. W.; He, M. *Curr. Topics Med. Chem.* **2005**, *5*, 1333–1350. (e) Bokor, É.; Kun, S.; Goyard, D.; Tóth, M.; Praly, J.-P.; Vidal, S.; Somsák, L. *Chem. Rev.* **2017**, *117*, 1687–1764.
3. (a) Giannis, A.; Sandhoff, K. *Tetrahedron Lett.* **1985**, *26*, 1479–1482. (b) Luengo, J. I.; Gleason, J. G. *Tetrahedron Lett.* **1992**, *33*, 6911–6914. (c) Uchiyama, T.; Vassilev, V. P.; Kajimoto, T.; Wong, W.; Lin, C.-C.; Huang, H.; Wong, C.-H. *J. Am. Chem. Soc.* **1995**, *117*, 5395–5396. (d) Uchiyama, T.; Woltering, T. J.; Wong, W.; Lin, C.-C.; Kajimoto, T.; Takebayashi, M.; Weitz-Schmidt, G.; Asakura, T.; Noda, M.; Wong, C.-H. *Bioorg. Med. Chem.* **1996**, *4*, 1149–1165. (e) Woltering, T. J.; Weitz-Schmidt, G.; Wong, C.-H. *Tetrahedron Lett.* **1996**, *37*, 9033–9036. (f) Hamzavi, R.; Dolle, F.; Tavitian, B.; Dahl, O.; Nielsen, P. E. *Bioconjugate Chem.* **2003**, *14*, 941–954. (g) Kolomiets, E.; Johansson, E. M. V.; Renaudet, O.; Darbre, T.; Reymond, J.-L. *Org. Lett.* **2007**, *9*, 1465–1468.

4. For original assignments of the NMR resonances of acetylated furanoses, see: (a) Euzen, R.; Ferrières, V.; Plusquellec, D. *J. Org. Chem.* **2005**, *70*, 847–855. (b) Hou, S.; Kováč, P. *Eur. J. Org. Chem.* **2008**, 1947–1952.
5. Nicolaou, K. C.; Hwang, C. K.; Duggan, M. E. *J. Am. Chem. Soc.* **1989**, *111*, 6682–6690.
6. Nicolaou, K. C.; Patron, A. P.; Ajito, K.; Richter, P. K.; Khatuya, H.; Bertinato, P.; Miller, R. A.; Tomaszewski, M. J. *Chem. Eur. J.* **1996**, *2*, 847–868.
7. Crystallization from isopropyl alcohol was previously described in: Parkan, K.; Werner, L.; Lövyová, Z.; Prchalová, E.; Kniežo L. *Carbohydr. Res.* **2010**, *345*, 352–362.

21 Synthesis and Characterization of 4-Methylphenyl 2,3,4,6-Tetra-O-Benzoyl-1-Thio-β-D-Galactopyranoside

*Matteo Panza, Michael P. Mannino, and Alexei V. Demchenko**
University of Missouri – St. Louis, One University Boulevard, St. Louis, Missouri, United States

Kedar N. Baryal [a]
Department of Chemistry, Michigan State University, East Lansing, Michigan, United States

CONTENTS

4-Methylphenyl 2,3,4,6-tetra-O-benzoyl-1-thio-β-D-galactopyranoside (**4**) has recently become a common glycosyl donor for the synthesis of various oligosaccharides.[1–12] Yet, a preparatively useful, high yielding procedure for the synthesis of this compound has not been reported, and characterization data are scattered or not reported

---

* Corresponding author: demchenkoa@umsl.edu
[a] Checker: kbaryal@chemistry.msu.edu

Carbohydrate Chemistry

97%

98% over two steps

for the pure substance.[13–18] Thioglycoside **4**, first reported by Kochetkov and coworkers in 1987,[13] is included in a library of glycosyl donors database for the development
of programmable one-pot oligosaccharide synthesis.[1] Presented herein is the synthesis and characterization of benzoylated thioglycoside **4** from commercially available
D-galactose pentaacetate **1**. Complete stereoselectivity and high yields in all steps
are the highlights of this synthetic sequence. Intermediate **2** can be obtained directly
from D-galactose by copper(II) triflate-catalyzed two-step conversion.[14]

EXPERIMENTAL

GENERAL METHODS

Reactions were performed using commercial reagents and solvents (Millipore-Sigma,
Acros or Fisher Scientific) and monitored by TLC on Kieselgel 60 F254 (EM Science).
1,2,3,4,6-Penta-*O*-acetyl-β-D-galactopyranose was purchased from Carbosynth.
Compounds were detected by UV light and by charring with 10% sulfuric acid in
methanol. Solvents were removed under reduced pressure at <40°C. Column chromatography was performed on silica gel 60 (70–230 mesh). Optical rotations were measured at "Jasco P-2000" polarimeter at 22°C. ^{1}H and ^{13}C NMR spectra were recorded
at 300 and 150 MHz, respectively, with a Bruker Avance spectrometer. The chemical
shifts are referenced to the signal of the residual CHCl$_3$ (δ_H = 7.26 ppm, δ_C = 77.16
ppm) for solutions in CDCl$_3$. HRMS were recorded with a JEOL MStation (JMS-700)
mass spectrometer. Melting points were measured using MSRS DigiMelt apparatus.

4-METHYLPHENYL 2,3,4,6-TETRA-*O*-ACETYL-1-THIO-β-D-GALACTOPYRANOSIDE (2)

Boron trifluoride diethyl etherate (1.58 mL, 12.8 mmol) was added dropwise at 0°C
to a solution of 1,2,3,4,6-penta-*O*-acetyl-β-D-galactopyranose (**1**, 1.0 g, 2.56 mmol)
and *p*-thiocresol (0.35 g, 2.82 mmol) in CH$_2$Cl$_2$ (10 mL), and the mixture was stirred
at 0°C for 1 h. The cooling was removed and the mixture was stirred for an additional 15 min at room temperature. The mixture was diluted with CH$_2$Cl$_2$ (~170 mL)
and washed with water (25 mL), sat. aq. NaHCO$_3$ (25 mL) and water (25 mL). The
organic phase was dried with MgSO$_4$, concentrated *in vacuo*, and the residue was
crystallized from CH$_2$Cl$_2$-hexane to afford the title compound (1.09 g, 94%), R_f 0.57

(1:1 hexanes–EtOAc); mp 115–116°C, reported 113–115°C[19]; [α]$_D$ +5.5 (c 1, CHCl$_3$), reported +4.4 (c 0.5, CHCl$_3$)[19]; ^{1}H NMR (CDCl$_3$) δ 1.97, 2.05, 2.10, 2.12 (4 s, 12H, 4 × COCH$_3$), 2.35 (s, 3H, CH$_3$), 3.91 (m, 1H, $J_{5,6a}$ = $J_{5,6b}$ 6.6 Hz, H-5), 4.07–4.23 (m, 2H, H-6a, H-6b), 4.64 (d, 1H, $J_{1,2}$ 10.0 Hz, H-1), 5.03 (dd, 1H, $J_{3,4}$ 3.3 Hz, H-3), 5.22 (dd, 1H, $J_{2,3}$ 10.0 Hz, H-2), 5.40 (br. d, 1H, H-4), 7.13 (d, 2H, J 7.9 Hz, aromatic), 7.41 (d, 2H, J 8.1 Hz, aromatic); ^{13}C NMR (CDCl$_3$) δ, 20.7, 20.8, 20.8, 21.0, 21.3 (5 C, 5 × CH$_3$), 61.7 (C-6), 67.3 (C-2), 67.4 (C-4), 72.1 (C-3), 74.5 (C-5), 87.1 (C-1), 128.7, 129.8, 133.3, 138.6 (6 C, aromatic), 169.6, 170.2, 170.3, 170.5 (4 C, 4 × C=O); HR FAB MS: m/z [M + Na$^+$] calcd for C$_{21}$H$_{26}$NaO$_9$S$^+$, 477.1190; found, 477.1195; Anal. calcd for C$_{21}$H$_{26}$O$_9$S: C, 55.50%; H, 5.77%; O, 31.68%; S, 7.05%, Found: C, 55.32%; H 5.77%. The reported NMR data[20] match those recorded for the title compound.

4-METHYLPHENYL 2,3,4,6-TETRA-O-BENZOYL-1-THIO-β-D-GALACTOPYRANOSIDE (4)

Compound 2 (1.09 g, 2.40 mmol) was dissolved in methanol (10 mL), the pH was adjusted to pH = 9 by the addition of a 1 M solution of NaOCH$_3$ in MeOH (~3 mL), and the mixture was stirred for 1.5 h at room temperature. After neutralization with Amberlite IR120 (H$^+$) prewashed and swelled with MeOH, the resin was filtered off and rinsed with methanol (5 × 5 mL). The combined filtrate (~25 mL) was concentrated in vacuo and dried in high vacuum. The residue containing crude compound p-tolyl 1-thio-β-D-galactopyranoside (3, 0.69 g, 2.40 mmol) was used for the next step without further purification.

Benzoyl chloride (1.67 mL, 14.4 mmol) was added dropwise at 0°C to a solution of compound 3 (0.69 g, 2.40 mmol) in pyridine (10 mL). The mixture was allowed to warm to room temperature and stirred for 16 h. The mixture was quenched by the addition of MeOH (2.0 mL), diluted with CH$_2$Cl$_2$ (120 mL) and washed with 1 M HCl (2 × 40 mL), sat. aq. NaHCO$_3$ (30 mL) and water (30 mL). The organic phase was separated, dried with MgSO$_4$, concentrated in vacuo and chromatography (hexane → 1:1 EtOAc–hexane) to afford the title compound 4 (1.58 g, 94%) as colorless foam, R_f 0.5 (7:3 hexane–EtOAc); [α]$_D$ +78 (c 1.0, CHCl$_3$), reported +83.1 (c 0.65, CHCl$_3$)[13]; ^{1}H NMR (CDCl$_3$) δ 2.38 (s, 3H, CH$_3$), 4.34–4.52 (m, 2H, H-5, H-6a), 4.69 (dd, 1H, $J_{5,6b}$ 6.5 Hz, $J_{6a,6b}$ 10.9 Hz, H-6b), 5.02 (d, 1H, $J_{1,2}$ 9.9 Hz, H-1), 5.64 (dd, 1H, $J_{3,4}$ 3.1 Hz, H-3), 5.78 (t, 1H, $J_{2,3}$ 9.9 Hz, H-2), 6.04 (br. d, 1H, H-4), 7.09–8.09 (m, 24H, aromatic); ^{13}C NMR (CDCl$_3$) δ 21.4 (CH$_3$), 62.6 (C-6), 68.0 (C-4), 68.4 (C-2), 73.1 (C-3), 75.1 (C-5), 86.1 (C-1), 127.4, 128.4, 128.5, 128.6, 128.8, 129.0, 129.4, 129.5, 129.7, 129.8, 129.9, 130.1, 133.4 (× 2), 133.7, 134.5, 138.7, 165.2, 165.5, 165.6, 166.1 (C aromatic); HR FAB MS: m/z [M + Na$^+$] calcd for C$_{41}$H$_{34}$NaO$_9$S$^+$, 725.1820; found, 725.1776. Anal. calcd for C$_{41}$H$_{34}$O$_9$S: C, 70.07%; H, 4.88%; O, 20.49%; S, 4.56%, found: C, 70.08%; H, 4.99%. NMR data matched, with minor differences, chemical shift, and J values reported by Kochetkov.[13]

ACKNOWLEDGMENTS

This work was supported by grants from the National Institute of General Medical Sciences (GM120673) and the National Science Foundation (USA, CHE-1800350). We thank Drs. Luo and Winter for HRMS determinations.

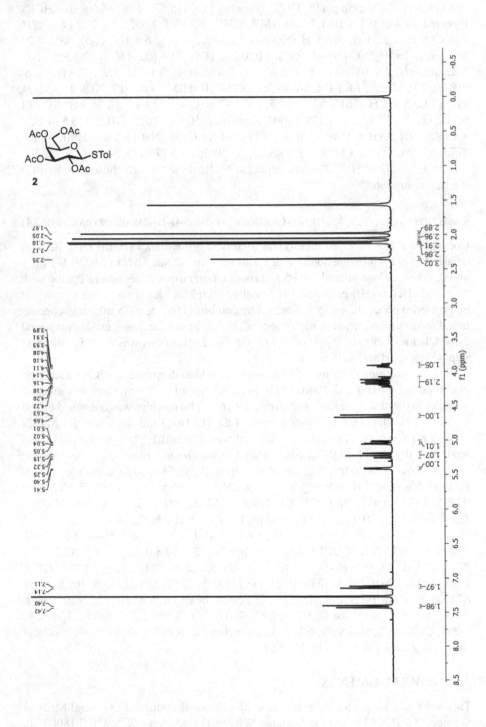

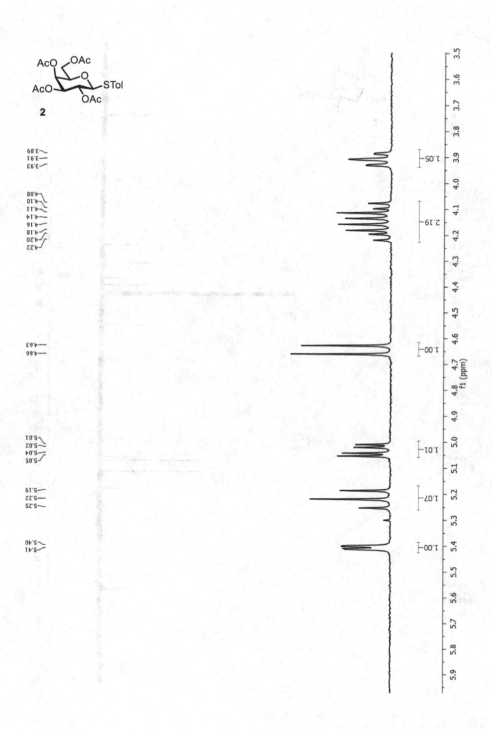

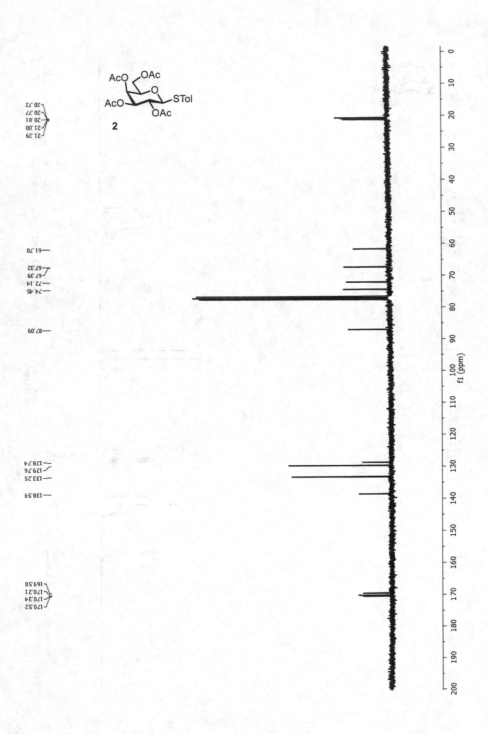

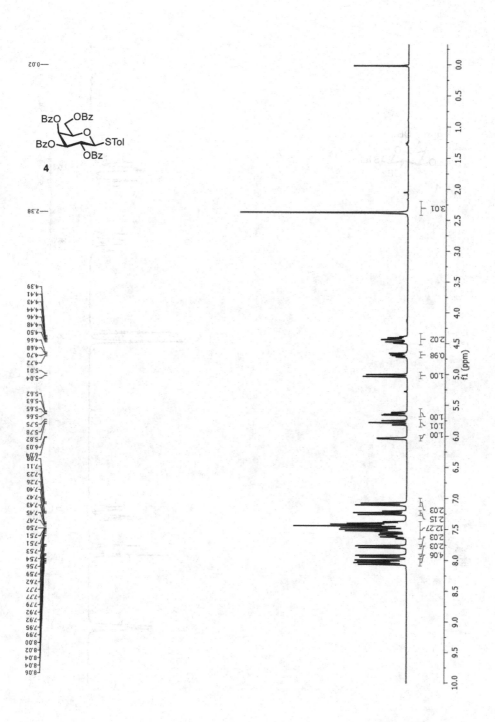

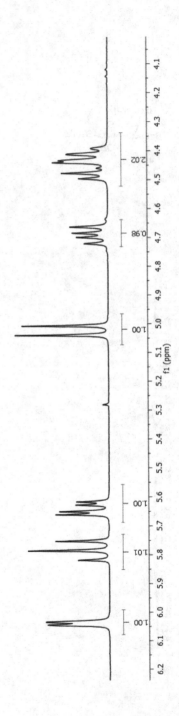

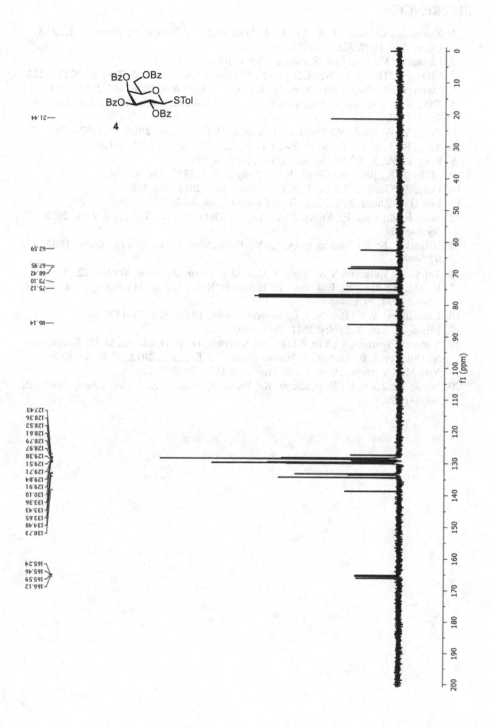

REFERENCES

1. Zhang, Z.; Ollmann, I. R.; Ye, X. S.; Wischnat, R.; Baasov, T.; Wong, C. H. *J. Am. Chem. Soc.* **1999**, *121*, 734–753.
2. Lahmann, M.; Oscarson, S. *Can. J. Chem.* **2002**, *80*, 889–893.
3. Huang, X.; Huang, L.; Wang, H.; Ye, X. S. *Angew. Chem. Int. Ed.* **2004**, *43*, 5221–5224.
4. Dasgupta, S.; Roy, B.; Mukhopadhyay, B. *Carbohydr. Res.* **2006**, *341*, 2708–2713.
5. Tanaka, N.; Ohnishi, F.; Uchihata, D.; Torii, S.; Nokami, J. *Tetrahedron Lett.* **2007**, *48*, 7383–7387.
6. Zeng, Y.; Wang, Z.; Whitfield, D.; Huang, X. *J. Org. Chem.* **2008**, *73*, 7952–7962.
7. Yang, B.; Jing, Y.; Huang, X. *Eur. J. Org. Chem.* **2010**, *2010*, 1290–1298.
8. Peng, P.; Ye, X.-S. *Org. Biomol. Chem.* **2011**, *9*, 616–622.
9. Maity, S. K.; Basu, N.; Ghosh, R. *Carbohydr. Res.* **2012**, *354*, 40–48.
10. Peng, P.; Xiong, D. C.; Ye, X. S. *Carbohydr. Res.* **2014**, *384*, 1–8.
11. Liu, G.-j.; Zhang, X.-t.; Xing, G.-w. *Chem. Commun.* **2015**, *51*, 12803–12806.
12. Mao, R.-Z.; Guo, F.; Xiong, D.-C.; Li, Q.; Duan, J.; Ye, X.-S. *Org. Lett.* **2015**, *17*, 5606–5609.
13. Nifant'ev, N. E.; Backinowsky, L. V.; Kochetkov, N. K. *Bioorg. Khim.* **1987**, *13*, 977–991.
14. Tai, C. A.; Kulkarni, S. S.; Hung, S. C. *J. Org. Chem.* **2003**, *68*, 8719–8722.
15. Kondo, H.; Aoki, S.; Ichikawa, Y.; Halcomb, R. L.; Ritzen, H.; Wong, C. H. *J. Org. Chem.* **1994**, *59*, 864–877.
16. Choudhury, A. K.; Roy, N. *J. Carbohydr. Chem.* **1997**, *16*, 1363–1371.
17. Misra, A.; Sau, A. *Synlett* **2011**, *2011*, 1905–1911.
18. Yang, B.; Yoshida, K.; Yin, Z.; Dai, H.; Kavunja, H.; El-Dakdouki, M. H.; Sungsuwan, S.; Dulaney, S. B.; Huang, X. *Angew. Chem. Int. Ed. Engl.* **2012**, *51*, 10185–10189.
19. Yde, M.; De Bruyne, C. K. *Carbohydr. Res.* **1973**, *26*, 227–229.
20. Wang, Z.; Zhou, L.; Ei-Boubbou, K.; Ye, X. S.; Huang, X. *J. Org. Chem.* **2007**, *72*, 6409–6420.

22 Alternative Synthesis of 1,2,4,6-Tetra-O-Acetyl-3-Deoxy-3-Fluoro-α,β-D-Glucopyranose

Jacob St-Gelais, Vincent Denavit, Danny Lainé, and Denis Giguère*
Département de Chimie, PROTEO, RQRM, Université Laval,
Québec City, Canada

Kevin Muru[a]
INRS-Institut Armand-Frappier, Université du Québec,
Laval, Canada

CONTENTS

Fluorinated carbohydrates are invaluable tools as, for example, mechanistic probes to decipher the mechanisms of glycosidases.[1] Also, the radiopharmaceutical field benefits tremendously from radiolabeled sugars used for positron emission tomography (PET).[2] As a consequence, the efficient preparation of such compounds is of great interest. As part of our ongoing program related to the preparation of fluorinated carbohydrates,[3] our attention was turned toward the preparation of 3-deoxy-3-fluoroglucose. This motif has been used as molecular probe[4] and its fluorine-18 counterpart has been used as potential tracer.[5]

The first synthesis of 3-deoxy-3-fluoroglucose was reported in 1966 and used an epoxide opening as a key step on 2,3-anhydro-4,6-di-O-methyl-α-D-allopyranoside.[6] Unfortunately, this method generated the desired 3-deoxy-3-fluoroglucose as the minor

---

* Corresponding author: denis.giguere@chm.ulaval.ca
[a] Checker: kevin.muru@iaf.inrs.ca

product of the three compounds formed. One year later, a new synthetic route was proposed for this compound by Foster *et al.* using 1,2:5,6-di-O-isopropylidene-α-D-allofuranose as starting material.[7] The latter is generated from an oxidation/reduction sequence from the corresponding glucose diacetonide. The preparation of 3-deoxy-3-fluoroglucose was based upon treatment of the 3-sulfonate of 1,2:5,6-di-O-isopropylidene-α-D-allofuranose with tetrabutylammonium fluoride. Optimization of this process was proposed in 1978 using diethylaminosulfur trifluoride (DAST) as deoxofluorinating agent on the same allofuranose scaffold.[8] The overall yield to generate 3-deoxy-3-fluoroglucose derivative from D-glucose is 15–18% (5 steps)[7] and a high yielding shorter synthetic route is thus needed. We proposed to use commercially available, inexpensive 1,6-anhydro-β-D-glucopyranose **1** as starting material. Also, our key deoxyfluorination step used Et$_3$N·3HF as practical fluorine source, thus avoiding highly reactive DAST reagent. To this end, we present the first synthesis of per-O-acetylated 3-deoxy-3-fluoro-glucopyranose (**4**) from levoglucosan **1** in four steps with an overall yield of 56%.

Levoglucosan **1** underwent facile *bis*-benzylation to afford compound **2** in 80% yield. Then, compound **3** was made indirectly via the preparation of the trifluoromethansulfonate followed by treatment with neat Et$_3$N·3HF. The nucleophilic fluorination proceeded with complete retention of configuration using microwave irradiation at 100°C for 2 h. The evidence for retention of configuration was shown using the ^{19}F NMR, with F-3 having distinctive coupling constants: dt, $^2J_{3F,3}$ 45.9 Hz, $^3J_{3F,2} = {}^3J_{3F,4}$ 17.5 Hz. Finally, the clear advantage of this methodology is the simultaneous acetolysis and benzyl ether hydrolysis under acidic conditions allowing formation of the desired 1,2,4,6-tetra-O-acetyl-3-deoxy-3-fluoro-D-glucopyranose **4** in 79% yield as an 9/1 (α/β) ratio.[9]

It is important to point out that the group of Kobayashi used a similar strategy to generate 3-deoxy-3-fluoro-(1→6)- α-D-glucopyranan.[10] They used a DAST-mediated deoxofluorination using DAST in benzene on intermediate **2**, but the yield (95%) could not be reproduced in our hands.

EXPERIMENTAL

GENERAL METHODS

Reactions in organic media were carried out under nitrogen atmosphere. ACS-grade solvents were used without further purification. Reactions were monitored by thin-layer chromatography (TLC) using silica gel 60 F$_{254}$-coated aluminum plates (SiliCycle). Visualization of the spots was affected by exposure to UV light and charring with a solution of 3 g of phenol in 5% H$_2$SO$_4$ in EtOH. Microwave irradiation was conducted in a Biotage Initiator Classic apparatus using 19 W of microwave power. Optical rotations were measured with a JASCO DIP-360 digital polarimeter. Nuclear magnetic resonance (NMR) spectra were recorded with an Agilent DD2 500 MHz spectrometer.

Proton (^{1}H) and carbon (^{13}C) chemical shifts (δ) are reported in part per millions (ppm) relative to the chemical shift of residual chloroform (7.26 ppm) as an internal standard. Fluorine (^{19}F) chemical shifts (δ) are reported in part per millions (ppm) relative to the chemical shift of hexafluorobenzene (162.29 ppm) as an internal standard. Coupling constants (J) are reported in Hertz (Hz), and the following abbreviations are used: singlet (s), doublet (d), doublet of doublets (dd), triplet (t), multiplet (m), and broad (br). Assignments of NMR signals were made by homonuclear correlation spectroscopy (COSY) and heteronuclear single quantum coherence two-dimensional spectroscopy. High-resolution mass spectra (HRMS) were measured with an Agilent 6210 LC time of flight mass spectrometer in electrospray mode. Elemental analysis was performed with a FLASH 2000 Analyzer (Thermo Scientific). Melting point was recorded on an OptiMelt Automated Melting Point System from Stanford Research Systems. 1,6-Anhydro-β-D-glucopyranose was purchased from Chem-Impex International, barium oxide, benzyl bromide, N,N-dimethylformamide (DMF), trifluoromethanesulfonic anhydride (Tf$_2$O), pyridine, dichloromethane (CH$_2$Cl$_2$), Et$_3$N·3HF, H$_2$SO$_4$, Ac$_2$O, and NaOAc were purchased from Sigma-Aldrich Chemical Co., Inc.

1,6-ANHYDRO-2,4-DI-O-BENZYL-β-D-GLUCOPYRANOSE (2)

To a solution of 1,6-anhydro-β-D-glucopyranose 1 (326 mg, 2.00 mmol) in DMF (8.0 mL) was added barium oxide (1.228 g, 8.03 mmol, 4 equiv). The mixture was stirred for 5 min, benzyl bromide (1.20 mL, 10.00 mmol, 5 equiv) was added, and the mixture was stirred at 60°C for 4 h. After this time, TLC analysis (1:1 hexanes–EtOAc) showed the appearance of a less-polar product (R_f 0.4). The mixture was cooled down to rt, methanol (10 mL) was added, and the mixture was stirred for 10 min. With the aid of EtOAc, the mixture was filtered through a Celite pad, the pad was rinsed with EtOAc, and the filtrate was concentrated under reduced pressure. Water (30 mL) was added to the residue, and the mixture was extracted with EtOAc (3 × 50 mL). The combined organic extracts were washed successively with saturated NaHCO$_3$ solution (1 × 30 mL) and brine (1 × 30 mL). The organic phase was dried over anhydrous MgSO$_4$, filtered, and concentrated under reduced pressure. Chromatography (4:1 → 3:2 hexane–EtOAc) provided the title compound as a white amorphous solid (495 mg, 72%). Recrystallization of a portion from 2:1 hexanes–EtOAc gave 2 as white needles. R_f 0.4 (1:1 hexanes–EtOAc); mp 103–105°C, lit.[11]: 106.5–107°C; [α]$_D$ −31.7 (c 0.7, CHCl$_3$), lit.[12]: [α]$_D$ −28.8 (c 1.0, CHCl$_3$); ^{1}H NMR (CDCl$_3$) δ 7.40–7.28 (m, 10H, Ar), 5.45 (t, 1H, $^3J_{1,2} = {}^4J_{1,3}$ 1.1 Hz, H-1), 4.73–4.66 (m, 4H, 2 × CH_2Ph), 4.58 (dq, 1H, $^3J_{5,6b}$ 5.3 Hz, $^3J_{5,6a} = {}^3J_{5,4} = {}^4J_{5,4}$ 1.1 Hz, H-5), 3.87 (tt, 1H, $^3J_{3,2} = {}^3J_{3,4}$ 4.0 Hz, $^4J_{3,2} = {}^3J_{3,4}$ 1.0 Hz, H-3), 3.82 (dd, 1H, $^2J_{6a,6b}$ 7.5 Hz, $^3J_{6a,5}$ 0.9 Hz, H-6a), 3.66 (ddd, 1H, $^2J_{6a,6b}$ 7.4 Hz, $^3J_{6b,5}$ 5.3 Hz, $^4J_{6b,4}$ 0.4 Hz, H-6b), 3.34 (ddt, 1H, $^3J_{4,3}$ 4.1 Hz, $^3J_{4,5}$ 1.3 Hz, $^4J_{4,2} = {}^4J_{4,6b}$ 0.6 Hz, H-4), 3.26 (ddt, 1H, $^3J_{2,3}$ 4.0 Hz, $^3J_{2,1}$ 1.2 Hz, $^4J_{2,4} = {}^4J_{2,OH}$ 0.5 Hz, H-2); ^{13}C NMR (CDCl$_3$) δ 138.0, 137.9, 128.67, 128.66, 128.09, 128.07, 128.06, 128.01 (10 C, Ar), 101.4 (1 C, C-1), 79.6 (1 C, C-4), 79.3 (1 C, C-2), 75.4 (1 C, C-5), 72.3, 71.9 (2 C, 2 × CH_2Ph), 70.6 (1 C, C-3), 66.7 (1 C, C-6); HRMS: m/z [M + Na]$^+$ calcd for C$_{20}$H$_{22}$O$_5$, 365.1359; found, 365.1359; Anal. calcd for C$_{20}$H$_{22}$O$_5$: C, 70.16; H, 6.48. Found: C, 70.45; H, 6.35.

1,6-ANHYDRO-2,4-DI-*O*-BENZYL-3-DEOXY-3-FLUORO-β-D-GLUCOPYRANOSE (3)

To a solution of compound 2 (840 mg, 2.454 mmol) in dry CH_2Cl_2/pyr. (8.0 mL/ 2.0 mL, 4:1) at 0°C was added, dropwise under argon, a solution of trifluoromethanesulfonic anhydride (Tf_2O) 1M in CH_2Cl_2 (5.0 mL, 5.0 mmol, 2 equiv). The yellow mixture was stirred for 15 min at 0°C. The mixture was allowed slowly (1 h) to reach rt, when TLC (1:1 hexanes–EtOAc) showed that the reaction was complete and that a less-polar product (R_f 0.6) was formed. The mixture was diluted with CH_2Cl_2 (40 mL) and washed successively with an aqueous saturated $NaHCO_3$ solution (1 × 20 mL), aqueous 1M HCl solution (1 × 20 mL), and brine (1 × 20 mL). The organic phase was dried over anhydrous $MgSO_4$, filtered, and concentrated under reduced pressure. The crude 1,6-anhydro-2,4-di-*O*-benzyl-3-*O*-trifluoromethanesulfonyl-β-D-glucopyranose was used for the next step without further purification.* It was dissolved in triethylamine trihydrofluoride ($Et_3N·3HF$, 6.0 mL, 37.0 mmol, 15 equiv) and irradiated in a microwave reactor at 100°C for 2 h. The mixture was diluted with water (20 mL) and extracted with CH_2Cl_2 (3 × 30 mL). The combined organic extracts were washed successively with a saturated $NaHCO_3$ solution (50 mL) and brine (30 mL), and the organic phase was dried over anhydrous $MgSO_4$, filtered, and concentrated. Chromatography (4:1 →7:3 hexanes–EtOAc) gave the title compound as a colorless oil (771 mg, 91%). R_f 0.6 (1:1 hexanes–EtOAc); $[α]_D$ –35.3 (*c* 0.7, $CHCl_3$), lit.[10]: $[α]_D$ –35.6 (*c* 1.0, $CHCl_3$); ^{1}H NMR ($CDCl_3$) δ 7.42–7.29 (m, 10*H*, Ar), 5.45 (d, 1*H*, $^3J_{1,2}$ 1.4 Hz, H-1), 4.77–4.59 (m, 6*H*, 2 × C*H*$_2$Ph, H-3, H-5), 3.81 (dt, 1*H*, $^2J_{6a,6b}$ 7.4 Hz, $^3J_{6a,5}$ = $^4J_{6a,4}$ 1.0 Hz, H-6a), 3.70 (ddd, 1*H*, $^2J_{6a,6b}$ 7.1 Hz, $^3J_{6b,5}$ 5.6 Hz, $^4J_{6b,4}$ 1.4 Hz, H-6b), 3.50–3.38 (m, 2*H*, H-2, H-4); ^{13}C NMR ($CDCl_3$) δ 137.5, 137.5, 128.7, 128.7, 128.2, 128.1, 128.0 (10 C, Ar), 100.6 (d, 1 C, $^3J_{1,3F}$ 3.0 Hz, C-1), 90.0 (d, 1 C, $^1J_{3,3F}$ 180.3 Hz, C-3), 76.1 (d, 1 C, $^2J_{4,3F}$ 25.6 Hz, C-4), 75.7 (d, 1 C, $^2J_{2,3F}$ 23.4 Hz, C-2), 74.5 (d, 1 C, $^3J_{5,3F}$ 2.9 Hz, C-5), 72.4, 71.8 (2 C, 2 × C*H*$_2$Ph), 65.8 (d, 1 C, $^4J_{6,3F}$ 2.4 Hz, C-6); ^{19}F NMR ($CDCl_3$) δ –183.6 (dt, 1 F, $^2J_{3F,3}$ 45.9 Hz, $^3J_{3F,2}$ = $^3J_{3F,4}$ 17.5 Hz, F-3); HRMS: *m/z* $[M + Na]^+$ calcd for $C_{20}H_{21}FO_4$, 367.1316; found, 367.1316.

1,2,4,6-TETRA-*O*-ACETYL-3-DEOXY-3-FLUORO-α,β-D-GLUCOPYRANOSE (4)

To a solution of compound 3 (356 mg, 1.033 mmol) in Ac_2O (3.0 mL, 31.0 mmol, 30 equiv) at 0°C was added sulfuric acid (H_2SO_4) (550 μL, 10.30 mmol, 10 equiv) dropwise under argon. The mixture was stirred for 23 h at 24°C, then sodium acetate (NaOAc) (1.736 g, 20.67 mmol, 20 equiv), was added and the mixture was stirred 30 min. Water (60 mL) was added and the mixture was extracted with EtOAc (3 × 60 mL). The combined organic extracts were washed successively with an aqueous saturated $NaHCO_3$ solution (1 × 100 mL) and brine (1 × 100 mL). The organic phase was dried with anhydrous $MgSO_4$, filtered, concentrated under reduced pressure, and chromatography (7:3 → 6:4 hexanes–EtOAc) gave the title compound as a white amorphous solid (293 mg, 81%, 9/1 α/β).† R_f 0.4 (1:1 hexanes–EtOAc); ^{1}H NMR ($CDCl_3$) δ 6.35 (t, 1*H*, $^3J_{1,2}$ = $^4J_{1,3F}$ 3.7 Hz, H-1α), 5.65 (dd, 1*H*, $^3J_{1,2}$ 8.3 Hz, $^4J_{1,3F}$ 0.4 Hz, H-1β), 5.29 (ddd, 1*H*, $^3J_{4,3F}$ 12.9 Hz, $^3J_{4,5}$ 10.3 Hz, $^3J_{4,3}$ 9.1 Hz, H-4α), 5.30–5.23

---

* The crude triflate can be stored in the freezer for 2 weeks without any noticeable decomposition.
† Attempt to crystallize selectively the α anomer was unsuccessful using various solvents.

(m, 2H, H-2β, H-4β), 5.18 (ddd, 1H, $^3J_{2,3F}$ 12.0 Hz, $^3J_{2,3}$ 9.8 Hz, $^3J_{2,1}$ 3.9 Hz, H-2α), 4.82 (dt, 1H, $^2J_{3,3F}$ 53.4 Hz, $^3J_{3,2} = {}^3J_{3,4}$ 9.4 Hz, H-3α), 4.60 (dt, 1H, $^2J_{3,3F}$ 51.9 Hz, $^3J_{3,2} = {}^3J_{3,4}$ 9.1 Hz, H-3β), 4.27 (dd, 1H, $^2J_{6a,6b}$ 12.6 Hz, $^3J_{6a,5}$ 4.5 Hz, H-6aβ), 4.25 (dd, 1H, $^2J_{6a,6b}$ 12.6 Hz, $^3J_{6a,5}$ 4.2 Hz, H-6aα), 4.13 (dt, 1H, $^2J_{6b,6a}$ 12.6 Hz, $^3J_{6b,5} = {}^5J_{6a,3F}$ 1.8 Hz, H-6bβ), 4.11 (dt, 1H, $^2J_{6b,6a}$ 12.6 Hz, $^3J_{6b,5} = {}^5J_{6a,3F}$ 1.9 Hz, H-6bα), 4.04 (dddt, 1H, $^3J_{5,4}$ 10.3 Hz, $^3J_{5,6a}$ 4.0 Hz, $^3J_{5,6b}$ 2.3 Hz, $^4J_{5,3F}$ 0.7 Hz, H-5α), 3.74 (dddd, 1H, $^3J_{5,4}$ 10.1 Hz, $^3J_{5,6a}$ 4.6 Hz, $^3J_{5,6b}$ 2.2 Hz, $^4J_{5,3F}$ 1.2 Hz, H-5β), 2.17 (s, 3H, COCH_3α), 2.13 (s, 3H, COCH_3α), 2.12 (s, 3H, COCH_3β), 2.11 (s, 3H, COCH_3β), 2.11 (s, 3H, COCH_3β), 2.10 (s, 3H, COCH_3α), 2.09 (s, 3H, COCH_3β), 2.09 (s, 3H, COCH_3α); ^{13}C NMR (CDCl$_3$) δ 170.8, 169.7, 169.3, 168.7 (4 C, 4 × COCH$_3$), 89.5 (d, 1 C, $^3J_{1,3F}$ 9.5 Hz, C-1), 89.2 (d, 1 C, $^1J_{3,3F}$ 189.9 Hz, C-3), 69.8 (d, 1 C, $^3J_{5,3F}$ 6.7 Hz, C-5), 69.7 (d, 1 C, $^2J_{2,3F}$ 17.9 Hz, C-2), 67.8 (d, 1 C, $^2J_{4,3F}$ 18.6 Hz, C-4), 61.4 (C-6), 21.0, 20.9, 20.8, 20.7 (4 C, 4 × COCH$_3$); ^{19}F NMR (CDCl$_3$) δ −195.97 (dt, 1 F, $^2J_{3F,3}$ 51.8 Hz, $^3J_{3F,2} = {}^3J_{3,4}$ 12.6 Hz, F-3β), −200.03 (dtd, 1 F, $^2J_{3F,3}$ 53.5 Hz, $^3J_{3F,2} = {}^3J_{3F,4}$ 12.4 Hz, $^4J_{3F,1}$ 3.9 Hz, F-3α); HRMS: m/z [M + Na]$^+$ calcd for C$_{14}$H$_{23}$FNO$_9{}^+$, 368.1351; found, 368.1360; Anal. calcd for C$_{14}$H$_{19}$FO$_9$: C, 48.00; H, 5.47. Found: C, 48.15; H, 5.50.

ACKNOWLEDGMENTS

This work was supported by the Natural Sciences and Engineering Research Council of Canada (NSERC), the Fonds de Recherche du Québec-Nature et Technologies, and the Université Laval. J. S.-G. thanks to Fonds Paul-Antoine-Giguère for a postgraduate fellowship.

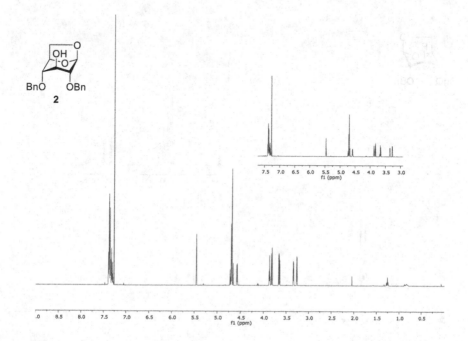

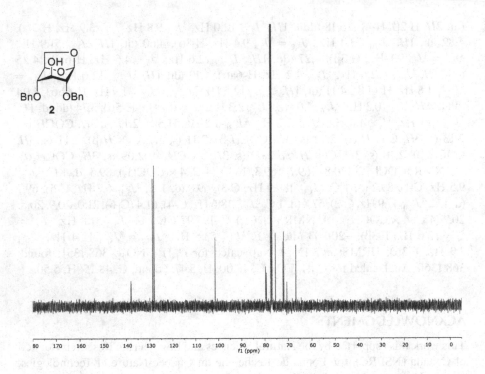

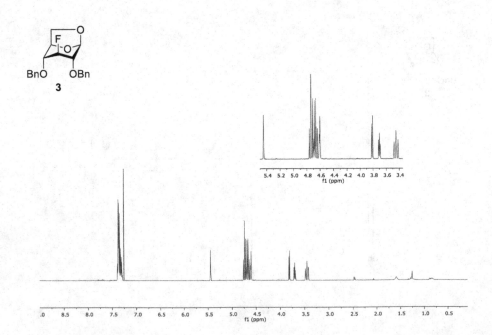

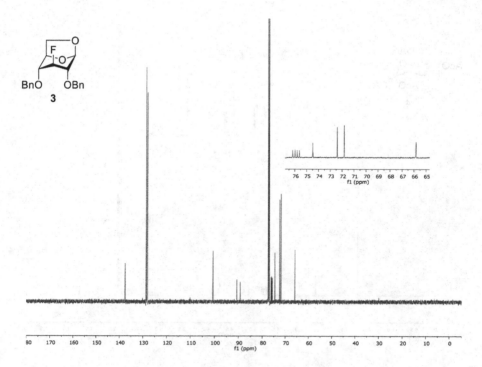

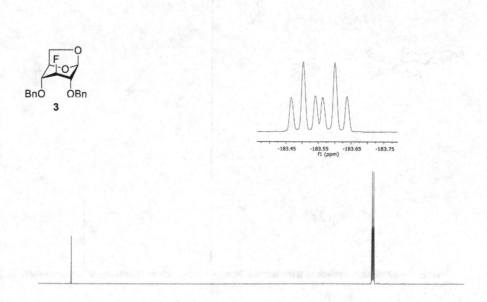

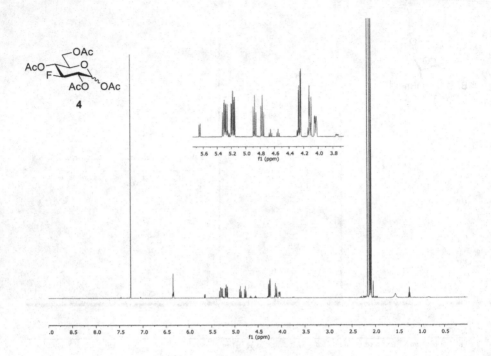

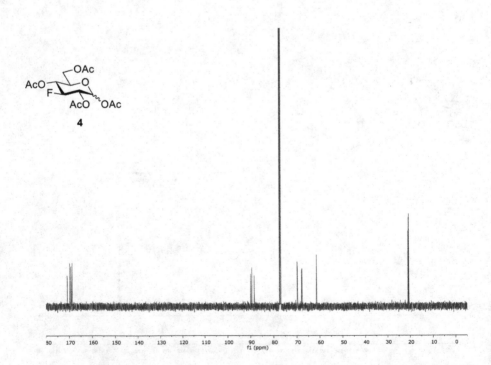

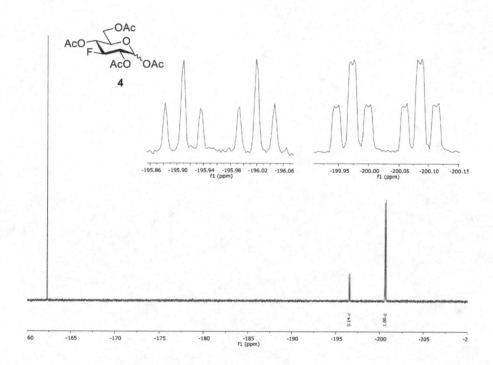

REFERENCES

1. (a) Namchuk, M.; Braun, C.; McCarter, J. D.; Withers, S. G. Fluorinated sugars as probes of glycosidase mechanism. In *ACS Symposium Series*, Eds.: Ojima, I.; McCarthy, J.; Welch, J. T. **1996**, *639*, 279–293; (b) Williams, S. J.; Withers, S. G. *Carbohydr. Res.* **2000**, *327*, 27–46.
2. Mankoff, D. A.; Dehdashti, F.; Shields, A. F. *Neoplasia*, **2000**, *2*, 71–88.
3. (a) Lainé, D.; Denavit, V.; Giguère, D. *J. Org. Chem.* **2017**, *82*, 4986–4992; (b) Denavit, V.; Lainé, D.; Le Heiget, G.; Giguère, D. Fluorine-containing carbohydrates: synthesis of 6-deoxy-6-fluoro-1,2:3,4-di-*O*-isopropylidene-α-D-galactopyranose. In, *Carbohydrate Chemistry: Proven Synthetic Methods*, Eds.: Vogel, C.; Murphy, P. **2017**, Vol. 4, pp 247–253.
4. (a) Halton, D. M.; Taylor, N. F.; Lopes, D. P. *J. Neurosc. Res.* **1980**, *5*, 241–252; (b) Riley, G. J.; Taylor, N. F. *Biochem. J.* **1973**, *135*, 773–777.
5. (a) Knust, E. J.; Machulla, H. J. *Radiochem. Radioanal. Lett.* **1983**, *59*, 7–14; (b) Gatley, S. J.; Shaughnessy, W. J. *Int. J. Appl. Radiat. Isot.* **1980**, *31*, 339–341; (c) Tewson, T. J.; Welch, M. J.; Raichc, M. E. *J. Label. Compd. Radiopharm.* **1979**, *16*, 10–11; (d) Tewson, T. J.; Welch, M. J.; Raichle, M. E. *J. Nucl. Med.* **1978**, *19*, 1339–1345.
6. Johansson, I.; Lindberg, B. *Carbohydr. Res.* **1966**, *1*, 467–473.
7. Foster, A. B.; Hems, R.; Webber, J. M. *Carbohydr. Res.* **1967**, *5*, 292–301.
8. Tewson, T. J.; Welch, M. J. *J. Org. Chem.* **1978**, *43*, 1090–1092.
9. (a) Weigel, T. M.; Liu, L.-d.; Liu, H.-w. *Biochemistry* **1992**, *31*, 2129–2139; (b) Foster, A. B.; Hems, R.; Hall, L. D. *Can. J. Chem.* **1970**, *48*, 3937–3945.
10. Kobayashi, K.; Kondo, T. *Macromolecules* **1997**, *30*, 6531–6535.
11. Zemplén, G.; Csuros, Z.; Angyal, S. *Chem. Ber.* **1937**, *70*, 1848–1856.
12. Iversen, T.; Bundle, D. R. *Can. J. Chem.* **1982**, *60*, 299–303.

23 Synthesis of Methyl 4-O-Benzoyl-2,3-O-Isopropylidene-α-D-Rhamnopyranoside: A Precursor to D-Perosamine

Mana Mohan Mukherjee, and Pavol Kováč*
NIDDK, LBC, National Institutes of Health,
Bethesda, Maryland, United States

Victoria Kohout[a]
Indiana University
Bloomington, Indiana, United States

CONTENTS

The chemistry of L-rhamnose (6-deoxy-L-mannose) was studied already during the early days of carbohydrate chemistry because of its presence in many plant polysaccharides. After the advent of conjugate vaccines from synthetic carbohydrate antigens, carbohydrate chemists became more interested in D-rhamnose, the enantiomer of L-rhamnose, and access to its derivatives because this monosaccharide is present in many bacterial polysaccharides. As D-rhamnose is a rare and expensive sugar, improvement in its synthesis is desirable. Access to 6-deoxysugars

* Corresponding author: kpn@helix.nih.gov
[a] Checker: vkohout@imail.iu.ed; under supervision of Nicola Pohl: npohl@indiana.edu

requires deoxygenation at HO-6 in a hexose which, when D-perosamine is the target, is normally done through hydrogenolysis of a 6-bromo or 6-iodo derivative of D-mannose.[1-3]

Our interest in D-rhamnose came about in connection with our efforts toward a conjugate vaccine for the disease caused by *Vibrio cholerae* O1. The protective antigen here, the O-specific polysaccharide, consists of a polymer of 3-deoxy-L-*glycero*-N-tetronylated 1,2-α-linked D-perosamine [perosamine: 4-amino-4-deoxy-D-rhamnose; 4-amino-4,6-dideoxy-D-mannose]. Perosamine is a constituent of many other bacterial polysaccharides.[1,4-8] Here, we describe deoxygenation at C-6 effected by the high-yielding redox rearrangement of benzylidene acetals,[9] in *n*-octane. This protocol gives access to 6-deoxy sugar **3** in one step from the acetal **2**, making it superior, in our opinion, to commonly used routes.

It is worth mentioning that several, and quite different, sets of physical constants have been reported for compound **2**.[10-13] Physical constants reported here for compound **2** are close to those reported by Madsen and Fraser-Reid.[10]

EXPERIMENTAL

GENERAL METHODS

Unless specified otherwise, all reagents and solvents were purchased from Sigma Chemical Company and used as supplied. Reactions were monitored by thin-layer chromatography (TLC) on silica gel 60 glass slides. Spots were visualized by charring with H_2SO_4 in EtOH (5% v/v) and/or UV light. Melting points were determined with a Kofler hot stage. Optical rotations were measured at ambient temperature with a Jasco P-2000 digital polarimeter. NMR spectra were measured at 25°C for solutions in $CDCl_3$, at 600 MHz for ^{1}H, and at 150 MHz for ^{13}C with a Bruker Avance spectrometer. Assignments of NMR signals were aided by 1D and 2D experiments ([1]H–[1]H homonuclear decoupling, APT, COSY, HSQC) run with the software supplied with the spectrometer. Chemical shifts were referenced to signals of tetramethylsilane (0 ppm) or $CDCl_3$ (77.00 ppm). Methyl 4,6-O-benzylidene-α-D-mannopyranoside (**1**) was prepared as reported[12] and had mp 147–148°C (CH_2Cl_2–benzene); $[\alpha]_D$ +63.8 (*c* 1.15, $CHCl_3$). Ref.[11,14] mp 146–147°C, $[\alpha]_D$ +64.3 (*c* 2.1, $CHCl_3$); Ref.[12] mp 141–143°C, $[\alpha]_D$ not reported; ^{1}H NMR ($CDCl_3$) δ 7.49–7.48 (m, 2*H*, Ar–H), 7.37–7.36 (m, 3*H*, Ar–H), 5.55 (s, 1*H*, PhCH), 4.73 (s, 1*H*, H-1), 4.26 (m, 1*H*, H-6a), 4.04 (m, 1*H*, H-3), 3.99 (bs, 1*H*, H-2), 3.90 (t, 1*H*, *J* 8.9 Hz, H-4), 3.84–3.77 (m, 2*H*, H-5, H-6b), 3.38 (s, 3*H*, OCH₃), 2.83 (d, 1*H*, *J* 3.1 Hz, 3-OH), 2.79 (d, 1*H*, *J* 2.9 Hz, 2-OH); [13]C ($CDCl_3$) δ 137.2, 129.3,

128.4, 126.3, 102.3 (PhCH), 101.2 (C-1), 78.9 (C-3), 70.7 (C-2), 68.8 (C-6), 68.6 (C-4), 62.9 (C-5), 55.1 (OCH$_3$). Solutions in organic solvents were dried with anhydrous MgSO$_4$ and concentrated at reduced pressure at > 40°C.

METHYL 4,6-*O*-BENZYLIDENE-2,3-*O*-ISOPROPYLIDENE-α-D-MANNOPYRANOSIDE (2)

2-Methoxypropene (2.4 mL, 25 mmol), followed by D-camphor-10-sulfonic acid (348 mg, 1.5 mmol), was added under argon to a solution of methyl 4,6-*O*-benzylidene-α-D-mannopyranoside[12] (1, 1.41 g, 5 mmol) in dry THF (40 mL), and the mixture was stirred at room temperature. After 2 h, TLC (3:1 hexane–EtOAc) showed complete consumption of starting material and formation of a single, less polar product (R_f = 0.7). The reaction was quenched with triethylamine (1 mmol, 0.12 mL) and the mixture was concentrated. A solution of the crude product in DCM was washed with saturated NaHCO$_3$ (250 mL), and the aqueous phase was backwashed with DCM (3× 25 mL). The combined organic phases were dried, concentrated, and chromatographed (80 g silica gel, 19:1→ 9:1 hexane–EtOAc). Excess 2-methoxypropene present in the crude mixture was eluted first, followed by the title product 2 (1.34 g, 83%, white solid), mp 112–113°C (hexane, twice), [α]$_D$ −19.7 (*c* 1.05, CHCl$_3$); Ref.[10] mp 106–108°C, [α]$_D$ −18.5 (*c* 1.0, CHCl$_3$). ^{1}H NMR (CDCl$_3$) δ 7.51–7.49 (m, 2*H*, Ar–H), 7.36–7.31 (m, 3*H*, Ar–H), 5.57 (s, 1*H*, PhCH), 4.96 (s, 1*H*, H-1), 4.34–4.29 (m, 2*H*, H-6a, H-3), 4.22 (d, 1*H*, $J_{2,3}$ 5.7 Hz, H-2), 3.79–3.74 (m, 1*H*, H-6b), 3.77–3.72 (m, 1*H*, H-5), 3.72–3.67 (m, 1*H*, H-4), 3.41 (s, 3*H*, OCH$_3$), 1.59 (s, 3*H*, CH$_3$), 1.37 (s, 3*H*, CH$_3$); ^{13}C (CDCl$_3$) δ 137.2, 129.0, 128.1, 126.3 (Ar–C), 109.6 (CMe$_2$), 101.9 (PhCH), 98.9 (C-1), 80.3 (C-4), 75.9 (C-2), 74.4 (C-3), 68.9 (C-6), 60.4 (C-5), 55.1 (OCH$_3$), 28.1 (CH$_3$), 26.1 (CH$_3$). ESI-MS: *m/z* [M + Na]$^+$ calcd for C$_{17}$H$_{22}$O$_6$Na 345.1314; found 345.1316; Anal. calcd for C$_{17}$H$_{22}$O$_6$: C, 63.34; H, 6.88. Found: C, 63.23; H, 6.88.

METHYL 4-*O*-BENZOYL-2,3-*O*-ISOPROPYLIDENE-α-D-RHAMNOPYRANOSIDE (3)

Di-*t*-butyl peroxide (DTBP) (0.4 mL, 2.5 mmol) followed by *tri*-isopropylsilanethiol (TIPS) (54 µL, 0.25 mmol) was added under argon to a solution of methyl 4,6-*O*-benzylidene-2,3-*O*-isopropylidene-α-D-mannopyranoside (2) (806 mg, 2.5 mmol) in dry *n*-octane (30 mL), and the mixture was refluxed. TLC (4:1 Hexane–EtOAc) after 2 h showed complete consumption of starting material and presence of a less polar product. After concentration, the residue was chromatographed (24 g silica gel, 19:1 → 9:1 hexane–EtOAc). The title product (687 mg, 86%, solid), when crystalized from hot hexane (twice, needles), showed mp 100–101°C, [α]$_D$ +5.1 (*c* 1.05, CHCl$_3$); lit.[15] mp 99–101°C (MeOH); [α]$_D$ +4.6 (*c* 3, CHCl$_3$). ^{1}H NMR (CDCl$_3$) δ 8.06–8.05 (d, 2*H*, *J* 7.6 Hz, Ar–H), 7.57 (t, 1*H*, *J* 7.6 Hz, Ar–H), 7.48–7.42 (t, 2*H*, *J* 7.6 Hz, Ar–H), 5.12 (dd, 1*H*, $J_{4,5}$ 10.1 Hz, $J_{4,3}$ 7.9 Hz, H-4), 4.95 (s, 1*H*, H-1), 4.33 (dd, 1*H*, $J_{3,4}$ 7.8 Hz, $J_{3,2}$ 5.5 Hz, H-3), 4.19 (d, 1*H*, $J_{2,3}$ 5.3 Hz, H-2), 3.87 (dq, 1*H*, $J_{5,4}$ 10.2 Hz, $J_{5,6}$ 6.4 Hz, H-5), 3.42 (s, 3*H*, OCH$_3$), 1.63 (s, 3*H*, CH$_3$), 1.36 (s, 3*H*, CH$_3$), 1.23 (d, 3*H*, $J_{6,5}$ 6.3 Hz, H-6); ^{13}C (CDCl$_3$) δ 165.8 (C=O), 133.2, 129.8, 128.4 (Ar–C), 109.9 (CMe$_2$), 98.1 (C-1), 76.0 (C-2), 75.9 (C-3), 75.1 (C-4), 64.0 (C-5), 55.0 (OCH$_3$), 27.8 (C-6), 26.4 (CH$_3$), 17.1 (CH$_3$).

ESI-MS: *m/z* [M + Na]⁺ calcd for $C_{17}H_{22}O_6Na$ 345.1314; found 345.1318; Anal. calcd for $C_{17}H_{22}O_6$: C, 63.34; H, 6.88. Found: C, 63.23; H, 6.88.

ACKNOWLEDGMENTS

This research was supported by the Intramural Research Program of the National Institutes of Health (NIH) and National Institute of Diabetes and Digestive and Kidney Diseases (DK059701).

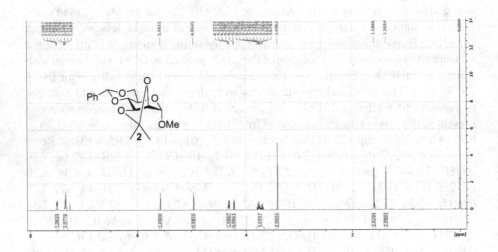

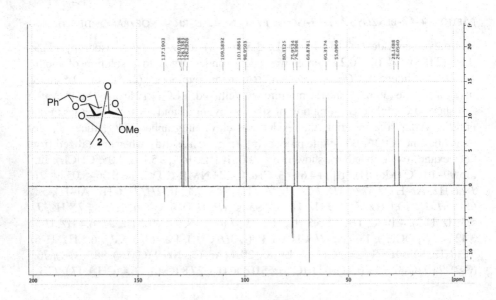

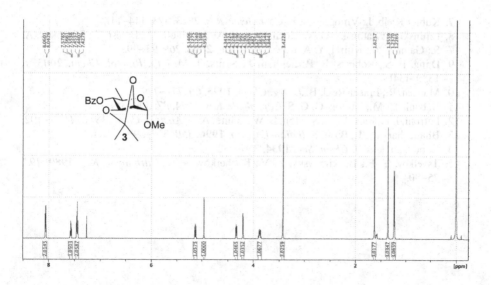

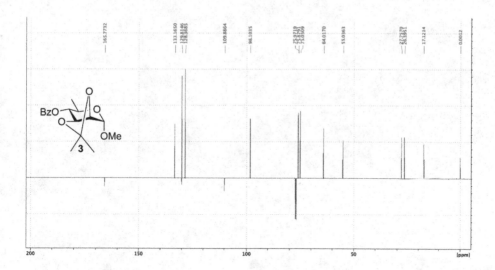

REFERENCES

1. Sarkar, K.; Roy, N. *J. Carbohydr. Chem.* **2006**, *25*, 53–68.
2. Eis, M. J.; Ganem, B. *Carbohydr. Res.* **1988**, *176*, 316–323.
3. Soliman, S. E.; Kovac, P. *Angew. Chem. Int. Ed. Engl.* **2016**, *55*, 12850–12853.
4. Perry, M. P.; Bundle, D. R. *Infect. Immun.* **1990**, *58*, 1391–1395.
5. Bundle, D. R.; Cherwonogrodzky, J. W.; Perry, M. P. *Infect. Immun.* **1988**, *56*, 1101–1106.
6. Kondo, S.; Sano, Y.; Isshiki, Y.; Hisatsune, K. *Microbiology (Reading, U. K.)*, **1996**, *142*, 2879–2885.

7. Kubler-Kielb, J.; Vinogradov, E. *Carbohydr. Res.* **2013**, *378*, 144–147.
8. Lipinski, T.; Zatonsky, G. V.; Kocharova, N. A.; Jaquinod, M.; Forest, E.; Shashkov, A. S.; Gamian, A.; Knirel, Y. A. *Eur. J. Biochem.* **2002**, *269*, 93–99.
9. Dang, H.-S.; Roberts, B. P.; Sekhon, J.; Smits, T. M. *Org. Biomol. Chem.* **2003**, *1*, 1330–1341.
10. Madsen, R.; Fraser-Reid, B. *J. Org. Chem.* **1995**, *60*, 772–779.
11. Bebault, G. M.; Dutton, G. G. S. *Carbohydr. Res.* **1974**, *37*, 309–319.
12. Patroni, J. J.; Stick, R. V.; Skelton, B. W.; White, A. H. *Austr. J. Chem.* **1988**, *41*, 91–102.
13. Bhattacharyya, T.; Basu, S. *Indian J. Chem.* **1996**, *35B*, 397–398.
14. Robertson, G. J. *J. Chem. Soc.* **1934**, 330–332.
15. Tsvetkov, Y. E.; Backinowsky, L. V.; Kochetkov, N. K. *Carbohydr. Res.* **1989**, *193*, 75–90.

24 Synthesis of Allyl and Dec-9-Enyl α-D-Mannopyranosides from D-Mannose

Mattia Vacchini[a], Laura Cipolla,
Roberto Guizzardi[b] and Barbara La Ferla*
Department of Biotechnology and Bioscience,
University of Milano Bicocca,
Milano, Italy

Martina Lahmann[b]
School of Natural Sciences, Bangor University,
Bangor, United Kingdom

CONTENTS

The term "click chemistry"[1a–c] is used for chemical transformations featured by high efficiency and selectivity (regio- and chemo-) and carried out particularly in nontoxic solvents, particularly in water. Major advantages are wide applicability, high yields, few byproducts, mild reaction conditions (quite often insensitive to aerobic conditions), readily available starting materials, and reagents. As a consequence, this methodology has found applications in many fields including polymer and material science, medicinal chemistry, molecular biology, and biotechnology.

A range of click reactions are used in these applications, such as the copper(I)-catalyzed azide–alkyne cycloaddition (CuAAC),[1b] the hetero-Diels–Alder reaction,[1c]

---

* Corresponding author: barbara.laferla@unimib.it
[a] Checker: m.lahmann@bangor.ac.uk
[b] These two authors contributed equally.

nucleophilic ring-opening of strained heterocyclic electrophiles,[1] or carbonyl transformation into oxime ethers and hydrazones.[2] Another click process that has gained renewed attention in recent time is the addition of thiols to alkenes,[3] which is currently called thiol–ene coupling (TEC). The reaction between an alkene and a thiol group can be induced thermally or photochemically through an anti-Markovnikov radical mechanism, which does not involve toxic transition metal catalysts. The TEC reaction can be successfully applied also to the synthesis of glycoconjugates.[4] Starting materials for the synthesis of glycoconjugates may be alkenyl glycosides, which can be later submitted to thiol-ene reaction. Key for TEC reactions is the easy availability of alkenyl glycosides, thus avoiding tedious protection/deprotection steps. Here, we applied a Fischer-type glycosylation reaction catalyzed by p-toluenesulfonic[5] acid to an unprotected monosaccharide D-mannose toward the synthesis of allyl and decenyl α-glycosides. The glycosylation reaction was performed with different amounts of the starting material (0.1 and 1 g, respectively, 0.553 and 5.53 mmol) and either allyl alcohol or decenyl alcohol as acceptor. Many protocols have already been published for the synthesis of allyl and decenyl mannopyranosides.[6–8] They usually start from protected sugars and involve intermediates, thus resulting in multistep processes. Experimentally, we observed high stereoselectivity toward the formation of α-glycosides in both cases described. We observed also that scaling up leads to a decrease in the reaction yield (product **1**). Compounds **1** and **2** were characterized as the corresponding acetates **3** and **4**.

EXPERIMENTAL

GENERAL METHODS

Reactions were performed with commercially available reagents and solvents, without further purification. D-Mannose (CAS no. 3458-28-4) was purchased from Sigma-Aldrich. Reactions were monitored by thin-layer chromatography (TLC) on silica gel 60F$_{254}$-coated glass plates (Merck) and visualized by charring with 10:45:45 (volumes) H$_2$SO$_4$–EtOH–H$_2$O or with (NH$_4$)$_6$Mo$_7$O$_{24}$ (21 g) and Ce(SO$_4$)$_2$ (1 g), in a solution of concentrated H$_2$SO$_4$ (31 mL) in water (500 mL). All reactions were terminated when differences in their course could no longer be observed (TLC). Solutions in organic solvents were dried over Na$_2$SO$_4$ and concentrated at <40°C. Flash column chromatography was performed on silica gel 230–400 mesh (Merck). NMR spectra were recorded at 400 MHz (^{1}H) and 100.6 MHz (^{13}C) with a Varian Mercury or with a Bruker Advance Neo instrument. Chemical shifts are reported in ppm referenced to residual solvent signal as an internal standard; J values are given in Hertz. For all the compounds, assignments of ^{1}H and ^{13}C NMR spectra were based on 2D proton–proton shift-correlation and 2D carbon–proton heteronuclear correlation spectra. Mass spectra were recorded on a QTRAP instrument (AB Sciex) equipped with an electrospray ion source. The samples were directly injected by a steel capillary, employing a spray voltage of 5.5 kV and a declustering potential of 5 V.

ALLYL α-D-MANNOPYRANOSIDE (1)

D-Mannose 0.1 or 1 g (Sigma-Aldrich, 0.56 and 5.56 mmol) was suspended in 5.5 and 6.5 mL of allyl alcohol, respectively, and pTsOH monohydrate (42 mg, 0.22 mmol and 422 mg, 2.22 mmol, 0.4 equiv, respectively) was added. The suspension was stirred at 100°C (oil bath) overnight. The clear solution was cooled to rt, triethylamine (50 or 500 μL, 0.33 or 3.3 mmol, 1.5 equiv) was added, and the solvent was evaporated under reduced pressure. The crude product was chromatographed (9:1 EtOAc–EtOH), affording the desired mannopyranoside **1** (94 mg, 0.43 mmol, 75% and 511 mg, 2.32 mmol, 40% yield, α with trace amount of β visible in the acetylated product **3**, R_f = 0.5 (5:1 EtOAc–EtOH) as a light-yellow oil. $[\alpha]_D^{20}$ + 70.0 (c 1.2, CH$_3$OH); lit.[9] $[\alpha]_D^{20}$ + 73.6 (c 1.0, CH$_3$OH); ^{1}H NMR (400 MHz, MeOD): δ 6.00–5.88 (m, 1H, CH$_2$= CH–CH$_2$), 5.29 (dd, 1H, J 17.2, 1.7 Hz, CH_2=CH–CH$_2$), 5.17 (dd, 1H, J 10.5, 1.7 Hz, CH_2= CH–CH$_2$), 4.79 (d, 1H, J 1.6 Hz H-1), 4.22 (dd, 1H, J 13.0, 5.5 Hz, –O–CH_2–CH=CH$_2$), 4.00 (dd, 1H, J 13.0, 5.5 Hz, –O–CH_2–CH=CH$_2$), 3.86–3.78 (m, 2H, H-2, H-6a), 3.74–3.67 (m, 2H, H-3, H-6b), 3.60 (t, 1H, J 9.5 Hz, H-4), 3.53 (ddd, 1H, J9.5, 5.7, 2.3 Hz, H-5). ^{13}C NMR (101 MHz, MeOD): δ 135.50 (CH$_2$=CH–CH$_2$), 117.25 (CH$_2$=CH– CH$_2$), 100.75 (C-1), 74.74 (C-5), 72.65 (C-3), 72.20 (C-2), 68.84 (O–CH$_2$–CH=CH$_2$), 68.67 (C-4), 62.95 (C-6). ESI–MS: m/z [M + H]$^+$ calcd for C$_9$H$_{16}$O$_6$, 220.0947; found, 220.1002.

DEC-9-ENYL α-D-MANNOPYRANOSIDE (2)

D-Mannose (500 mg, 2.78 mmol) was suspended in 5 mL of 9-decen-1-ol, pTsOH (211 mg, 1.108 mmol, 0.4 equiv) was added, and the suspension was stirred at 100°C overnight. The suspension (the starting material did not dissolve completely) was cooled to rt and triethylamine (230 μL, 1.662 mmol, 1.5 equiv) was added. After concentration, the crude product was chromatographed (9.9:0.1 → 9.5:0.5 EtOAc–EtOH), affording the desired, amorphous mannopyranoside **2** (435 mg, 1.67 mmol, 60%), mainly α but with trace amount of β, R_f = 0.5 (5:1 EtOAc–EtOH). Crystallization from MeOH, EtOH, and CHCl$_3$ was unsuccessful. $[\alpha]_D^{20}$ + 42.9 (c 1.1, CH$_3$OH); lit. $[\alpha]_D^{20}$ + 56 (c 0.5, CH$_3$OH),[6] $[\alpha]_D^{20}$ + 50 (c 0.9, CH$_3$OH).[6] ^{1}H NMR (400 MHz, MeOD): δ 5.81 (ddt, 1H, J17.0, 10.2, 6.7 Hz, CH$_2$=CH–CH$_2$), 4.98 (dd, 1H, J 17.1, 1.5 Hz, CH_2=CH– CH$_2$), 4.91 (dd, 1H, J 10.2, 2.2 Hz, CH_2=CH–CH$_2$), 4.73 (d, 1H, J 0.9 Hz, H-1), 3.82 (dd, 1H, J 11.8, 2.2 Hz, H-6a), 3.78 (dd, 1H, J 3.0, 1.6 Hz, H-2), 3.77–3.66 (m, 3H, –O–CH_2–, H-3, H-6b), 3.61 (t, 1H, J 9.5 Hz, H-4), 3.55–3.49 (m, 1H, H-5), 3.41 (dt, 1H, J 9.6, 6.3 Hz, –O–CH_2–), 2.05 (q, 2H, J 6.9 Hz, CH$_2$=CH–CH_2–), 1.64–1.53

(m, 2*H*, O–CH$_2$–CH$_2$–), 1.45–1.28 (m, 10*H*, –(CH$_2$)$_5$–); ^{13}C NMR (101 MHz, MeOD): δ 140.12 (CH$_2$=CH–CH$_2$), 114.71 (CH$_2$=CH–CH$_2$), 101.53 (C-1), 74.55 (C-5), 72.67 (C-3), 72.28 (C-2), 68.61 (C-4), 68.56 (–O–CH$_2$–(CH$_2$)$_7$), 62.91 (C-6), 34.87 (CH$_2$=CH–CH$_2$–(CH$_2$)$_7$–O), 30.60 (O–CH$_2$–CH$_2$–(CH$_2$)$_6$), 30.54, 30.51, 30.17, 30.10, 27.33 (O–CH$_2$–CH$_2$–(CH$_2$)$_5$–CH$_2$–CH=CH$_2$). ESI–MS: *m/z* [M + H]$^+$ calcd for C$_{16}$H$_{30}$O$_6$: 318.2042; observed 318.1263; monoisotopic *m/z* = 341.0400 (Na$^+$ adduct), 357.0900 (K$^+$ adduct). Anal. calcd for C$_{16}$H$_{30}$O$_6$: C, 60.35; H, 9.50; found C, 60.42; H, 9.49.

ALLYL 2,3,4,6-TETRA-*O*-ACETYL-α-D-MANNOPYRANOSIDE (3)

Compound **1** 120 mg (0.55 mmol) was dissolved in a mixture of pyridine (4 mL) and Ac$_2$O (2 mL), and the solution was stirred at rt overnight. MeOH (1 mL) was added, and the solvent was evaporated under vacuum. The crude product was suspended in AcOEt (15 mL) and the solution was washed successively with 5% HCl (2 × 15 mL) and H$_2$O (15 mL). The organic phase was dried and the solvent was removed under reduced pressure affording the desired mannopyranoside **3** (192 mg, 0.50 mmol, 91% yield), as a light-yellow oil. [α]$_D^{20}$ + 48.3 (*c* 1.0, CHCl$_3$); lit. [α]$_D^{20}$ + 46.9 (*c* 1.0, CHCl$_3$).[9] ^{1}H NMR (400 MHz, CDCl$_3$): δ 5.96–5.84 (m, 1*H*, C*H*=CH$_2$), 5.36 (dd, 1*H*, *J* 10.0, 3.4 Hz, H-3), 5.33–5.20 (m, 4*H*, H-2, H-4, CH=C*H*$_2$), 4.86 (d, 1*H*, *J* 1.8 Hz, H-1), 4.28 (dd, 1*H*, *J* 12.2, 5.3 Hz, C*H*$_2$CH=CH$_2$), 4.18 (dd, 1*H*, *J* 12.8, 5.3 Hz, H-6a), 4.10 (dd, 1*H*, *J* 12.2, 2.2 Hz, C*H*$_2$CH=CH$_2$), 4.06–3.97 (m, 2*H*, H-5, H-6b), 2.14 (s, 3*H*, C*H*$_3$CO), 2.10 (s, 3*H*, CH$_3$CO), 2.03 (s, 3*H*, CH$_3$CO), 1.98 (s, 3*H*, CH$_3$CO). ^{13}C NMR (101 MHz, CDCl$_3$): δ 170.78, 170.19, 170.02, 169.88 (4 C=O), 133.04 (*C*H=CH$_2$), 118.58 (CH=*C*H$_2$), 96.72 (C-1), 69.76, 69.21, 68.67, 66.33 (C-2, C-3, C-4, C-5), 68.79 (*C*H$_2$CH=CH$_2$), 62.60 (C-6), 21.00, 20.85, 20.81, 20.81 (4 *C*H$_3$CO) ESI–MS: *m/z* [M + H]$^+$ calcd for C$_{17}$H$_{24}$O$_{10}$ 388.1369, observed monoisotopic *m/z* = 411.1272 (Na$^+$ adduct). All characterization data are in accordance with the literature data.[10,11]

DEC-9-ENYL 2,3,4,6-TETRA-*O*-ACETYL-α-D-MANNOPYRANOSIDE (4)

A solution of compound **2** (100 mg, 0.31 mmol) in 2:1 pyridine–Ac$_2$O (6 mL) was stirred at rt overnight. The full conversion was confirmed by TLC. MeOH (1 mL) was added, and the solvent was evaporated under vacuum. The crude product was chromatographed (7.3:2.7 petroleum ether–EtOAc), affording the desired man-nopyranoside **4** (140 mg, 0.29 mmol, 93% yield) as a light yellow oil, [α]$_D^{20}$ +40.1 (*c* 1.1 g/mL, CHCl$_3$); lit.[6] [α]$_D^{20}$ +38.4 (*c* 0.9, CHCl$_3$). ^{1}H NMR (400 MHz, CDCl$_3$): δ 5.80 (ddt, 1*H*, *J* 16.9, 10.2, 6.7 Hz, C*H*=CH$_2$), 5.34 (dd, 1*H*, *J* 10.0, 3.4 Hz, H-3), 5.26 (t, 1*H*, *J* 10.0 Hz, H-4), 5.22 (dd, 1*H*, *J* 3.2, 1.6 Hz, H-2), 4.98 (dd, 1*H*, *J* 17.1, 1.5 Hz, CH=C*H*$_2$), 4.92 (d, 1*H*, *J* 10.2 Hz, CH=C*H*$_2$), 4.79 (bs, 1*H*, H-1), 4.27 (dd, 1*H*, *J* 12.2, 5.3 Hz, H-6a), 4.09 (dd, 1*H*, *J* 12.2, 2.2 Hz, H-6b), 3.97 (ddd, 1*H*, *J* 9.5, 5.2, 2.2 Hz, H-5), 3.66 (dt, 1*H*, *J* 9.4, 6.8 Hz, OCH$_2$), 3.43 (dt, 1*H*, *J* 9.5, 6.6 Hz, OCH$_2$), 2.14 (s, 3*H*, CH$_3$CO), 2.09 (s, 3*H*, CH$_3$CO), 2.04–2.00 (m, 1*H*, –C*H*$_2$–CH=CH$_2$) 2.03 (s, 3*H*, CH$_3$CO), 1.98 (s, 3*H*, CH$_3$CO), 1.59–1.56 (m, 2*H*, OCH$_2$C*H*$_2$), 1.41–1.23 (m, 11*H*, –(CH$_2$)$_5$–CH$_2$–CH=CH$_2$). ^{13}C NMR (101 MHz, CDCl$_3$): δ 170.79, 170.24, 170.03, 169.89 (4 C=O), 139.28 (*C*H=CH$_2$), 114.27 (CH=*C*H$_2$), 97.69 (C-1), 69.87, 69.28, 68.50 (C-2, C-3, C-4), 68.68 (OCH$_2$), 66.41 (C-5), 62.66 (C-6), 33.88 (CH$_2$=CH–CH$_2$–(CH$_2$)$_7$–O),

29.47, 29.41, 29.37, 29.16, 29.00, 26.17 (6 CH$_2$), 21.03, 20.86, 20.82, 20.82 (4 CH_3CO). ESI–MS: *m/z* [M + H]$^+$ calcd for C$_{24}$H$_{38}$O$_{10}$ 486.2465, observed monoisotopic *m/z* = 509.2368 (Na$^+$ adduct), 525.2108 (K$^+$ adduct). Anal. calcd for C$_{24}$H$_{38}$O$_{10}$ C, 59.24; H, 7.87; found C, 59.48; H, 7.85.

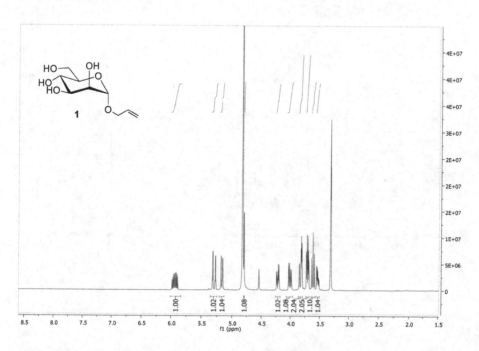

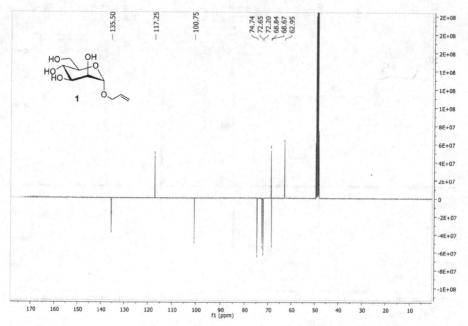

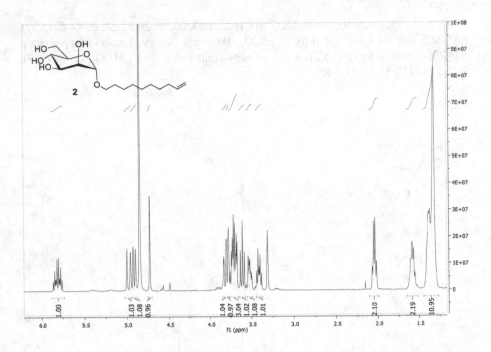

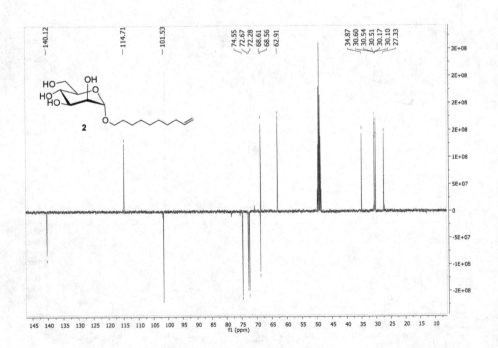

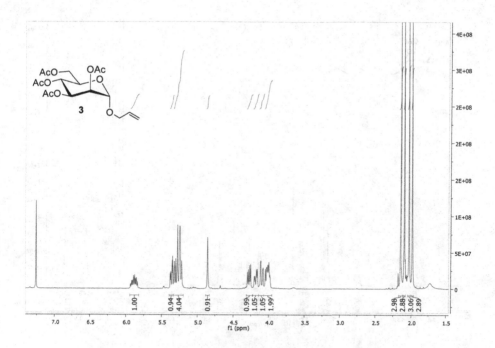

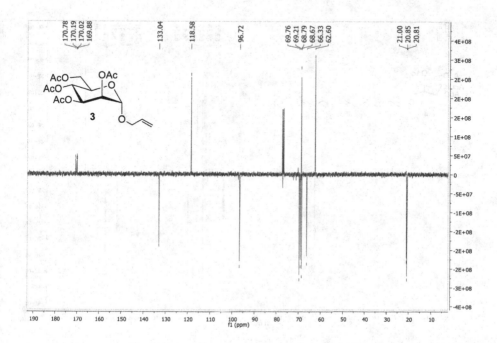

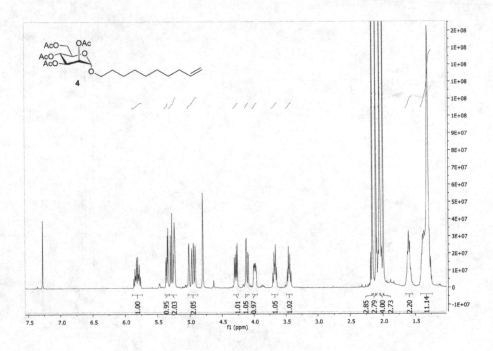

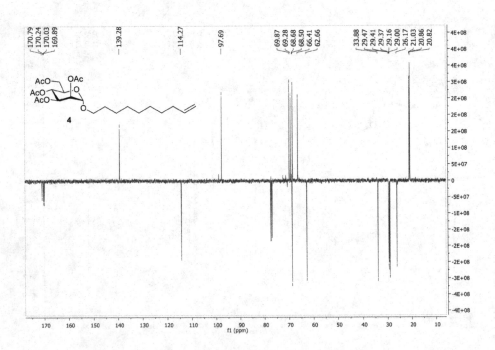

REFERENCES

1. (a) Kolb, H.C.; Finn, M.G.; Sharpless, K.B. *Angew. Chem. Int. Ed. Engl.* **2001**, *40*, 2004–2021. (b) Liang, L.; Astruc, D. *Coordination Chem. Rev.* **2011**, *255*, 2933–2945. (c) Corey, E. J. *Angew. Chem. Int. Ed.* **2002**, *41*, 1650–1667.
2. Rostovtsev, V.V.; Green, L.G.; Fokin, V.V.; Sharpless, K.B. *Angew. Chem. Int. Ed.* **2002**, *41*, 2596–2599.
3. Dondoni, A. *Angew. Chem. Int. Ed.* **2008**, *47*, 8995–8997.
4. Buskas, T.; Söderberg, E.; Konradsson, P.; Fraser-Reid, B. *J. Org. Chem.* **2000**, *65*, 958–963.
5. Seeberger, P.H.; Stallforth, P.; De Liberto, G.; Cavallari, M. Carbohydrate-Glycolipid Conjugate Vaccines, Patent: US 20150238597 A1.
6. Baldoni, L.; Marino, C. *Carbohydr. Res.* **2013**, *374*, 75–81.
7. Horejsí, V.; Kocourek, J. *Biochim. Biophys. Acta* **1973**, *297*, 346–351.
8. Nikolaev, A.V.; Rutherford, T.J.; Ferguson M.A.J; Brimacombe, J.S. *J. Chem. Soc., Perkin Trans.* **1995**, 1, 1977–1987.
9. Poláková, M.; Roslund, M.U.; Ekholm, F.S.; Saloranta, T.; Leino, R. *Eur. J. Org. Chem.* **2009**, 870–888.
10. (a) León, E. I.; Martín, A.; Pérez-Martín, I.; Quintanal, L. M.; Suárez, E. *Eur. J. Org. Chem.* **2012**, *20*, 3818–3829. (b) Martín, A.; Quintanal, L. M.; Suárez, E. *Tetrahedron Lett.* **2007**, *48*, 5507–5511.
11. Page, D.; Roy, R. *Glycoconj. J.* **1997**, *14*, 345–356.

25 Synthesis of 2,2,2-Trifluoroethyl Glucopyranoside and Mannopyranoside via Classical Fischer Glycosylation

*Afraz Subratti, and Nigel K. Jalsa**
The University of the West Indies,
St. Augustine, Trinidad and Tobago

Bhoomendra Bhongade[a]
Department of Pharmaceutical Chemistry,
RAK College of Pharmaceutical Sciences,
RAK Medical and Health Sciences University,
Ras Al Khaimah, United Arab Emirates

CONTENTS

2,2,2-Trifluoroethyl glucopyranosides have found several applications, including selective tags for proteins,[1] and probes for studying various protein binding and catalytic sites.[2–4] These derivatives have been commonly synthesized via multistep routes from free sugars by Mitsunobu conditions,[5] solvolysis,[6] insertion of a glycosylidene

---

* Corresponding author: nigel.jalsa@sta.uwi.edu
[a] Checker: bhoomendra@rakmhsu.ac.ae

(i) TFE, AcCl, reflux, 24 h
(ii) Ac_2O, pyr, rt, 12 h

1: R_1 = OH; R_2 = H
2: R_1 = H; R_2 = OH

3: R_1 = OAc; R_2 = H (51 %)
(3a = α anomer; 3b = β anomer)
4: R_1 = H; R_2 = OAc (59 %)

carbene,[7] or glycosylation with an acetate donor.[8,9] The most recent approach necessitated the employment of microwave conditions using trifluoroethanol (TFE) and acetyl chloride.[1] Attempts at the classical Fischer glycosylation were identified as being unsuccessful due to a combination of sugar insolubility and weak nucleophilicity and low boiling point of the solvent.

We report herein the successful glycosylation of glucose and mannose employing TFE as the solvent and in situ-generated HCl as the promoter at reflux temperature. Upon removal and recovery of the solvent, the crude product was acetylated with acetic anhydride and pyridine to facilitate purification. Yields of 51% for the glucose derivative (3) and 59% for the mannose derivative (4) are attractive considering the essentially one-pot process. We attribute the success of this simple method to the larger than usual solvent-to-sugar ratio employed.

EXPERIMENTAL

GENERAL METHODS

All chemicals used were reagent grade and used as purchased, unless otherwise stated. Reaction mixtures were stirred magnetically. Gravity column chromatography was carried out using high purity silica gel (porosity: 60 Å, particle size: 63–200 μm, bulk density: 0.5 g/mL, pH range: 6.0–8.0, residual water: <7.0%). Silica gel plates on aluminum backing (silica G TLC plates, w/UV 254) were used for thin-layer chromatography (TLC), with ammonium molybdate (ammonium molybdate [VI] tetrahydrate [25 g] in 1 M H_2SO_4 [500 mL]) for visualization. Characterization of compounds and confirmation of the structure was done by (i) high-resolution mass spectrometry (HRMS) by Bruker Daltonics micrOTOF-Q instrument via electron spray ionization, (ii) optical rotation ($[\alpha]_D$) measurement (Bellingham & Stanley ADP 220 polarimeter at 25.0°C); and (iii) nuclear magnetic resonance (NMR) spectroscopy with Bruker 300- or 600-MHz spectrometers. Chemical shifts are reported in ppm and coupling constants in Hertz multiplicities are stated as follows; s (singlet), *appt* s (apparent singlet), d (doublet), dd (doublet of doublets), *appt* dd (apparent doublet of doublets), ddd (doublet of doublet of doublets), dt (doublet of triplets), *appt* dt (apparent doublet of triplets), t (triplet), *pt* (pseudo triplet), td (triplet of doublets), q (quartet), m (multiplet), etc. Combustion analysis was performed with a PerkinElmer 2400 Series II CHNS/O. Solutions in organic solvents were dried with anhydrous Na_2SO_4 and concentrated at reduced pressure and <40°C.

GENERAL SYNTHETIC PROTOCOL

D-Glucose (1) or D-mannose (2) (1000 mg, 5.551 mmol) were suspended in freshly distilled TFE (50 mL), and acetyl chloride (1.97 mL, 27.753 mmol, 5.0 equiv) was slowly added with stirring and the mixture was refluxed for 24 h. Minor starting material was observed at point; however, prolonged reaction time (>24 h) did not result in any more conversion. The TFE was then removed under reduced pressure and the crude product was dissolved in pyridine (15 mL). Acetic anhydride (10 mL) was added and the solution was stirred at room temperature for 12 h. After concentration, a solution of the residue in CH_2Cl_2 (50 mL) was washed successively with 1 M HCl (3 × 50 mL) and water (3 × 50 mL). The organic layer was dried, concentrated, and column chromatography (7:3 petroleum ether–EtOAc) gave mixture of glycosides 3a (R_f 0.3), and 3b (R_f 0.2) in an approximate ratio of α:β = 3:1 (NMR) as a pale yellow solid (combined yield from 1, 1211 mg, 51%); and 4 (α exclusively) as a pale yellow solid (from 2, 1429 mg, 59%) R_f (0.5, 1:1 petroleum ether–EtOAc).

TRIFLUOROETHYL 2,3,4,6-TETRA-O-ACETYL-α- (3A)
AND β-D-GLUCOPYRANOSIDE (3B)

For characterization, the mixture of anomers was re-chromatographed by gradient elution from a column of silica gel. (petroleum ether → 9:1 petroleum ether–EtOAc → 8:2 petroleum ether–EtOAc) → 7:3 (petroleum ether–EtOAc). Elution with the latter solvent system was continued until complete elution of both anomers was achieved.

α-Anomer (3a): $[\alpha]_D$ −15.0 (c 0.80, $CDCl_3$); mp 81–83°C (from ethanol). ^1H NMR ($CDCl_3$) δ 5.49 (t, 1H, $J_{2,3}$ 9.9 Hz, $J_{3,4}$ 9.9 Hz, H-3), 5.21 (d, 1H, $J_{1,2}$ 3.8 Hz, H-1), 5.09 (t, 1H, $J_{3,4}$ 9.9 Hz, $J_{4,5}$ 9.9 Hz, H-4), 4.88 (dd, 1H, $J_{1,2}$ 3.8 Hz, $J_{2,3}$ 9.9 Hz H-2), 4.27 (dd, 1H, $J_{5,6}$ 4.5 Hz, $J_{6,6'}$ 12.4 Hz, H-6), 4.12 (d, 1H, $J_{5,6'}$ 2.3 Hz, $J_{6,6'}$ 12.4 Hz, H-6′), 3.91–4.06 (m, 3H, H-5, CH_2CF_3), 2.10 (s, 3H, $COCH_3$), 2.08 (s, 3H, $COCH_3$), 2.04 (s, 3H, $COCH_3$), 2.03 (s, 3H, $COCH_3$); ^{13}C NMR ($CDCl_3$) δ 170.5 (1 C, $COCH_3$), 170.2 (1 C, $COCH_3$), 169.9 (1 C, $COCH_3$), 169.5 (1 C, $COCH_3$), 123.4 (1 C, J_{C-F} 278.5 Hz, CF_3), 96.7 (1 C, J_{C1-H1} 175.1 Hz, C-1), 70.3 (1 C, C-2), 69.5 (1 C, C-5), 68.1 (1 C, C-3), 68.0 (1 C, C-4), 65.6 (1 C, J_{C-F} 35.5, CH_2CF_3), 61.5 (1 C, C-6), 20.6 (2 C, −$COCH_3$), 20.4 (1 C, $COCH_3$), 20.3 (1 C, $COCH_3$); HRMS: m/z [M + Na]$^+$ calcd for $C_{16}H_{21}F_3O_{10}Na$, 453.0985; found: 453.0996. Anal. calcd for $C_{16}H_{21}F_3O_{10}$: C, 44.66; H, 4.92; O, 37.18. Found: C, 44.71; H, 4.98; O, 37.25.

β-Anomer (3b): $[\alpha]_D$ + 51.4 (c 1.1, $CDCl_3$); mp 79–81°C (from ethanol). ^1H NMR ($CDCl_3$) δ 5.23 (t, 1H, $J_{2,3}$ 9.4 Hz, $J_{3,4}$ 9.4 Hz, H-3), 5.10 (t, 1H, $J_{3,4}$ 9.4 Hz, $J_{4,5}$ 9.4 Hz, H-4), 5.04 (t, 1H, $J_{1,2}$ 9.4 Hz, $J_{2,3}$ 9.4 Hz, II-2), 4.66 (d, 1H, $J_{1,2}$ 9.4 Hz, H-1), 4.27 (dd, 1H, $J_{5,6}$ 4.6 Hz, $J_{6,6'}$ 12.4 Hz, H-6), 4.12 (dd, 1H, $J_{5,6'}$ 2.8 Hz, $J_{6,6'}$ 12.4 Hz, H-6′), 3.90–4.03 (m, 2H, CH_2CF_3), 3.74 (ddd, 1H, $J_{4,5}$ 9.4 Hz, $J_{5,6}$ 4.6 Hz, $J_{5,6'}$ 2.8 Hz, H-5), 2.10 (s, 3H, $COCH_3$), 2.05 (s, 3H, $COCH_3$), 2.03 (s, 3H, $COCH_3$), 2.02 (s, 3H, $COCH_3$); ^{13}C NMR ($CDCl_3$) δ 170.6 (1 C, $COCH_3$), 170.1 (1 C, $COCH_3$), 169.33 (1 C, $COCH_3$), 169.29 (1 C, $COCH_3$), 123.4 (1 C, J_{C-F} 278.5 Hz, CF_3), 100.7 (1 C, J_{C1-H1} 163.8 Hz, C-1), 72.3 (1 C, C-5), 72.2 (1 C, C-3), 70.7 (1 C, C-2), 68.1 (1 C, C-4), 66.0 (1 C, J_{C-F} 35.5 Hz, CH_2CF_3) 61.6 (1 C, C-6), 20.7 (1 C, $COCH_3$), 20.5 (2 C, $COCH_3$), 20.4 (1 C, $COCH_3$). HRMS: m/z [M + Na]$^+$ calcd for $C_{16}H_{21}F_3O_{10}Na$, 453.0985;

found: 453.0992. Anal. calcd for $C_{16}H_{21}F_3O_{10}$: C, 44.66; H, 4.92; O, 37.18. Found: C, 44.74; H, 5.09; O, 37.32.

TRIFLUOROETHYL 2,3,4,6-TETRA-*O*-ACETYL-α-D-MANNOPYRANOSIDE (4)

$[\alpha]_D$ +22.2 (*c* 0.9, CDCl$_3$); mp 71–73°C (from ethanol). ^{1}H NMR (CDCl$_3$) δ 5.32–5.29 (m, 3*H*, H-2, H-3, H-4), 4.92 (*appt* s, 1*H*, H-1), 4.28 (dd, 1*H*, $J_{5,6}$ 5.4 Hz, $J_{6,6'}$ 12.3 Hz, H-6), 4.13 (dd, 1*H*, $J_{5,6'}$ 2.4 Hz, $J_{6,6'}$ 12.3 Hz, H-6'), 4.01–3.98 (m, 3*H*, H-5, C*H*$_2$CF$_3$), 2.17 (s, 3*H*, COC*H*$_3$), 2.12 (s, 3*H*, COC*H*$_3$), 2.06 (s, 3*H*, COC*H*$_3$), 2.01 (s, 3*H*, COC*H*$_3$); ^{13}C NMR (CDCl$_3$) δ 170.6 (1 C, *C*OCH$_3$), 169.9 (1 C, *C*OCH$_3$), 169.8 (1 C, *C*OCH$_3$), 169.7 (1 C, *C*OCH$_3$), 122.3 (1 C, *C*F$_3$), 98.0 (1 C, J_{C1-H1} 173.6 Hz, C-1), 69.4 (1 C, C-5), 68.9 (1 C, C-3), 68.5 (1 C, C-2), 65.8 (1 C, C-4), 65.0 (1 C, J_{C-F} 36.2 Hz, *C*H$_2$CF$_3$), 62.2 (1 C, C-6), 20.8 (1 C, COC*H*$_3$), 20.69 (2 C, COC*H*$_3$), 20.65 (1 C, COC*H*$_3$). HRMS: *m/z* [M + Na]$^+$ calcd for $C_{16}H_{21}F_3O_{10}$Na, 453.0985; found, 453.0994. Anal. calcd for $C_{16}H_{21}F_3O_{10}$: C, 44.66; H, 4.92; O, 37.18. Found: C, 44.79; H, 5.01; O, 37.33.

ACKNOWLEDGMENTS

The Authors thank The University of the West Indies for financial support.

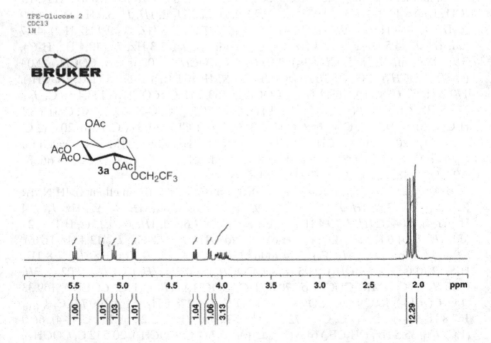

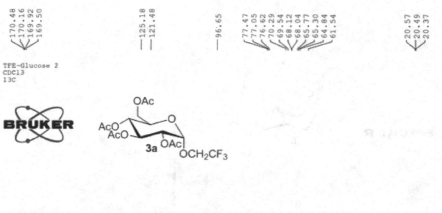

TFE-Glucose 2
CDC13
13C

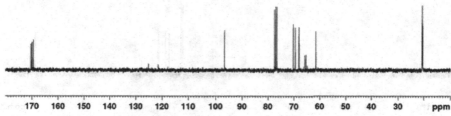

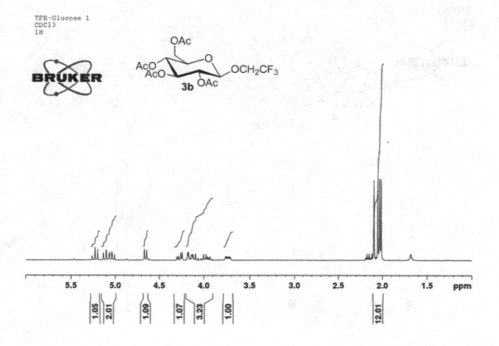

TFE-Glucose 1
CDC13
1H

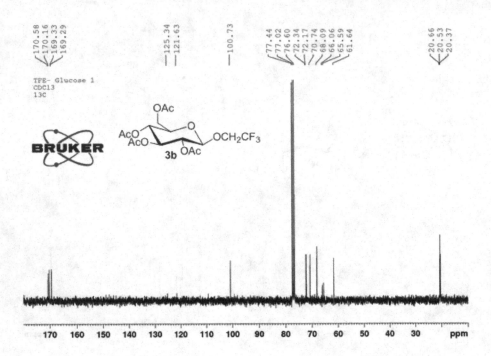

170.58, 170.16, 169.33, 169.29 · 125.34, 121.63 · 100.73 · 77.44, 77.02, 76.60, 72.34, 72.17, 70.74, 70.09, 68.09, 66.06, 65.59, 61.64 · 20.66, 20.53, 20.37

TFE- Glucose 1
CDCl3
13C

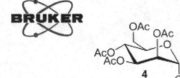

3b

170 160 150 140 130 120 110 100 90 80 70 60 50 40 30 ppm

TFE–Mannose
CDCl3
1H

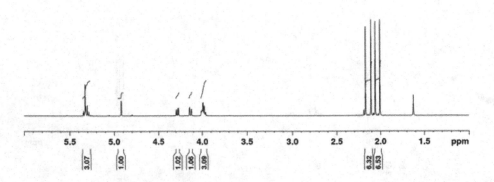

4

5.5 5.0 4.5 4.0 3.5 3.0 2.5 2.0 1.5 ppm

3.07 · 1.00 · 1.02 · 1.06 · 3.09 · 6.32 · 6.53

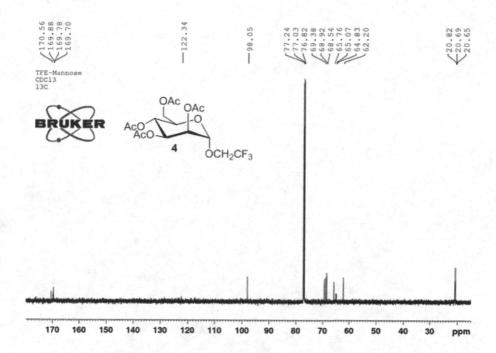

REFERENCES

1. Fröhlich, R. F. G.; Schrank, E.; Zangger K. *Carbohydr. Res.* **2012**, *361*, 100–104.
2. Richard, J. P.; Westerfeld, J. G.; Lin, S. *Biochemistry* **1995**, *34*, 11703–11712.
3. Poretz, R. D.; Goldstein, I. J. *Biochemistry* **1970**, *9*, 2890–2896.
4. Richard, J. P.; Huber, R. E.; Lin, S.; Heo, C.; Amyes, T. L. *Biochemistry* **1996**, *35*, 12377–12386.
5. Gueyrard, D.; Rollin, P.; Nga, T. T. T.; Ourévitch, M.; Bégué, J.-P. Bonnet-Delpon, D. *Carbohydr. Res.* **1999**, *318*, 171–179.
6. Sinnott, M. L.; Jencks, W. P. *J. Am. Chem. Soc.* **1980**, *102*, 2026–2032.
7. Vasella, A.; Briner, K.; Soundararajan, N.; Platz, M. S. *J. Org. Chem.* **1991**, *56*, 4741–4744.
8. Beaver, M. G.; Woerpel, K. A. *J. Org. Chem.* **2010**, *75*, 1107–1118.
9. Xue, J. L.; Cecioni, S.; He, L.; Vidal, S.; Praly, J.-P. *Carbohydr. Res.* **2009**, *344*, 1646–1653.

26 Synthesis of 2-{2-[2-(N-Tert-Butyloxycarbonyl) Ethoxy]Ethoxy}Ethyl β-D-Glucopyranoside

Hélène B. Pfister, and Pavol Kováč*
NIDDK, LBC, National Institutes of Health,
Bethesda, Maryland, United States

Marek Baráth[a]
Institute of Chemistry, Slovak Academy of Sciences
Bratislava, Slovakia

CONTENTS

Azidoalkyl glycosides are useful intermediates in the synthesis of their corresponding aminoalkyl glycoside counterparts, as well as in the preparation of glycoconjugates through azide-alkyne cycloadditions.[1] We previously described in this series a synthesis of 6-azidohexyl 2,3,4,6-tetra-*O*-acetyl-β-D-glucopyranoside, the per-*O*-acetate of one of the simplest compounds in this class of substances.[2] However, after further functionalization of this derivative, its hydrogenolysis yielded significant amount of a side-product resulting from the reductive dimerization of the azidoalkyl linker.[3] We now describe the synthesis of 2-[2-(2-azidoethoxy)ethoxy]

---

* Corresponding author: e-mail: kpn@helix.nih.gov
[a] Checker: chemmbar@savba.sk

ethyl 2,3,4,6-tetra-*O*-acetyl-β-D-glucopyranoside (**2**), wherein the azidohexyl linker is replaced by an ethylene glycol derivative. Similar to our previous experience,[4–5] and observation of others,[6–8] hydrogenolysis of the azide function in this linker was uneventful. Ethylene glycol-type linkers have other advantages over their aliphatic chain counterparts. They are known to be more water soluble and show lower immunogenicity and nonspecific binding character.[9] Also, as pointed out,[10] these spacer arms are considered to be superior to the poly(methylene) types, which have a strong tendency to coil in aqueous milieu, thus lessening their effective lengths.

Herein, we describe synthesis of azide **2** and its protected amine derivative **4** from 1,2,3,4,6-penta-*O*-acetyl-β-D-glucopyranose (**1**). Compound **2** was previously described[11–13] but not fully characterized. Syntheses of compounds **3** and **4** are reported here for the first time. Derivatives **2** and **4** can be used as amine-equipped building blocks for further modifications. Glucopyranoside **2** was prepared in good yield by BF$_3$·OEt$_2$-catalyzed reaction of 2-[2-(2-azidoethoxy)ethoxy]ethanol with acetate **1**. Product **2** was further converted into the Boc-protected amine **4** by Zemplén deacetylation,[14] followed by the one-pot conversion of the azido group into the Boc-protected amine, applying hydrogenolysis in the presence of di-*tert*-butyl dicarbonate (Boc$_2$O).

EXPERIMENTAL

GENERAL METHODS

High-performance liquid chromatography (HPLC) grade solvents were used, and reactions requiring anhydrous conditions were carried out under nitrogen or argon. Hydrogenation/hydrogenolysis was done under atmospheric pressure. The palladium-on-charcoal catalyst used was purchased from Engelhard Industries (Escat 103, Lot# FC96303). Reactions were monitored by thin-layer chromatography (TLC) on silica gel 60 glass slides. Spots were visualized by charring with H$_2$SO$_4$ in EtOH (5% v/v). Optical rotations were measured at ambient temperature with a Jasco P-2000 digital polarimeter. NMR spectra were measured at 25°C, at 600 MHz for ^{1}H, and at 150 MHz for ^{13}C with a Bruker Avance spectrometer. Assignments of NMR signals were aided by 1D and 2D experiments (APT, COSY, and HSQC) run with the software supplied with the spectrometer. With CDCl$_3$ as solvent, ^{1}H and ^{13}C chemical shifts were referenced to signals of tetramethylsilane (0 ppm) and CDCl$_3$ (77.00 ppm). With D$_2$O as solvent, ^{1}H and ^{13}C chemical shifts were referenced to signals of acetone (external, 2.22 and 30.89 ppm for protons and ^{13}C, respectively). With MeOD as solvent, ^{1}H and ^{13}C chemical shifts were referenced to signals of MeOD (3.31 and 49.00 ppm, respectively). Samples for combustion analysis were prepared as described.[15] Liquid chromatography-electron spray-ionization mass spectrometry (ESI–MS) was performed with a Hewlett-Packard 1100 MSD spectrometer. Solutions in organic solvents were dried with anhydrous Na_2SO_4 and concentrated at reduced pressure/<40°C.

2-[2-(2-Azidoethoxy)ethoxy]ethyl 2,3,4,6-tetra-
O-acetyl-β-D-glucopyranoside (2)

BF_3·OEt_2 (6.0 mL, 48.7 mmol) was slowly added at 0°C to a solution of 1,2,3,4,6-penta-O-acetyl-β-D-glucopyranose (**1**, 4.75 g, 12.2 mmol) and 2-[2-(2-azidoethoxy)ethoxy] ethanol (4.36 g, 24.9 mmol) in anhydrous DCM (200 mL). The mixture was stirred at room temperature under argon for 46 h when TLC (4:1 toluene–acetone) showed that the starting material was virtually consumed, and that one major product was formed. Several more polar side products were also present. The mixture was neutralized with Et_3N and concentrated. Flash chromatography (toluene-acetone 95:5 → 85:15) gave glycoside **2** (4.16 g, 8.23 mmol, 67%, colorless oil), which failed to crystallize from common solvents; $[\alpha]_D^{22}$ –10.4 (c 1.0, $CHCl_3$); ^1H NMR ($CDCl_3$) δ 5.21 (t, 1*H*, $J_{2,3}$= $J_{3,4}$ 9.5 Hz, H-3), 5.09 (dd, 1*H*, $J_{3,4}$ 9.5 Hz, $J_{4,5}$ 9.9 Hz, H-4), 5.00 (dd, 1*H*, $J_{1,2}$ 8.0 Hz, $J_{2,3}$ 9.6 Hz, H-2), 4.62 (d, 1*H*, $J_{1,2}$ 8.0 Hz, H-1), 4.26 (dd, 1*H*, $J_{5,6a}$ 4.8 Hz, $J_{6a,6b}$ 12.3 Hz, H-6a), 4.14 (dd, 1*H*, $J_{5,6b}$ 2.4 Hz, $J_{6a,6b}$ 12.3 Hz, H-6b), 3.95 (m, 1*H*, CH_2 linker), 3.75 (m, 1*H*, CH_2-linker), 3.71 (ddd, 1*H*, $J_{4,5}$ 10.0 Hz, $J_{5,6a}$ 4.7 Hz, $J_{5,6b}$ 2.4 Hz, H-5), 3.68-3.63 (m, 8*H*, CH_2-linker), 3.41 (t, 2*H*, J 5.1 Hz, CH_2-linker), 2.09, 2.05, 2.03, 2.01 (4 s, 12*H*, OCOCH_3); ^{13}C NMR ($CDCl_3$) δ 170.7, 170.3, 169.4, 169.4 (4 × O*C*OCH$_3$), 100.8 (C-1), 72.8 (C-3), 71.7 (C-5), 71.2 (C-2), 70.7, 70.7, 70.4, 70.0, 69.0 (5 × CH$_2$-linker), 68.4 (C-4), 61.9 (C-6), 50.7 (CH$_2$-linker), 20.7, 20.7, 20.6, 20.6 (3 × OCO*C*H$_3$); ESI-HRMS: m/z [M + Na]+ calcd for $C_{20}H_{31}N_3O_{12}Na$, 528.1805; found, 528.1808. Anal. calcd for $C_{20}H_{31}N_3O_{12}$: C, 47.52; H, 6.18; N, 8.31. Found: C, 47.64; H, 6.28; N, 8.29.

2-[2-(2-Azidoethoxy)ethoxy]ethyl β-D-glucopyranoside (3)

To a solution of acetate **2** (522 mg, 1.03 mmol) in MeOH (6 mL) was added methanolic solution of NaOMe (1 M) until the pH was consistently 10 (≈ 150 μL). The reaction was stirred at room temperature for 1 h, when TLC showed complete conversion (9:1 DCM–MeOH) and formation of a more polar product. The mixture was diluted with MeOH (≈ 5 mL) and neutralized with Amberlite IR-120 H+ resin. The suspension was filtered and the filtrate concentrated, to give compound **3** (357 mg) in nearly theoretical yield as colorless oil. For analysis, a small amount was eluted from a column of silica gel (DCM → 85:15 DCM–MeOH) to give pure **3**; $[\alpha]_D^{22}$ –17.4 (c 1.0, CH_3OH); ^1H NMR (D_2O) δ 4.48 (d, 1*H*, $J_{1,2}$ 8.0 Hz, H-1), 4.05 (m, 1*H*, CH_2-linker), 3.90 (dd, 1*H*, $J_{5,6a}$ 2.2 Hz, $J_{6a,6b}$ 12.3 Hz, H-6a), 3.83 (m, 1*H*, CH_2-linker), 3.75 (t, 2*H*, J 4.2 Hz, CH_2-linker), 3.72-3.71 (m, overlapped, 6*H*, CH_2-linker), 3.70 (dd, 1*H*, overlapped, $J_{5,6b}$ 6.0 Hz, $J_{6a,6b}$ 12.4 Hz, H-6b), 3.50 (t, 2*H*, overlapped, J 3.9 Hz, CH_2-linker), 3.48 (t, 1*H*, overlapped, $J_{2,3}$ = $J_{3,4}$ 9.3 Hz, H-3), 3.44 (ddd, 1*H*, $J_{4,5}$ 9.7 Hz, $J_{5,6a}$ 2.2 Hz, $J_{5,6b}$ 6.0 Hz, H-5), 3.37 (dd, 1*H*, $J_{3,4}$ 9.2 Hz, $J_{4,5}$ 9.7 Hz, H-4), 3.28 (dd, 1*H*, $J_{1,2}$ 8.0 Hz, $J_{2,3}$ 9.4 Hz, H-2); ^{13}C NMR (D_2O) δ 102.0 (C-1), 75.6 (C-5), 75.4 (C-3), 72.8 (C-2), 69.4 (CH$_2$-linker), 69.3 (C-4), 69. 3, 69.2, 68.9, 68.4 (4 × CH$_2$-linker), 60.5 (C-6), 49.8 (CH$_2$-linker); ESI-HRMS: m/z [M + Na]+ calcd for $C_{12}H_{23}N_3O_8Na$

360.1383; found 360.1376. Anal. calcd for $C_{12}H_{23}N_3O_8$: C, 42.73; H, 6.87; N, 12.46. Found: C, 42.62; H, 6.75; N, 12.30.

2-{2-[2-(N-TERT-BUTYLOXYCARBONYL)ETHOXY]ETHOXY}ETHYL β-D-GLUCOPYRANOSIDE (4)

A suspension of Pd/C (75 mg) in a small amount of MeOH was added under a stream of N_2 to a solution of azide **3** (357 mg, 1.03 mmol) and Boc_2O (510 mg, 2.06 mmol) in MeOH (7 mL). The suspension was degassed, flushed with H_2 three times, and stirred under H_2 at room temperature. After 1 day, TLC (85:15 CHCl₃–MeOH) showed complete conversion. The suspension was filtered over Celite and the Celite was thoroughly washed with MeOH. After concentration, chromatography gave the desired **4** (370 mg, 0.90 mmol, 87%) as colorless oil. $[\alpha]_D^{23}$ −13.3 (c 1.0, H_2O); ¹H NMR (MeOD) δ 4.33 (d, 1*H*, $J_{1,2}$ 7.8 Hz, H-1), 4.02 (m, 1*H*, CH₂-linker), 3.86 (dd, 1*H*, $J_{5,6a}$ 1.9 Hz, $J_{6a,6b}$ 12.1 Hz, H-6a), 3.73–3.61 (m, 7*H*, CH₂-linker), 3.66 (m, 1*H*, overlapped, H-6b), 3.51 (t, 2*H*, J 5.6 Hz, CH₂-linker), 3.36 (m, 1*H*, H-3), 3.28–3.27 (m, 2*H*, H-4, H-5), 3.22 (t, 2*H*, J 5.5 Hz, overlapped, CH₂-linker), 3.20 (dd, 1*H*, $J_{1,2}$ 7.9 Hz, $J_{2,3}$ 9.2 Hz, H-2), 1.44 (s, 9*H*, C(CH₃)₃); ¹³C NMR (MeOD) δ 158.5 (NHCOO), 104.5 (C-1), 80.1 (C(CH₃)₃), 78.0 (C-5), 77.9 (C-3), 75.1 (C-2), 71.6 (C-4), 71,5, 71.5, 71.2, 71.1, 69.6 (5×CH₂-linker), 62.8 (C-6), 41.3 (CH₂-linker), 28.8 (C(CH₃)₃); ESI-HRMS: m/z [M + Na]⁺ calcd for $C_{17}H_{33}NO_{10}Na$ 434.2002; found 434.1996. Anal. calcd for $C_{17}H_{33}NO_{10}$: C, 49.63; H, 8.08; N, 3.40. Found: C, 49.34; H, 8.10; N, 3.46.

ACKNOWLEDGMENTS

This research was supported by the Intramural Research Program of the National Institutes of Health (NIH), and the National Institute of Diabetes and Digestive and Kidney Diseases (NIDDK).

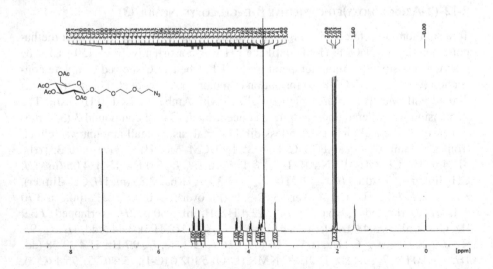

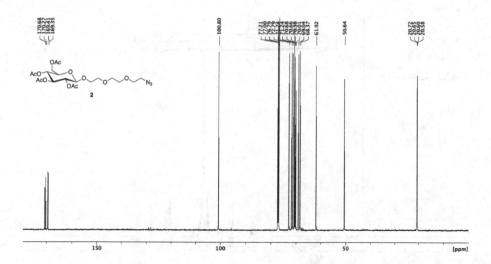

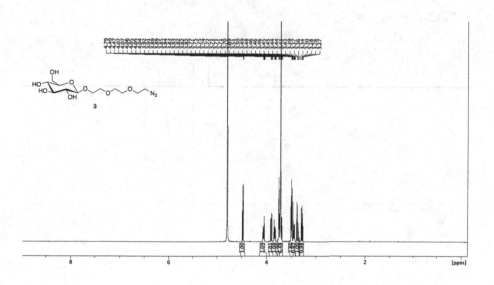

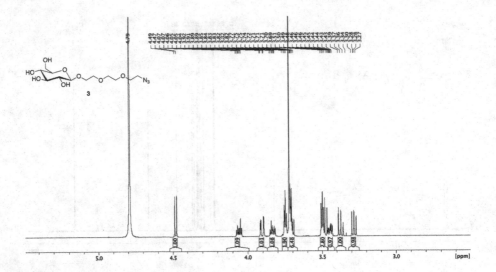

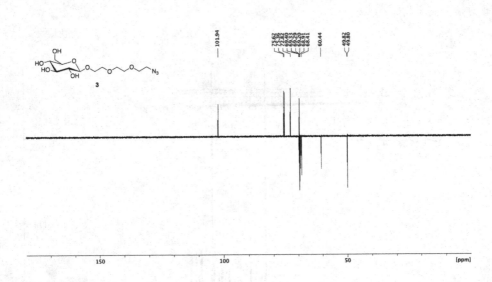

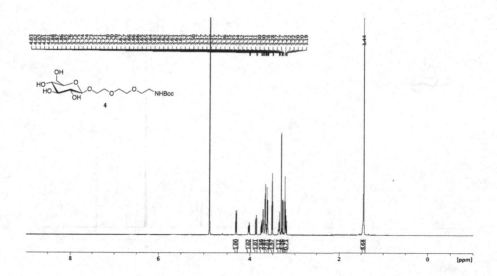

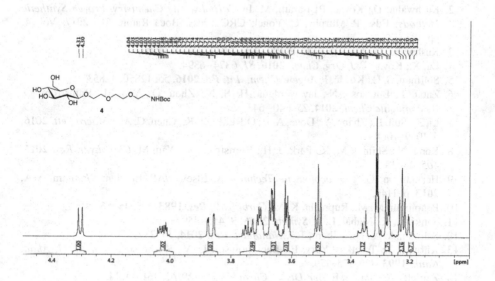

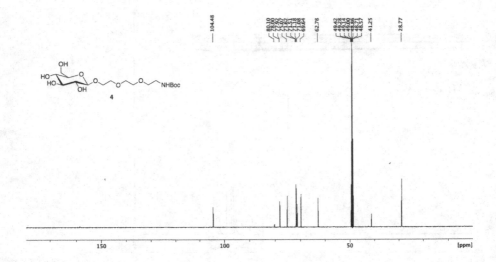

REFERENCES

1. Tiwari, V. K.; Mishra, B. B.; Mishra, K. B.; Mishra, N.; Singh, A. S.; Chen, X. *Chem. Rev.* **2016**, *116*, 3086–3240.
2. Kushwaha, D.; Kováč, P.; Barath, M. In *Carbohydrate Chemistry. Proven Synthetic Methods*; Eds.: P. Murphy, C. Vogel; CRC Press: Boca Raton, FL, **2017**; Vol. 4, pp. 183–188.
3. Kushwaha, D.; Xu, P.; Kováč, P. *RSC Adv.* **2017**, *7*, 7591–7603.
4. Lu, X.; Kováč, P. *J. Org. Chem.* **2016**, *81*, 6374–6394.
5. Soliman, S. E.; Kováč, P. *Angew. Chem., Int. Ed.* **2016**, *55*, 12850–12853.
6. Zhou, J.; Butchosa, N.; Jayawardena, H. S. N.; Zhou, Q.; Yan, M.; Ramström, O. *Bioconjugate Chem.* **2014**, *25*, 640–643.
7. Li, Z.; Sun, L.; Zhang, Y.; Dove, A. P.; O'Reilly, R. K.; Chen, G. *ACS Macro Lett.* **2016**, *5*, 1059–1064.
8. Kong, N.; Shimpi, M. R.; Park, J. H.; Ramström, O.; Yan, M. *Carbohydr. Res.* **2015**, *405*, 33–38.
9. Hermanson, G. T. *Bioconjugate Techniques*, Elsevier/AP, London; Waltham, MA, **2013**, p. 1146.
10. Ponpipom, M. M.; Ruprecht, K. M. *Carbohydr. Res.* **1983**, *113*, 45–56.
11. Sanki, A. K.; Mahal, L. K. *Synlett*, **2006**, *3*, 455–459.
12. Salman, A. A.; Heidelberg, T. *J. Nanopart. Res.* **2014**, *16*, 2399.
13. Kitov, P. I.; Tsvetkov Yury, E.; Backinowski, L. V.; Kochetkov, N. K. *Dokl. Akad. Nauk.* **1993**, *329*, 175–179.
14. Zemplén, G.; Pacsu, E. *Ber. Dtsch. Chem. Ges.* **1929**, *62*, 1613–1614.
15. Kováč, P. In *Carbohydrate Chemistry: Proven Synthetic Methods*; Eds.: G. Van Der Marel, J. Codee; CRC Press-Taylor & Francis Group: Boca Raton, FL, **2014**; Vol. 2, pp. XI–XXV.

27 Synthesis of 2-Propynyl 2-Acetamido-3,4,6-Tri-O-Acetyl-2-Deoxy-1-Thio-β-D-Glucopyranoside, 2-Propynyl 3,4,6-Tri-O-Acetyl-2-Deoxy-2-Phthalimido-1-Thio-β-D-Glucopyranoside and Their 2-(2-Propynyloxy-ethoxy)ethyl Analogs

*Alejandro E. Cristófalo, Hugo O. Montenegro, María Emilia Cano, and María Laura Uhrig**
Universidad de Buenos Aires, Facultad de Ciencias
Exactas y Naturales, Departamento de Química Orgánica,
Pabellón 2, Ciudad Universitaria,
Buenos Aires, Argentina
Consejo Nacional de Investigaciones Científicas y
Técnicas (CONICET)-UBA, Centro de Investigación en
Hidratos de Carbono (CIHIDECAR),
Buenos Aires, Argentina

Juan P. Colomer[a]
Universidad Nacional de Córdoba, Instituto de
Investigaciones en Físico-Química de Córdoba –
(UNC – INFIQC – CONICET), Facultad de Ciencias
Químicas, Departamento de Química Orgánica,
Edificio de Ciencias 2, Ciudad Universitaria –
Córdoba, Argentina

---

* Corresponding author: mluhrig@qo.fcen.uba.ar
[a] Checker: juanpablo@fcq.unc.edu.ar

CONTENTS

The popular CuAAC (Cu(I)-catalyzed azide-alkyne 1,3-dipolar cycloaddition), also referred to as the *click* reaction,[1-3] has been adopted by carbohydrate chemists for the construction of neoglycoconjugates.[4-6] Thus, the synthesis of alkynyl- and azide-sugar precursors[7-9] through readily available and high-yielding methods has become an important topic.[10-12] One option for the synthesis of neoglycoconjugates is to employ thiosugars as recognition elements, which are more flexible isosteric structures than the naturally occurring O-glycosides.[13,14] Still, they are recognized by carbohydrate-binding proteins without the loss of affinity, with the advantage of their increased resistance to hydrolysis.[15]

Among the most abundant monosaccharides, N-acetyl glucosamine is found as component of N- and O-linked glycans in glycoproteins, and also as fundamental component of glycosaminoglycans. It is well known that the presence of the 2-N-acetamido group diminishes the reactivity of the anomeric position toward glycosylation. The phthalimido group is a useful alternative for the preparation of hexosamine building blocks, as it prevents the participation of the NH group in many side reactions.

In 2003, Ibatullin *et al.* reported the one-pot transformation of monosaccharide peracetates into thioglycosides by treatment with BF$_3$.OEt$_2$ in the presence of thiourea. By addition of a base (TEA) and an electrophile, thioglycosides of glucose, galactose, maltose, and cellobiose were obtained in 55–85% yields.[16] Thiodisaccharides were also obtained by this methodology.[17] This approach was revisited a few years later by Tiwari.[18] A traditional alternative method to access thioglycosides involves the preparation of anomeric halides,[19–21] conversion thereof to the corresponding isothiouronium salts,[22] followed by reaction with a base in the presence of an electrophile.[23,24]

Here, we describe the synthesis of precursors of alkynyl-functionalized β-N-protected thioglucosamine 3–6 by a one-pot procedure starting from either N-acetyl or N-phthaloyl glucosamine peracetates 1 and 2. Our methodology involves the isothiouronium salt intermediate which decomposes to the thioaldose. The latter reacts with the highly electrophilic propargyl bromide to give the desired products 3 and 4, respectively. Alternative alkynyl-functionalized electrophiles derived from diethylene glycol can be used to obtain the corresponding thioglycosides 5 and 6, respectively, which are suitable for conjugation with biologically active compounds. This one-pot method, employing N-acetyl or N-phthaloyl glucosamine peracetates 1 and 2 as starting materials, is a convenient alternative to obtain alkynyl-functionalized β-thioglucosamine analogs.

EXPERIMENTAL

GENERAL METHODS

Acetonitrile was refluxed with CaH_2 and distilled before use. Solvents used for chromatography were also distilled before use. Thin-layer chromatography (TLC) was performed on silica gel 60 F_{254} plates (Merck). Visualization of the spots was effected by charring with 5% (v/v) sulfuric acid in EtOH, containing 0.5% p-anisaldehyde. Column chromatography was performed on silica gel 60 (230–400 mesh, Merck), by elution with the solvent systems indicated in each case. Propargyl bromide (80 wt.% in toluene) was purchased from Aldrich. Thiourea was recrystallized from ethanol. 2-Acetamido-1,3,4,6-tetra-O-acetyl-2-deoxy-β-D-glucopyranose (1) and 1,3,4,6-tetra-O-acetyl-2-deoxy-2-phthalimido-D-glucopyranose (2) were prepared as described.[25,26] 3-(2-(2-Iodoethoxy)ethoxy)prop-1-yne was prepared from 2-(2-chloroethoxy)ethanol, by reaction with NaH/propargyl bromide and further substitution of the chlorine atom by NaI in butanone.[27] Strata C18-E cartridges were purchased from Phenomenex. [1]H and ^{13}C nuclear magnetic resonance (NMR) spectra were recorded at 25°C at 500 and 125.7 MHz, respectively, using a Bruker Avance II 500 spectrometer. For ^{1}H, ^{13}C chemical shifts are reported in parts per million relative to tetramethylsilane or a residual solvent peak ($CHCl_3$: [1]H: δ 7.26 ppm, [13]C: δ 77.2 ppm). Assignments of ^{1}H and ^{13}C were assisted by 2D ^{1}H COSY and 2D ^{1}H–^{13}C HSQC experiments. High-resolution mass spectra (HRMS) were obtained by electrospray ionization (ESI) and Q-TOF in a Bruker micrOTOF-Q II spectrometer. Optical rotations were measured at 20°C in a 1-dm cell with a PerkinElmer 343 polarimeter. Elemental analyses were measured in a Carlo Erba EA 1108 apparatus. Melting points were measured in a Fisher-Johns apparatus.

2-Propynyl 2-acetamido-3,4,6-tri-O-acetyl-2-deoxy-1-thio-β-D-glucopyranoside (3)

BF$_3$·OEt$_2$ (806 μL, 6.42 mmol) was added under Argon atmosphere to a suspension of 2-acetamido-1,3,4,6-tetra-O-acetyl-2-deoxy-β-D-glucopyranose 1 (1.00 g, 2.57 mmol) and thiourea (878 mg, 11.56 mmol) in anhydrous acetonitrile (5.50 mL), and the mixture was refluxed for 3 h. Thiourea (78 mg, 1.03 mmol) followed by BF$_3$·OEt$_2$ (129 μL, 1.03 mmol) was added at room temperature to the yellow solution, and the mixture was refluxed for one additional hour, when TLC (EtOAc) showed disappearance of the starting sugar (R_f 0.5) and appearance of a brownish spot at the origin, corresponding to the isothiouronium salt. The solution was cooled to 20°C and triethylamine (1.42 mL, 10.28 mmol) followed by propargyl bromide (315 μL, 2.82 mmol) was added.* After stirring for 18 h at 20°C, analysis by TLC showed the presence of the title product (major, R_f 0.5, EtOAc) and its α isomer (R_f 0.6).

The solution was concentrated under vacuum; the residue was suspended in EtOAc (100 mL) and extracted with water (1 × 50 mL). The organic phase was dried (MgSO$_4$), concentrated under vacuum, and chromatography (1:1 → 1:4 hexane–EtOAc) gave first the α anomer as a syrup (106 mg, 10%); [α]$_D^{20}$ +148.9 (c 1.0, EtOAc); ^{1}H NMR (CDCl$_3$) δ 5.75 (d, 1H, $J_{2,NH}$ 8.6 Hz, NH), 5.67 (d, 1H, $J_{1,2}$ 5.4 Hz, H-1), 5.15 (dd, 1H, $J_{3,4}$ 9.3, $J_{4,5}$ 9.8 Hz, H-4), 5.05 (dd, 1H, $J_{3,4}$ 9.3, $J_{2,3}$ 11.2 Hz, H-3), 4.56 (ddd, 1H, $J_{1,2}$ 5.4, $J_{2,3}$ 11.2, $J_{2,NH}$ 8.6 Hz, H-2), 4.32 (ddd, 1H, $J_{5,6a}$ 4.3, $J_{5,6b}$ 2.1, $J_{4,5}$ 9.8 Hz, H-5), 4.28 (dd, 1H, $J_{5,6a}$ 4.3, $J_{6a,6b}$ 12.2 Hz, H-6a), 4.07 (dd, 1H, $J_{5,6b}$ 2.1, $J_{6a,6b}$ 12.2 Hz, H-6b), 3.40, 3.24 (2 dd, 1H each, $J_{CH2,C≡CH}$ 2.6, $J_{CH2Sgem}$ 16.7 Hz, CH$_2$S), 2.25 (t, 1H, $J_{CH2,C≡CH}$ 2.6 Hz, C≡CH), 2.09, 2.03, 1.95 (4 s, 12H, 4 × CH$_3$CO); ^{13}C NMR (CDCl$_3$) δ 171.8, 170.8, 170.2, 169.4 (4 × C=O), 83.8 (C-1), 78.8 (HC≡C), 72.2 (HC≡C), 71.4 (C-3), 69.0 (C-5), 68.1 (C-4), 61.9 (C-6), 52.4 (C-2), 23.4, 20.9, 20.8, 20.7 (4 × CH$_3$CO), 18.3 (CH$_2$S); ESI-MS: m/z [M + Na]$^+$ calcd for C$_{17}$H$_{23}$NNaO$_8$S, 424.1037; found, 424.1028.

Eluted next was the desired β product 3 (386 mg, 39%), mp 170–171°C (from water); [α]$_D^{20}$ −103.1 (c 1, CHCl$_3$). Ref.[28] [α]$_D^{20}$ −45 (c 1, CHCl$_3$) for the amorphous material; ^{1}H NMR (CDCl$_3$) δ 5.87 (d, 1H, $J_{2,NH}$ 9.3 Hz, NH), 5.18 (dd, 1H, $J_{3,4}$ 9.5, $J_{2,3}$ 10.1 Hz, H-3), 5.09 (dd, 1H, $J_{3,4}$ 9.5, $J_{4,5}$ 10.0 Hz, H-4), 4.80 (d, 1H, $J_{1,2}$ 10.5 Hz, H-1), 4.22 (dd, 1H, $J_{5,6a}$ 5.1, $J_{6a,6b}$ 12.4 Hz, H-6a), 4.16 (ddd, 1H, $J_{2,3}$ 10.1, $J_{1,2}$ 10.5, $J_{2,NH}$ 9.3 Hz, H-2), 4.12 (dd, 1H, $J_{5,6b}$ 2.3, $J_{6a,6b}$ 12.4 Hz, H-6b), 3.70 (ddd, 1H, $J_{5,6b}$ 2.3, $J_{5,6a}$ 5.1, $J_{4,5}$ 10.0 Hz, H-5), 3.55, 3.27 (2 dd, 1H each, $J_{CH2,C≡CH}$ 2.5, $J_{CH2Sgem}$ 16.5 Hz, CH$_2$S), 2.26 (t, 1H, $J_{CH2,C≡CH}$ 2.5 Hz, C≡CH), 2.06, 2.01, 1.94 (4 s, 12H, 4 × CH$_3$CO); ^{13}C NMR (CDCl$_3$) δ 171.2, 170.8, 170.4, 169.4 (4 × C=O), 83.1 (C-1), 79.3 (HC≡C), 76.0 (C-5), 73.9 (C-3), 71.8 (HC≡C), 68.5 (C-4), 62.2 (C-6), 53.0 (C-2), 23.3, 20.8, 20.7 (4 × CH$_3$CO), 17.5 (CH$_2$S). ESI-MS: m/z [M + Na]$^+$ calcd for C$_{17}$H$_{23}$NNaO$_8$S, 424.1037; found, 424.1029. Anal. calcd for C$_{17}$H$_{23}$NO$_8$S: C, 50.86; H, 5.77; N, 3.49. Found: C, 50.83; H, 5.64; N, 3.83.

---

* Upon addition of TEA, the solution became brownish and turned green when propargylbromide was added.

2-(2-Propynyloxethoxy)ethyl 2-acetamido-3,4,6-tri-*O*-acetyl-2-deoxy-1-thio-β-D-glucopyranoside (5)

2-Acetamido-1,3,4,6-tetra-*O*-acetyl-2-deoxy-β-D-glucopyranose **1** (1.00 g, 2.57 mmol) was treated as described for the preparation of **3**, except 3-(2-(2-iodoethoxy) ethoxy)prop-1-yne (716 mg, 2.82 mmol) was used instead of propargylbromide. Chromatography (1:1 → 3:7 hexane–EtOAc) gave first the α anomer as a syrup (R_f 0.6, EtOAc) (151 mg, 12%); $[\alpha]_D^{20}$ +8.9 (*c* 1, CHCl₃); ¹H NMR (CDCl₃) δ 5.81 (d, 1*H*, $J_{2,NH}$ 8.8 Hz, N*H*), 5.46 (d, 1*H*, $J_{1,2}$ 5.4 Hz, H-1), 5.11 (dd, 1*H*, $J_{3,4}$ 9.3, $J_{4,5}$ 9.5 Hz, H-4), 5.06 (dd, 1*H*, $J_{3,4}$ 9.3, $J_{2,3}$ 10.5 Hz, H-3), 4.50 (ddd, 1*H*, $J_{1,2}$ 5.4, $J_{2,NH}$ 8.8, $J_{2,3}$ 10.5, Hz, H-2), 4.35 (ddd, 1*H*, $J_{5,6b}$ 2.3, $J_{5,6a}$ 4.4, $J_{4,5}$ 9.5 Hz, H-5), 4.26 (dd, 1*H*, $J_{5,6a}$ 4.4, $J_{6a,6b}$ 12.4 Hz, H-6a), 4.19 (d, 2*H*, $J_{CH2,C≡CH}$ 2.4 Hz, C*H₂*C≡CH), 4.06 (dd, 1*H*, $J_{5,6b}$ 2.3, $J_{6a,6b}$ 12.4 Hz, H-6b), 3.72–3.61 (m, 6*H*, 3 × OC*H₂*), 3.86, 3.75 (2 m, 1*H* each, C*H₂*S), 2.44 (t, 1*H*, $J_{CH2,C≡CH}$ 2.4 Hz, C≡C*H*), 2.08, 2.02, 2.01, 1.95 (4 s, 12*H*, 4 × C*H₃*CO); ¹³C NMR (CDCl₃) δ 171.7, 170.8, 170.1, 169.4 (4 × *C*O), 84.8 (C-1), 79.6 (H*C*≡C), 74.8 (HC≡*C*), 71.4 (C-3), 70.4, 70.3, 69.1 (*C*H₂O), 68.5 (C-5), 68.2 (C-4), 62.1 (C-6), 58.5 (*C*H₂C≡CH), 52.5 (C-2), 30.8 (*C*H₂S), 23.3, 20.9, 20.8, 20.7 (4 × *C*H₃CO); ESI-MS: *m/z* [M + H]⁺ calcd for C₂₁H₃₂NO₁₀S, 490.1741; found, 490.1750; [M + Na]⁺ calcd for C₂₁H₃₁NNaO₁₀S, 512.1561; found, 512.1573.

Syrupy† β anomer **5** was eluted next (R_f 0.5, EtOAc). Further purification was performed by passing through a C18 cartridge (0:1 → 1:1 MeOH–H₂O), whereupon 420 mg, 33% of pure **5** was obtained, $[\alpha]_D^{20}$ –12.7 (*c* 1, EtOAc). ¹H NMR (CDCl₃) δ 6.56 (d, 1*H*, $J_{2,NH}$ 9.4 Hz, N*H*), 5.11–5.06 (m, 2*H*, H-3, H-4), 4.95 (d, 1*H*, $J_{1,2}$ 10.7 Hz, H-1), 4.39, 4.33 (2 dd, 1*H* each, $J_{CH2,C≡CH}$ 2.3 Hz, J_{gem} 16.2 Hz, C*H₂*C≡CH), 4.23 (dd, 1*H*, $J_{5,6a}$ 4.7 Hz, $J_{6a,6b}$ 12.3 Hz, H-6a), 4.08–4.16 (m, 2*H*, H-6b, H-2), 3.91–3.55 (m, 7*H*, H-5, 3 × OC*H₂*), 3.03 (ddd, 1*H*, $J_{CHaS,CH2}$ 3.2, $J_{CHaS,CH2}$ 10.2, $J_{CHaS,CHbS}$ 14.0 Hz, C*H*aS), 2.65 (dt, 1*H*, $J_{CHbS,CH2}$ 3.3, $J_{CHaS,CHbS}$ 14.0 Hz, C*H*bS), 2.50 (t, 1*H*, $J_{CH2,C≡CH}$ 2.3 Hz, C≡C*H*), 2.08, 2.01, 1.96 (4 s, 12*H*, 4 × C*H₃*CO); ¹³C NMR (CDCl₃) δ 171.1, 171.0, 170.5, 169.4 (4 × *C*=O), 86.4 (C-1), 79.6 (H*C*≡C), 75.8 (HC≡*C*), 75.5 (C-5), 74.6 (C-3), 73.4, 70.4, 69.3 (3 × *C*H₂O), 68.5 (C-4), 62.5 (C-6), 58.5 (O*C*H₂C≡CH), 53.4 (C-2), 30.7 (*C*H₂S), 23.3, 20.9, 20.8 (4 × *C*H₃CO). ESI-MS: *m/z* [M + Na]⁺ calcd for C₂₁H₃₁NNaO₁₀S, 512.1561; found, 512.1561. Anal. calcd for C₂₁H₃₁NO₁₀S: C, 51.52; H, 6.38; N, 2.86; S, 6.55. Found: C, 51.15; H, 6.20; N, 3.06; S, 6.50.

2-Propynyl 3,4,6-tri-*O*-acetyl-2-deoxy-2-phthalimido-1-thio-β-D-glucopyranoside (4)

BF₃·OEt₂ (1.84 mL, 14.63 mmol) was added at 20°C under Ar atmosphere to a suspension of 1,3,4,6-tetra-*O*-acetyl-2-deoxy-2-phthalimido-D-glucopyranose **2** (1.00 g, 2.09 mmol) and thiourea (730 mg, 9.61 mmol) in anhydrous acetonitrile (4.50 mL) and the mixture was refluxed at 82°C for 12 h. Analysis by TLC (1:1 hexane–EtOAc) of the yellow solution formed showed disappearance of the starting sugar (R_f 0.4) and appearance of a brownish spot at the origin, corresponding to the isothiouronium

---

† Attempt to crystallize the material from H₂O, EtOH/H₂O, EtOH, *n*-BuOH, *i*-PrOH and *n*-hexane were unsuccessful.

salt. Triethylamine (2.30 mL, 16.72 mmol) followed by propargyl bromide (303 µL, 2.72 mmol) was added (see footnote "*") and the stirring was continued for 18 h at 20°C, when TLC (1:1 hexane–EtOAc) showed the presence of **4** (R_f 0.5, 1:1 hexane–EtOAc). After concentration at reduced pressure, the residue was suspended in EtOAc (100 mL) and extracted with water (3 × 50 mL). The organic phase was dried (MgSO$_4$), concentrated under vacuum and chromatography (9:1 → 6:4 hexane–EtOAc) gave amorphous **4** (510 mg, 50%)) (see footnote "†"); $[\alpha]_D^{20}$ +18.2 (*c* 1, CHCl$_3$); ^{1}H NMR (CDCl$_3$) δ 7.86, 7.74 (2 m, 2*H* each, H-aromatic), 5.88 (dd, 1*H*, $J_{3,4}$ 9.1, $J_{2,3}$ 10.5 Hz, H-3), 5.70 (d, 1*H*, $J_{1,2}$ 10.5 Hz, H-1), 5.20 (dd, 1*H*, $J_{3,4}$ 9.1, $J_{4,5}$ 10.2 Hz, H-4), 4.43 (t, 1*H*, $J_{1,2}$≈$J_{2,3}$ 10.5 Hz, H-2), 4.33 (dd, 1*H*, $J_{5,6a}$ 4.8, $J_{6a,6b}$ 12.3 Hz, H-6a), 4.19 (dd, 1*H*, $J_{5,6b}$ 2.2, $J_{6a,6b}$ 12.3 Hz, H-6b), 3.92 (ddd, 1*H*, $J_{5,6b}$ 2.2, $J_{5,6a}$ 4.8, $J_{4,5}$ 10.2 Hz, H-5), 3.53, 3.26 (2 dd, 1*H* each, $J_{CH2,C\equiv CH}$ 2.6, $J_{CH2Sgem}$ 16.4 Hz, CH$_2$S), 2.22 (t, 1*H*, $J_{CH2,C\equiv CH}$ 2.6 Hz, C≡C*H*), 2.11, 2.03, 1.86 (3 s, 9*H*, 3 × CH$_3$CO); ^{13}C NMR (CDCl$_3$) δ 170.9, 170.2, 169.6, 167.7, 167.3 (5 × *C*=O), 134.6, 134.5, 131.7, 131.3, 123.9 (C-aromatic), 80.2 (C-1), 78.6 (H*C*≡C), 76.1 (C-5), 72.3 (C-3), 71.5 (H*C*≡C), 68.9 (C-4), 62.2 (C-6), 53.6 (C-2), 20.9, 20.8, 20.6 (3 × *C*H$_3$CO), 17.9 (*C*H$_2$). ESI-MS: *m/z* [M + Na]$^+$ calcd for C$_{23}$H$_{23}$NNaO$_9$S, 512.0986; found, 512.0962. Anal. calcd for C$_{23}$H$_{23}$NO$_9$S: C, 56.43; H, 4.74; N, 2.86; S, 6.55. Found: C, 56.30; H, 4.70; N, 2.90; S, 6.40.

2-(2-PROPYNYLOXETHOXY)ETHYL 3,4,6-TRI-*O*-ACETYL-2-PHTHALIMIDO-2-DEOXY-1-THIO-β-D-GLUCOPYRANOSIDE (6)

The procedure described before for the synthesis of compound **5** was followed, but using 3-(2-(2-iodoethoxy)ethoxy)prop-1-yne (690 mg, 2.72 mmol) instead of propargylbromide. Chromatography (9:1 → 3:2 hexane-EtOAc) gave syrupy (see footnote "†") **6** (483 mg, 40%); R_f 0.4 (1:1 hexane–EtOAc); $[\alpha]_D^{20}$ +18.5 (*c* 1, CHCl$_3$); ^{1}H NMR (CDCl$_3$) δ 7.85, 7.74 (2 m, 2*H* each, H-aromatic), 5.82 (dd, 1*H*, $J_{3,4}$ 9.2, $J_{2,3}$ 10.2 Hz, H-3), 5.55 (d, 1*H*, $J_{1,2}$ 10.6 Hz, H-1), 5.17 (dd, 1*H*, $J_{3,4}$ 9.2, $J_{4,5}$ 10.2 Hz, H-4), 4.37 (dd, 1*H*, $J_{2,3}$ 10.2, $J_{1,2}$ 10.6 Hz, H-2), 4.30 (dd, 1*H*, $J_{5,6a}$ 4.8, $J_{6a,6b}$ 12.3 Hz, H-6a), 4.17 (dd, 1*H*, $J_{5,6b}$ 2.3, $J_{6a,6b}$ 12.3 Hz, H-6b), 4.17 (d, 2*H*, $J_{CH2,C\equiv CH}$ 2.6 Hz, CH$_2$C≡CH), 3.90 (ddd, 1*H*, $J_{5,6b}$ 2.3, $J_{5,6a}$ 4.8, $J_{4,5}$ 10.2 Hz, H-5), 3.68–3.58 (m, 6*H*, 3 × OCH$_2$), 2.93, 2.75 (2 m, 1*H* each, CH$_2$S), 2.43 (t, 1*H*, $J_{CH2,C\equiv CH}$ 2.6 Hz, CH$_2$C≡C*H*), 2.11, 2.03, 1.86 (3 s, 9*H*, 3 × CH$_3$CO); ^{13}C NMR (CDCl$_3$) δ 170.9, 170.2, 169.6, 167.8, 167.3 (5 × *C*=O), 134.6, 134.4, 131.7, 131.3, 123.9, 123.8 (C-aromatic), 81.4 (C-1), 79.7 (H*C*≡C), 76.1 (C-5), 74.8 (H*C*≡C), 71.6 (C-3), 71.1, 70.3, 69.1 (3 × *C*H$_2$O), 69.1 (C-4), 62.4 (C-6), 58.5 (O*C*H$_2$C≡CH), 53.9 (C-2), 29.7 (*C*H$_2$S), 21.0, 20.8, 20.6 (3 × *C*H$_3$CO). ESI-MS: *m/z* [M + H]$^+$ calcd for C$_{27}$H$_{32}$NNaO$_{11}$S, 578.1691; found, 578.1675. Anal. calcd for C$_{27}$H$_{31}$NO$_{11}$S: C, 56.14; H, 5.41; N, 2.42; S, 5.55. Found: C, 56.20; H, 5.40; N, 2.50; S, 5.40.

ACKNOWLEDGMENTS

Support for this work from the National Research Council CONICET and Universidad de Buenos Aires, Argentina, is gratefully acknowledged. AEC and HOM are fellows from CONICET. MLU is a research member of CONICET.

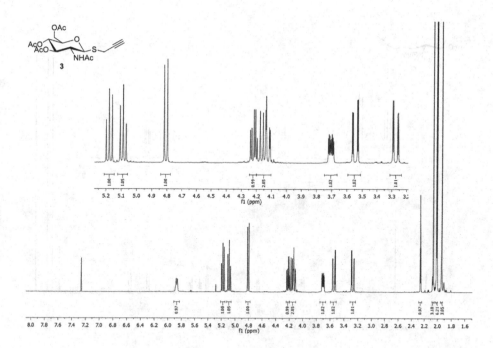

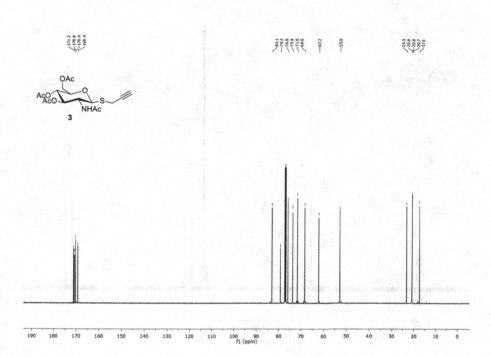

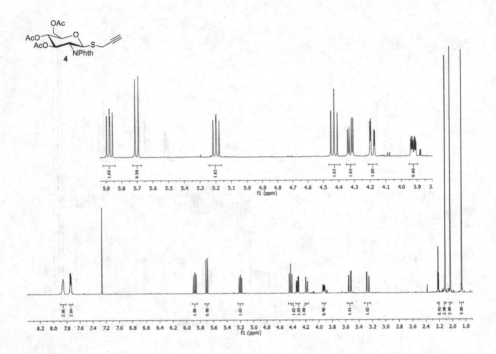

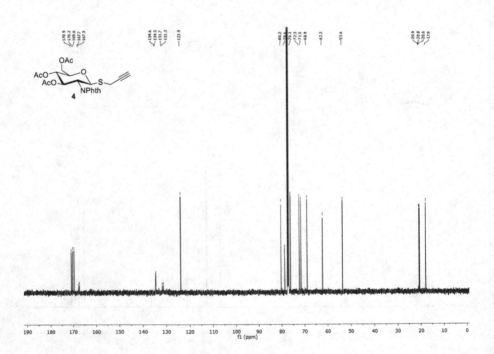

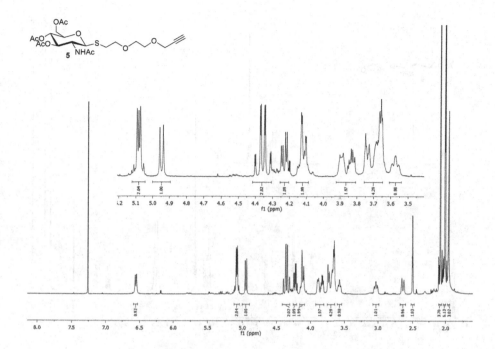

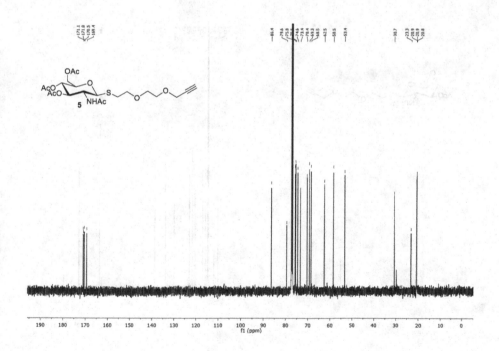

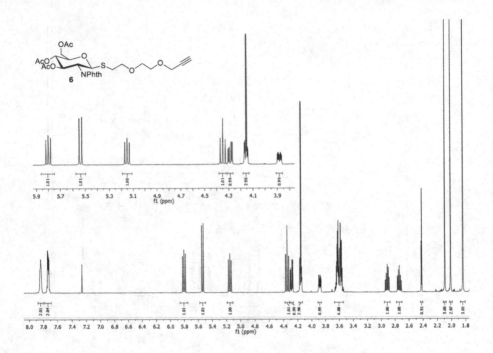

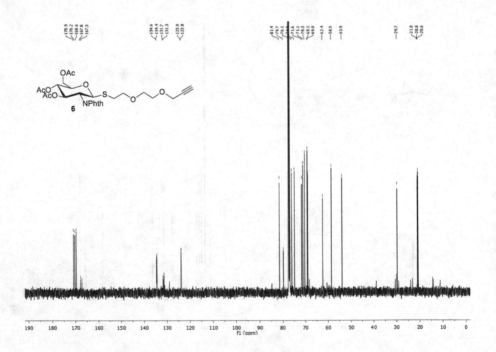

REFERENCES

1. Huisgen, R. *Angew. Chem. Int. Ed.* **1963**, *2* (10), 565–598.
2. Rostovtsev, V. V.; Green, L. G.; Fokin, V. V.; Sharpless, K. B. *Angew. Chemie – Int. Ed.* **2002**, *41* (14), 2596–2599.
3. Tornøe, C. W.; Christensen, C.; Meldal, M. *J. Org. Chem.* **2002**, *67* (9), 3057–3064.
4. Chabre, Y. M.; Roy, R. Design and creativity in synthesis of multivalent neoglycoconjugates. *Adv. Carbohydr. Chem. Biochem.* **2010**, *63*, 165–393.
5. Miller, N.; Williams, G. M.; Brimble, M. A. *Int. J. Pept. Res. Ther.* **2010**, *16* (3), 125–132.
6. Luttheroth, K. R.; Harris. P. W. R.; Wright, T. H., Kaur. H.; Sparrow, K.; Yang, S.-H.; Cooper, G. J. S.; Brimble, M. A. *Org. Biomol. Chem.* **2017**, *15*, 5602–5608.
7. Gouin, G. G.; Bultel, L.; Falentin, C.; Kovensky, J. *Eur. J. Org. Chem.* **2007**, 1160–1167.
8 Tatina, M. B.; Kusunuru, A. K.; Yousuf, S. K.; Mukherjee, D. *Org. Biomol. Chem.* **2014**, *12*, 7900–7903.
9. Sethi, K. P.; Kartha, R. *Carbohydr. Res.* **2016**, *434*, 132–135.
10. Uhrig, M. L.; Kovensky, J. Clicking sugars onto sugars: oligosaccharide analogs and glycoclusters on carbohydrate scaffolds; In *Click Chemistry in Glycoscience, New Developments and Strategies;* Witczak, Z. W. and Bielski, R. Eds.; John Wiley and Sons: Hoboken, NJ, **2013**, 107–142.
11. Kushwaha, D.; Dwivedi, P.; Kuanar, S. K.; Tiwari, V. K. *Curr. Org. Synth.* **2013**, *10* (1), 90–135.
12. Ortiz Mellet, C.; Méndez-Ardoy, A.; García Fernández, J. M. Click multivalent glycomaterials: glycoclusters, glycodendrimers, glycopolymers, hybrid glycomaterials and glycosurfaces; In Click Chemistry in Glycoscience, *New Developments and Strategies;* Witczak, Z. W. and Bielski, R. Eds.; John Wiley and Sons: Hoboken, NJ, **2013**, 143–182.
13. García-Herrero, A.; Montero, E.; Muñoz, J. L.; Espinosa, J. F.; Vian, A.; García, J. L.; Asensio, J. L.; Cañada, F. J.; Jiménez-Barbero, J. *J. Am. Chem. Soc.* **2002**, *124* (17), 4804–4810.
14. Calle, L.; Roldós, V.; Cañada, F. J.; Uhrig, M. L.; Cagnoni, A. J.; Manzano, V. E.; Varela, O.; Jiménez-Barbero, J. *Chem. - A Eur. J.* **2013**, *19* (13), 4262–4270.
15. Driguez, H. *Top. Curr. Chem.* **1997**, *187*, 85–116.
16. Ibatullin, F. M.; Shabalin, K. A.; Jänis, J. V.; Shavva, A. G. *Tetrahedron Lett.* **2003**, *44* (43), 7961–7964.
17. Ibatullin, F. M.; Shabalin, K. A.; Jänis, J. V.; Selivanov, S. I. *Tetrahedron Lett.* **2001**, *42* (27), 4565–4567.
18. Tiwari, P.; Agnihotri, G.; Misra, A. K. *J. Carbohydr. Chem.* **2005**, *24* (7), 723–732.
19. Horton, D.; Johnson, A. L.; McKusick, B. C. *Org. Syn. Coll.* **1973**, *5*, 1.
20. McLaren, J. M.; Stick, R. V.; Webb, S. *Aust. J. Chem.* **1977**, *30*, 2689–2693.
21. Floyd, N.; Vijayakrishnan, B.; Koeppe, J. R.; Davis, B. G. *Angew. Chem. Int. Ed.*, **2009**, *48*, 7798–7802.
22. Horton, D.; Wolfrom, M. L. *J. Org. Chem.* **1961**, *27* (1911), 1794–1800.
23. Ghosh, S.; Tiwari, P.; Pandey, S.; Misra, A. K.; Chaturvedi, V.; Gaikwad, A.; Bhatnagar, S.; Sinha, S. *Bioorg. Med. Chem. Lett.* **2008**, *18*, 4002–4005.
24. Cano, M. E.; Agustí, R.; Cagnoni, A. J.; Tesoriero, M. F.; Kovensky, J.; Uhrig, M. L.; Lederkremer, R. M. *Beilstein J. Org. Chem.* **2014**, *10*, 3073–3086.
25. Hernández-Torres, J. M.; Liew, S.-T.; Achkar, J.; Wei, A. *Synthesis (Stuttg).* **2002**, 487–490.
26. Debenham, J.; Rodebaugh, R.; Fraser-Reid, B. *Liebigs Ann.* **1997**, 791–802.
27. Thiebault, N.; Lesur, D.; Godé, P.; Moreau, V.; Djedaïni-Pilard, F. *Carbohydr. Res.* **2008**, *343*, 2719–2728.
28. Chen, Q.; Cui, Y.; Cao, J.; Han, B. H. *Polymer (Guildf).* **2011**, *52* (2), 383–390.

28 Synthesis of 1'-(4'-Thio-β-D-Ribofuranosyl) Uracil

Mieke Guinan, and Gavin J. Miller*
Lennard-Jones Laboratory, School of Chemical and Physical
Sciences, Keele University,
Keele, United Kingdom

Dylan Lynch[a]
School of Chemistry, Trinity College Dublin,
University of Dublin,
Dublin, Ireland

Mark Smith
Rioscience LLC,
Sunnyvale, California, United States

CONTENTS

4'-Thionucleosides, where the furanose ring oxygen is replaced with sulfur, have attracted significant interest due to their biological activity and metabolic stability.[1,2] In recent years, a growing number of reports have detailed the chemical synthesis and evaluation of 4'-thio nucleoside analogs,[3–7] alongside methodologies for the provision of thiosugar-containing scaffolds.[8] As part of our nucleoside analog research program, we required a convenient and scalable access to the title compound **3**, starting from **1**.

The original report for synthesizing **3** was made by Imbach and coworkers in their synthesis of 4'-thio-β-D-oligoribonucleotides and utilized **1** as the starting material,[9] recording a 38% yield for the two-step process. The synthetic procedures and full characterization of **2** and **3** were not detailed in this publication, and we

---

* Corresponding author: g.j.miller@keele.ac.uk
[a] Checker: under supervision of Eoin Scanlan: SCANLAE@tcd.ie

present this information herein. Beginning with anomeric acetate **1**, we completed a Vorbrüggen[10,11] reaction to install the uracil base in 76% yield. We found that this reaction required some optimization to give consistently good yields across a range of scales, paying particular attention to purification of the material. A related synthesis of 2′,3′,5′tri-*O*-benzoyl-β-D-ribofuranosyl uracil reported by Shirozu involved refluxing the same reagents (protected ribose derivative, silylated uracil, and TMSOTf) for 3 h in acetonitrile.[12] In our hands, this same reaction using **1** as a substrate produced only low yields of **2**, with the solution very quickly turning a dark brown color. By reducing the temperature to just below reflux (75°C), the final isolated yield improved, with the reaction staying orange throughout. Crude ¹H NMR showed the β-anomer formed preferentially, indicated by a 10:1 β:α ratio. Purification of this crude material was best completed using a two-step process; first passing through a short silica plug (to remove TLC baseline impurities), followed by suspension of the resultant solid in boiling petroleum ether (10 volumes) and adding EtOAc (5 volumes) dropwise. This procedure produced **2** as a fine white suspension, which could easily be collected by filtration and left the remaining impurities (including the small amount of α-anomer) in solution. This method delivered the desired β-anomer (J_{H1-H2} 6.8 Hz) in excellent yields (76%) and proved to be consistent on scale-up from 1.0 to 10.0 g. Completing the second purification step with room temperature solvent produced lower yields of the precipitated product **2**. NMR analysis confirmed the correct structure for **2**, notably with the presence of signals for the alkene in the nucleobase (H-6′ and H-5′ assigned at 7.74 and 5.56 ppm, respectively), with vicinal coupling to each other (J 8.2 Hz). HSQC NMR showed these protons correlated to signals at 139.7 and 103.8 ppm, respectively, in the ¹³C NMR spectrum.

Finally, global deprotection was achieved by stirring **2** with 4.5-mol equivalents of NH_3 in MeOH (7.0M solution) for 72 h to obtain **3** and analytical data for this compound matched those previously reported.[13] In summary, we have developed a reliable and scalable process for the conversion of **1** into thiouridine **3** in 71% yield over two steps.

EXPERIMENTAL

GENERAL METHODS

All reagents and solvents which were available commercially were purchased from Acros, Alfa Aesar, Fisher Scientific, or Sigma Aldrich. All reactions in nonaqueous solvents were conducted in flame-dried glassware under a nitrogen atmosphere with a magnetic stirring device. Solvents were purified by passing through activated alumina columns and used directly from a Pure Solv-MD solvent purification system and were transferred under nitrogen. Reactions requiring low temperatures used the following cooling baths: −78°C (dry ice/acetone), −30°C (dry ice/acetone), −15°C (NaCl/ice/water) and 0°C (ice/water). Infrared spectra were recorded neat on a PerkinElmer Spectrum 100 FT-IR spectrometer; selected absorbencies (v_{max}) are reported in cm^{-1}. ^{1}H NMR spectra were recorded at 400 MHz and ^{13}C spectra at 100 MHz, respectively, using a Bruker AVIII400 spectrometer. ^{1}H NMR signals were assigned with the aid of gDQCOSY. ^{13}C NMR signals were assigned with the aid of gHSQCAD. Coupling constants are reported in Hertz. Chemical shifts (δ, in ppm) are standardized against the deuterated solvent peak. NMR data were analyzed using Nucleomatica iNMR software. ^{1}H NMR splitting patterns were assigned as follows: s (singlet), d (doublet), t (triplet), dd (doublet of doublets), ddd (doublet of doublet of doublets), or m (multiplet and/or multiple resonances). Reactions were followed by TLC using Merck silica gel 60F254 analytical plates (aluminum support) and were developed using standard visualizing agents: short wave UV radiation (245 nm) and 5% sulfuric acid in methanol/Δ. Purification *via* flash column chromatography was conducted using silica gel 60 (0.043–0.063 mm). Melting points were recorded using open glass capillaries on a Gallenkamp melting points apparatus and are uncorrected. Optical activities were recorded on automatic polarimeter Rudolph autopol I (concentration in g/100 mL). MS and HRMS (ESI) were obtained on Waters (Xevo, G2-XS Tof) or Waters Micromass LCT spectrometers using a methanol mobile phase. High-resolution (ESI) spectra were obtained on a Xevo, G2-XS Tof mass spectrometer. HRMS was obtained using a lock mass to adjust the calibrated mass scale. Methyl dichlorophosphate and benzyl alcohol (anhydr. Sureseal) were purchased from Sigma-Aldrich and used as received.

2',3',5'-O-BENZOYL,1'-(4'-THIO-β-ᴅ-RIBOFURANOSYL) URACIL (2)

Uracil (2.90 g, 26.0 mmol, 1.4 equiv) was suspended in anhydrous pyridine (19.4 mL, 240 mmol), and the flask charged with hexamethyldisilazane (39.8 mL, 190 mmol, 9.9 equiv) and the mixture refluxed for 3 h under N_2. The solvent was removed *in vacuo* and the flask immediately fitted with a septum to obtain crude 2,4-O-silylated uracil as a colorless oil. Some cloudiness was observed in this oil due to exposure to air during the release of pressure from the rotary evaporator. The flask was fitted with a septum, flushed with N_2, and the oil transferred *via* syringe to a flask containing a solution of **1** (10.0 g, 19.2 mmol) in anhydrous MeCN (100 mL), rinsing the uracil-containing flask with anhydrous MeCN (4 × 25 mL). The solution was cooled to 0°C, trimethylsilyl triflate (2.7 mL, 15.0 mmol, 0.8 equiv) was added dropwise, and the solution stirred at 0°C for 5 min. The solution was heated to 75°C for 72 h, until disappearance of the starting material ($R_f = 0.85$, EtOAc–hexane, 1:1) and two new major spots ($R_f = 0.4$ and 0.3, 1:1 EtOAc–hexane) were observed. The reaction was quenched at 0°C with Et$_3$N (1.8 mL) and stirred for 15 min at this temperature. The solvent was reduced to <10 mL *in vacuo* and the residue re-diluted with EtOAc (300 mL). The organic phase was washed with saturated aqueous NaHCO$_3$ (3 × 100 mL), brine (100 mL), dried over Na$_2$SO$_4$, filtered and concentrated *in vacuo* to give crude **2** as an orange-brown oil. Impurities observed on the baseline of the TLC plate (3:7 EtOAc–petroleum ether) were removed by flash chromatography, making a dry load of crude **2** with silica gel (100 g) and eluting with 3:7 EtOAc–petroleum ether (4.0 L). Solvent was removed from the product-containing fractions *in vacuo*, and this material suspended in petroleum ether (10 volumes, 120 mL). This mixture was then heated to 80°C and EtOAc (5 volumes, 60 mL) was added dropwise over 10 min, forming a white precipitate. The hot suspension was filtered through a sintered funnel, washing with room temperature petroleum ether (50 mL) to yield **2** as a white solid (8.12 g, 14.2 mmol, 76%). R_f 0.41 (EtOAc/hexane, 1:1), mp 207–209°C, $[\alpha]_D^{27}$ −80.0 (*c* 1.0, CHCl₃); ¹H NMR (400 MHz, CDCl₃) δ 8.13 (dd, 2*H*, *J* 7.1 1.4 Hz, ArH), 8.04 (dd, 2*H*, *J* 8.3, 1.2 Hz, ArH), 7.93 (dd, 2*H*, *J* 8.3, 1.2 Hz, ArH), 7.74 (d, 1*H*, *J* 8.2 Hz, H-6), 7.65–7.34 (m, 7*H*, ArH), 7.2–7.12 (m, 2*H*, ArH), 6.69 (d, 1*H*, *J* 6.8 Hz, H-1′), 5.98–5.96 (m, 2*H*, H-2′, H-3′), 5.56 (d, 1*H*, *J* 8.2, 1.2 Hz, H-5), 4.85 (dd, 1*H*, *J* 12.0, 5.6 Hz, H-5′a), 4.71 (dd, 1*H*, *J* 12.0, 4.7 Hz, H-5′b), 4.06 (td, 1*H*, *J* 5.1, 2.5 Hz, H-4′); 13C NMR (101 MHz, CDCl₃) δ 166.1 (C=O), 165.4 (C=O), 165.2 (C=O), 162.1 (C=O, U), 150.4 (C=O, U), 139.7 (C₅), 133.9 (Ar-C), 133.8 (Ar-C), 130.0 (Ar-C), 130.0 (Ar-C), 129.8 (Ar-C), 129.1 (Ar-C), 128.9 (Ar-C), 128.7 (Ar-C), 128.6 (Ar-C), 128.3 (Ar-C), 103.9 (C-6), 76.7 (C-2′ or C-3′), 74.3 (C-2′ or C-3′) 64.5 (C-5′), 61.8 (C-1′), 48.1 (C-4′); ESI HRMS *m/z* found: (M + H)⁺ 573.1351, C₃₀H₂₄N₂O₈S, requires (M + H)⁺ 573.1326; Anal. calcd for C₃₀H₂₄N₂O₈S: C, 62.93; H, 4.23; N, 4.89; S, 5.60; found C, 63.18; H, 4.30; N. 5.09; S, 5.61.

1′-(4′-THIO-β-D-RIBOFURANOSYL) URACIL (3)

2′,3′,5′-Tri-*O*-benzoyl,1′-(4′-thio-β-D-ribofuranosyl) uracil **2** (8.12 g, 14.2 mmol, 1.0 equiv) was suspended in MeOH (95 mL) and cooled to 0°C. The flask was charged with a 7.0 N solution of NH₃ in MeOH (18.2 mL, 128 mmol, 9.0 equiv) at 0°C, warmed to 40°C, and stirred for 72 h until complete disappearance of the starting material (R_f = 0.4, 1:1 EtOAc–hexane) and **3** (R_f = 0.3, 1:5 MeOH–CH₂Cl₂) was observable by TLC. The solvent was removed *in vacuo* to yield an orange solid that was triturated with CH₂Cl₂ (200 mL), and the precipitate filtered through a sintered funnel, washing with CH₂Cl₂ (30 mL) and acetone (15 mL) to afford **3** as a beige solid (3.55 g, 13.6 mmol, 96%). R_f = 0.3 (1:5 MeOH–CH₂Cl₂); ¹H NMR (400 MHz, D₂O) δ 8.22 (d, 1*H*, *J* 8.1 Hz, H-6), 5.98 (d, 1*H*, *J* 5.7 Hz, H-1′), 5.93 (d, 1*H*, *J* 8.1 Hz, H-5), 4.36 (dd, 1*H*, *J* 5.4, 4.0 Hz, H-2′), 4.22 (t, 1*H*, *J* 4.1 Hz, H-3′), 3.88 (m, 2*H*, H-5′a, H-5′b), 3.50 (m, 1*H*, H-4′); ¹H NMR (400 MHz, d₆-DMSO) δ 11.16 (s, 1*H*, N*H*), 8.01 (d, 1*H*, *J* 8.1 Hz, H-6), 5.91 (d, 1*H*, *J* 7.4 Hz, H-1′), 5.70 (d, 1*H*, *J* 8.0 Hz, H-5), 4.28–4.08 (m, 1H, H-2′), 4.04 (s, 1 H, H-3′), 3.61 (m, 2*H*, H-5′a, H-5′b), 3.19 (d, 1*H*, *J* 11.8 Hz, H-4′), ¹³C NMR (101 MHz, D₂O) δ 166.1 (C=O, U), 152.3 (C=O, U), 143.0 (C6), 102.4 (C5), 77.4 (C2′), 73.4 (C3′), 64.3 (C1′), 62.4 (C5′), 52.2 (C4′); NMR data were in good agreement with literature values in (CD₃)₂SO.[13] ESI HRMS *m/z* found: (M + H)⁺ 261.0530 C₉H₁₂N₂O₅S, requires (M+H)⁺ 261.0540.

ACKNOWLEDGMENTS

Keele University and Rioscience LLC are thanked for a studentship to M.G. We also thank the EPSRC UK National Mass Spectrometry Facility (NMSF) at Swansea University.

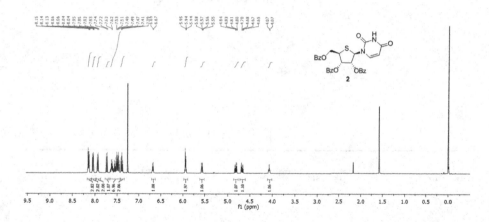

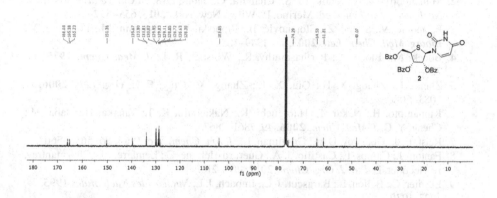

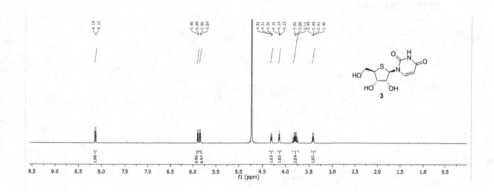

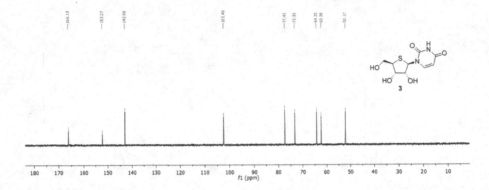

REFERENCES

1. Yokoyama, M. *Synthesis* **2000**, 12, 1637–1655.
2. Mulamoottil, V. A.; Majik, M. S.; Chandra, G.; Jeong L. S. In *Chemical Synthesis of Nucleoside Analogues*, ed. Merino, P., Wiley, New York, **2013**, 655–697.
3. Pejanović, V.; Stokić, Z.; Stojanović, B.; Piperski, V.; Popsavinm, M.; Popsavin, V., *Bioorg. Med. Chem. Lett.* **2003**, *13*, 1849–1852.
4. Bobek, M.; Bloch, A.; Parthasarathy, R.; Whistler, R. L. *J. Med. Chem.* **1975**, *18*, 784–787.
5. Zheng, F.; Zhang, X. H.; Qiu, X. L.; Zhang, X.; Qing, F. L. *Org. Lett.* **2006**, *8*, 6083–6086.
6. Kumamoto, H.; Nakai, T.; Haraguchi, K.; Nakamura, K. T.; Tanaka, H.; Baba, M.; Cheng, Y. C. *J. Med. Chem.* **2006**, *49*, 7861–7867.
7. Reist, E. J.; Gueffroy, D. E.; Goodman, L., *J. Am. Chem. Soc.* **1964**, *86*, 5658–5663.
8. Fanton, J.; Camps, F.; Castillo, J. A.; Guérard-Hélaine, C.; Lemaire, M.; Charmantray, F.; Hecquet, L. *Eur. J. Org. Chem.* **2012**, 203–210.
9. Leydier, C.; Bellon, L.; Barascut, J. L.; Imbach, J. L. *Nucleosides Nucleotides* **1995**, *14*, 1027–1030.
10. Vorbrüggen, H.; Krolikiewicz, K.; Bennua, B. *Chem. Ber.* **1981**, *114*, 1234–1255.
11. Vorbrüggen, H.; Bennua, B. *Tetrahedr. Lett.* **1978**, *19*, 1339–1342.
12. Shirouzu, H.; Morita, H.; Tsukamoto, M. *Tetrahedron* **2014**, *70*, 3635–3639.
13. Nishizono, N.; Baba, R.; Nakamura, C.; Oda, K.; Machida, M. *Org. Biomol. Chem.* **2003**, *1*, 3692–3697.

29 Synthesis of an Orthogonally Protected L-Idose Derivative Using Hydroboration/Oxidation

Fruzsina Demeter
Department of Pharmaceutical Chemistry,
University of Debrecen,
Debrecen, Hungary,
MTA-DE Molecular Recognition and Interaction Research
Group, University of Debrecen,
Debrecen, Hungary,
Doctoral School of Chemistry, University of Debrecen,
Debrecen, Hungary

Gordon Jacob Boehlich[a]
Universität Hamburg, Institut für Pharmazie,
Hamburg, Germany

Anikó Borbás[*]
Department of Pharmaceutical Chemistry,
University of Debrecen,
Debrecen, Hungary

Mihály Herczeg[*]
University of Debrecen,
Debrecen, Hungary,
Research Group for Oligosaccharide Chemistry
of the Hungarian Academy of Sciences,
Debrecen, Hungary

CONTENTS

[*] Corresponding authors: herczeg.mihaly@science.unideb.hu; borbas.aniko@pharm.unideb.hu
[a] Checker: under supervision of N. Schützenmeister: nina.schuetzenmeister@chemie.uni-hamburg.de

Heparin is a linear polyanionic polysaccharide, consisting of repeating (1 → 4) linked D-glucosamine and hexuronic acid units. The hexuronic acid units can be either D-glucuronic acid or its C-5 epimer L-iduronic acid.[1–4] Heparin plays important biological roles, but the most studied effect is the long-known anticoagulant activity.[4,5] The crucial step in the synthesis of heparin-analog oligosaccharides is the efficient preparation of the L-iduronic acid unit. L-Idose is an expensive commodity because it does not occur in nature in free form. Its existing syntheses are cumbersome and inefficient.[6–9] Idose derivatives obtained by these routes are generally not applicable directly to heparin/heparan sulfates syntheses, and further multistep transformation is required to generate the properly functionalized glycosyl donors.[10–13] Ikegami and coworkers reported a generally applicable method to L-pyranosides utilizing the diastereoselective hydroboration/oxidation.[14] They examined the C-5 epimerization of α-*O*-glycosides of different configuration (5-enopyranosides derived from glucosides, galactosides, and mannosides), whereas our research group applied the hydroboration/oxidation approach for the direct synthesis of protected 1-thioidosides.[15,16]

Here we present an efficient synthesis of an orthogonally protected thioidoside **4**, which, after an oxidative 4,6-acetal ring closure, can be utilized as a glycosyl donor for stereoselective synthesis of heparin-analog oligosaccharides. We demonstrated recently that the 4,6-arylmethylene acetal of thioidosides ensures full 1,2-*trans* selectivity in the absence of a participating group at C-2.[15,16] The protective group strategy provides access to heparin analogs where the benzyl group at position 2 masks the hydroxyl to be sulfated, and the benzoyl group at position 3 protects the hydroxyl group to be liberated in the final product.

It has been proven that the α-anomeric configuration is a prerequisite of the high L-*ido*-selectivity in the hydroboration step.[13–15] Hence, the synthesis of the planned L-idose derivative started from the orthogonally protected α-configured thioglucoside **1**.[17] Nucleophilic iodination of **1** with I$_2$ afforded compound **2**, followed by elimination of hydrogen iodide with DBU to give 6-deoxy-α-D-*xylo*-hex-5-enopyranoside **3**. Using BH$_3$·THF complex in the hydroboration and H$_2$O$_2$/NaHCO$_3$ reagent in the subsequent oxidation step, the desired L-idose derivative **4** was obtained in good yield (53% over two steps). The standard oxidation conditions (i.e., H$_2$O$_2$, NaOH, or NaHCO$_3$) reported for *O*-glycosides proved to be well suited for thioglycosides as oxidation of the anomeric thioacetal functionality was not observed. The resulting L-idose deriva-tive **4** can be converted to a fully protected derivative by an oxidative ring closure and utilized for the synthesis of heparinoids as an α-selective L-idosyl donor.[16]

EXPERIMENTAL

GENERAL METHODS

Optical rotations were measured at room temperature on a PerkinElmer 241 auto-matic polarimeter. TLC was performed on Kieselgel 60 F_{254} (Merck) silica gel plates with visualization by charring with 5% H_2SO_4 in EtOH. Column chroma-tography was performed on silica gel 60 (Merck 0.063–0.200 mm) and Sephadex LH-20 (Sigma-Aldrich, bead size: 25–100 mm). Solutions in organic solvents were dried over $MgSO_4$ and concentrated under vacuum. ^1H and ^{13}C NMR spectroscopy (^1H: 400 MHz; ^{13}C: 100.28 MHz) were performed with Bruker DRX-400 spectrom-eters at 25°C. Chemical shifts are referenced to $SiMe_4$ or sodium 3-(trimethylsilyl)-1-propanesulfonate (DSS, $d = 0.00$ ppm for ^1H nuclei) and to residual solvent signals ($CDCl_3$: $\delta 77.00$ ppm, CD_3OD: $\delta 49.15$ ppm for ^{13}C nuclei). Signal-nuclei assignments were based on 2D NMR measurements (COSY and HSQC). HRMS measurements were carried out with a maXis II UHR ESI-QTOF MS instrument (Bruker) in posi-tive ionization mode. The following parameters were applied for the electrospray ion source: capillary voltage: 3.6 kV; end plate offset: 500 V; nebulizer pressure: 0.5 bar; dry gas temperature: 200°C and dry gas flow rate: 4.0 L/min. Constant background correction was applied for each spectrum, the background was recorded before each sample by injecting the blank sample matrix (solvent). Na-formate calibrant was injected after each sample which enabled internal calibration during data evalua-tion. Mass spectra were recorded by otofControl version 4.1 (build: 3.5, Bruker) and processed by Compass DataAnalysis version 4.4 (build: 200.55.2969). 1-M $BH_3\cdot THF$ complex was purchased from Acros Organics.

ETHYL 3-*O*-BENZOYL-2-*O*-BENZYL-6-DEOXY-6-IODO-4-*O*-(2-NAPHTHYL)METHYL-1-THIO-α-D-GLUCOPYRANOSIDE (2)

To a solution of compound **1**[17] (879 mg, 1.575 mmol) in dry toluene (12.7 mL), tri-phenylphosphine (620 mg, 2.363 mmol, 1.5 equiv), imidazole (322 mg, 4.725 mmol, 3.0 equiv) and iodine (560 mg, 2.205 mmol, 1.4 equiv) were added. The mixture was stirred at 75°C for 30 min and cooled to room temperature. A solution of $NaHCO_3$ (529 mg) in water (6.9 mL) was added with stirring and, after 5 min, 10% aq. $Na_2S_2O_3$ (12.6 mL) was added. The mixture was diluted with EtOAc (200 mL) and washed with H_2O (2 × 50 mL). The organic layer was separated, dried, and concentrated, and chromatography (7:3 *n*-hexane–acetone) gave **2** as an amorphous foam. After dissolving the material in ethyl acetate and removal of the solvent a crystalline solid was obtained. Yield, 821 mg (78%); mp 92–94°C (from EtOAc–*n*-hexane); $[\alpha]_D^{25}$ +89.3 (*c* 0.14, $CHCl_3$); R_f 0.50 (7:3 *n*-hexane–acetone); ^1H NMR ($CDCl_3$, 400 MHz) δ 7.99–7.13 (m, 17*H*, arom), 5.76 (t, 1*H*, $J_{2,3}$, $J_{3,4}$ 9.4 Hz, H-3), 5.48 (d, 1*H*, $J_{1,2}$ 5.6 Hz, H-1), 4.77 (s, 2*H*, NAP-C*H*$_2$), 4.63 (d, 1*H*, J_{gem} 12.6 Hz, Bn-C*H*$_{2a}$), 4.46 (d, 1*H*, J_{gem}

12.6 Hz, Bn-CH_{2b}), 4.03–3.99 (m, 1H, H-5), 3.84 (dd, 1H, $J_{1,2}$ 5.6 Hz, $J_{2,3}$ 9.8 Hz, H-2), 3.60 (t, 1H, $J_{3,4}$, $J_{4,5}$ 9.2 Hz, H-4), 3.52 (dd, 1H, J 2.8 Hz, J 10.8 Hz, H-6a), 3.46 (dd, 1H, J 5.1 Hz, J 10.8 Hz, H-6b), 2.68-2.57 (m, 2H, SCH_2CH$_3$), 1.29 (t, 3H, J 7.4 Hz, SCH$_2$CH_3) ppm; ^{13}C NMR (CDCl$_3$, 100 MHz) δ 165.5 (C$_q$ Bz), 137.4, 134.9, 133.1, 130.0 (5 C, 5 × C$_q$ arom), 133.2–126.0 (17 C, arom), 82.7 (C-1), 80.2 (C-4), 76.5 (C-2), 74.9, 71.8 (2 C, NAP-CH$_2$, Bn-CH$_2$), 74.3 (C-3), 69.2 (C-5), 23.9 (SCH$_2$CH$_3$), 14.8 (SCH$_2$CH$_3$), 8.1 (C-6) ppm; MS (UHR ESI-QTOF): m/z [M + Na]$^+$ calcd for C$_{33}$H$_{33}$INaO$_5$S: 691.0986; found: 691.0979. Anal. calcd for C$_{33}$H$_{33}$IO$_5$S: C, 59.28; H, 4.98; S, 4.80. Found: C, 59.08; H, 4.98; S, 4.61.

ETHYL 3-*O*-BENZOYL-2-*O*-BENZYL-4-*O*-(2-NAPHTHYL) METHYL-1-THIO-β-L-IDOPYRANOSIDE (4)

To a solution of compound **2** (668 mg, 1.000 mmol) in dry THF (23.00 mL), DBU (374 µL, 2.500 mmol) was added and the mixture was stirred at reflux for 20 h. EtOAc (175 mL) was added,* and the mixture was washed successively with 0.1 M aq. H$_2$SO$_4$ (30 mL), saturated aq. Na$_2$S$_2$O$_3$ (30 mL), saturated aq. NaHCO$_3$ (30 mL) and H$_2$O (2 × 30 mL) until neutral pH. The organic layer was dried and concentrated, and chromatography (3:1 *n*-hexane–EtOAc) gave **3** as colorless crystals.† Yield 342 mg (63%); mp 72–74°C (from EtOAc–*n*-hexane); [α]$_D^{25}$ +60.5 (*c* 0.18, CHCl$_3$); R_f 0.48 (3:1 *n*-hexane–EtOAc); ^{1}H NMR (CDCl$_3$, 400 MHz) δ 7.88–7.12 (m, 17H, arom), 5.66 (t, 1H, $J_{2,3}$, $J_{3,4}$ 9.0 Hz, H-3), 5.42 (d, 1H, $J_{1,2}$ 5.0 Hz, H-1), 5.06 (s, 1H, H-6a), 4.91 (d, 1H, J_{gem} 12.2 Hz, NAP-CH_{2a}), 4.83 (s, 1H, H-6b), 4.66–4.61 (m, 2H, NAP-CH_{2b}, Bn-CH_{2a}), 4.52 (d, 1H, J_{gem} 12.4 Hz, Bn-CH_{2b}), 4.03 (d, 1H, $J_{3,4}$ 9.0 Hz, H-4), 3.89 (dd, 1H, $J_{1,2}$ 5.0 Hz, $J_{2,3}$ 8.9 Hz, H-2), 2.70–2.61 (m, 2H, SCH_2CH$_3$), 1.30 (t, 3H, J 7.4 Hz, SCH$_2$CH_3) ppm; ^{13}C NMR (CDCl$_3$, 100 MHz) δ 165.3 (C$_q$ Bz), 152.6 (C-5), 137.3, 135.0, 133.1, 133.0 (5 C, 5 × C$_q$ arom), 133.1–125.9 (17 C, arom), 98.7 (C-6), 83.6 (C-1), 76.4 (C-2), 76.3 (C-4), 73.2 (C-3), 73.1, 72.0 (2 C, NAP-CH$_2$, Bn-CH$_2$), 23.9 (SCH$_2$CH$_3$), 14.5 (SCH$_2$CH$_3$) ppm; MS (UHR ESI-QTOF): m/z [M + Na]$^+$ calcd for C$_{33}$H$_{32}$NaO$_5$S: 563.1863; found: 563.1861.

To a solution of compound **3** (300 mg, 0.555 mmol) in dry THF (1.43 mL), 1-M BH$_3$·THF complex (5.55 mL, 5.55 mmol, 10 equiv) was added. The mixture was stirred at 0°C for 1.5 h, and 30% H$_2$O$_2$ (1.43 mL) followed by saturated aq. NaHCO$_3$ (3.0 mL) were added at 0°C. The mixture was stirred for 1 h and allowed to warm up to room temperature. EtOAc (100 mL) was added, and the mixture was washed successively with saturated aq. NH$_4$Cl (2 × 25 mL), H$_2$O (25 mL), and brine (25 mL) until neutral pH. The organic layer was dried and concentrated, and chromatography (95:5 CH$_2$Cl$_2$–acetone) gave **4** as colorless syrup. Yield, 164 mg (53%); [α]$_D^{25}$

---

* When methyl-*tert*-butyl ether was used instead of ethyl acetate, phase separation was much faster.
† The product crystallizes only after purification by column chromatography.

+36.6 (*c* 0.12, CHCl$_3$); R_f 0.50 (19:1 CH$_2$Cl$_2$–acetone); ^{1}H NMR (CDCl$_3$, 400 MHz) δ 8.01–7.25 (m, 17*H*, arom), 5.65 (s, 1*H*, H-3), 4.96 (d, 1*H*, J_{gem} 12.5 Hz, NAP-C*H*$_{2a}$), 4.95 (s, 1*H*, H-1), 4.84 (d, 1*H*, J_{gem} 12.0 Hz, Bn-C*H*$_{2a}$), 4.75 (d, 1*H*, J_{gem} 12.0 Hz, Bn-C*H*$_{2b}$), 4.65 (d, 1*H*, J_{gem} 12.2 Hz, NAP-C*H*$_{2b}$), 4.06 (dd, 1*H*, *J* 7.9 Hz, *J* 11.6 Hz, H-6a), 3.92–3.89 (m, 1*H*, H-5), 3.68 (s, 1 *H*, H-2), 3.56–3.54 (m, 1*H*, H-6b), 3.50 (s, 1*H*, H-4), 2.80–2.70 (m, 2*H*, SC*H*$_2$CH$_3$), 2.13 (s, 1*H*, H-6-O*H*), 1.30 (t, 3*H*, *J* 7.4 Hz, SCH$_2$C*H*$_3$) ppm; ^{13}C NMR (CDCl$_3$, 100 MHz) δ 165.0 (C$_q$ Bz), 137.5, 134.9, 133.1, 133.0 129.7 (5 C, 5 × C$_q$ arom), 133.7–126.1 (17 C, arom), 84.2 (C-1), 77.9 (C-5), 75.4 (C-2), 73.3, 71.8 (2 C, NAP-CH$_2$, Bn-CH$_2$), 71.1 (C-4), 65.6 (C-3), 62.6 (C-6), 25.9 (SCH$_2$CH$_3$), 15.4 (SCH$_2$CH$_3$) ppm; MS (UHR ESI-QTOF): *m/z* [M + Na]$^+$ calcd for C$_{33}$H$_{34}$NaO$_6$S: 581.1968; found: 581.1962. Anal. calcd for C$_{33}$H$_{34}$O$_6$S: C, 70.95; H, 6.13; S, 5.74. Found: C, 71.29; H, 6.13; S, 5.49.

ACKNOWLEDGMENTS

The authors gratefully acknowledge financial support for this research from the Premium Postdoctoral Program of HAS (PPD 461038) and from the EU and co-financed by the European Regional Development Fund under the project GINOP-2.3.2-15-2016-00008.

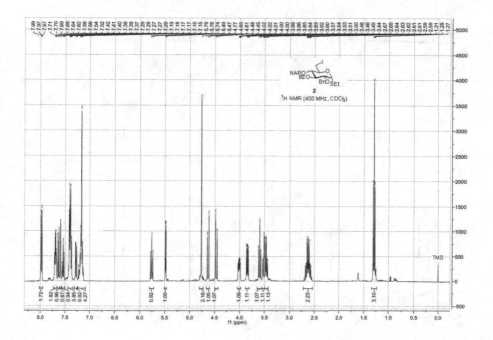

Carbohydrate Chemistry

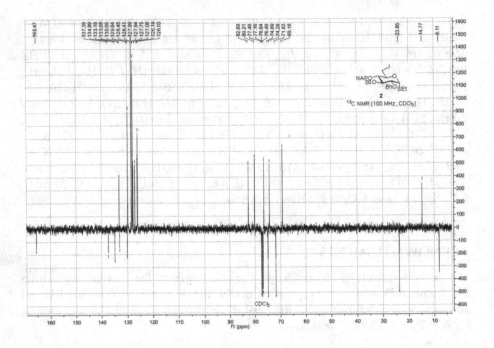

^{13}C NMR (100 MHz, CDCl$_3$)

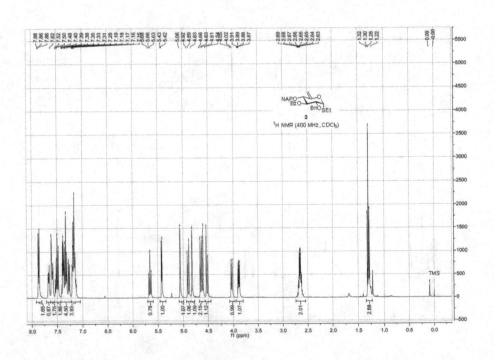

^{1}H NMR (400 MHz, CDCl$_3$)

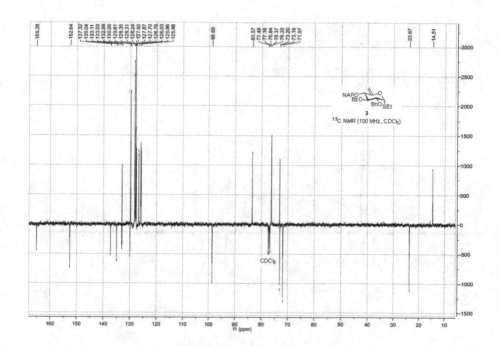

NAPO
BzO
BnO
SEt

3

^{13}C NMR (100 MHz, CDCl$_3$)

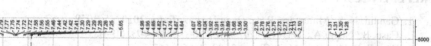

CDCl$_3$

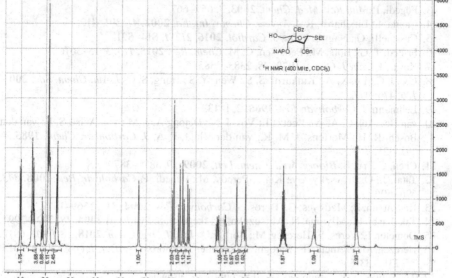

HO
OBz
SEt
NAPO
OBn

4

^{1}H NMR (400 MHz, CDCl$_3$)

TMS

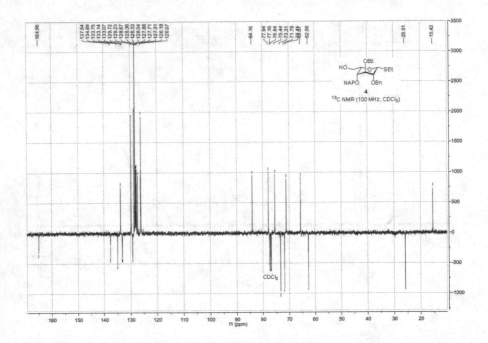

^{13}C NMR (100 MHz, CDCl$_3$)

REFERENCES

1. Gandhi, N. S.; Manecera, R. L. *Drug Discov. Today* **2010**, *15*, 1058–1067.
2. Casu, B.; Naggi A.; Torri G. *Carbohydr. Res.* **2014**, *403*, 60–68.
3. Fügedi, P. *Mini Rev. Med. Chem.* **2003**, *3*, 659–667.
4. Capila, I.; Lindhardt, R. J. *Angew. Chem. - Int. Ed.* **2002**, *41*, 391–412.
5. Cassinelli, G.; Naggi, A. *Int. J. Cardiol.* **2016**, *212S1*, S14–S21.
6. Frihed, T. G.; Bols, M.; Pedersen, C. M. *Chem. Rev.* **2015**, *115*, 3615–3676.
7. Kovar, J. *Can. J. Chem.* **1970**, *48*, 2383–2385.
8. Lee, J. C.; Lu, X. A.; Kulkarni, S. S.; Wen, Y. S.; Hung, S. C. *J. Am. Chem. Soc.* **2004**, *126*, 476–477.
9. Lehmann, J. *Carbohydr. Res.* **1966**, *2*, 1–13.
10. van Boeckel, C. A. A.; Beetz, T.; Vos, J. N.; Dejong A. J. M.; van Aelst, S. F.; van den Bosch, R. H.; Mertens, J. M. R.; van der Vlugt, F. A. *J. Carbohydr. Chem.* **1985**, *4*, 293–321.
11. Chen, C.; Yu, B. *Bioorg. Med. Chem. Lett.* **2009**, *19*, 3875–3879.
12. Tatai, J.; Osztrovszky, Gy.; Kajtár-Peredy, M.; Fügedi, P. *Carbohydr. Res.* **2008**, *343*, 596–606.
13. Kuszmann, J.; Medgyes, G.; Boros, S. *Carbohydr. Res.* **2004**, *339*, 1569–1579.
14. Takahashi, H.; Miyama, N.; Mitsuzuka, H.; Ikegami, S. *Synthesis* **2004**, *18*, 2991–2994.
15. Demeter, F.; Veres, F.; Herczeg, M.; Borbás, A. *Eur. J. Org. Chem.* **2018**, *48*, 6901–6912.
16. Herczeg, M.; Demeter, F.; Balogh, T.; Kelemen, V.; Borbás, A. *Eur. J. Org. Chem.* **2018**, *25*, 3312–3316.
17. The synthesis of compound **1** will be published elsewhere.

30 4-O-Acetyl-2-Azido-3,6-di-O-Benzyl-2-Deoxy-α/β-D-Glucopyranose

Yuko Yoneda*, and Yuichi Mikota
Faculty of Agriculture, Shizuoka University,
Shizuoka, Japan

Takaaki Goto, and Toshinari Kawada*
Graduate School of Life and Environmental Sciences,
Kyoto Prefectural University,
Kyoto, Japan

Martin Thonhofer[a]
Institute of Organic Chemistry,
Graz University of Technology,
Graz, Austria

CONTENTS

* Corresponding authors: yoneda.yuko@shizuoka.ac.jp; kawada@kpu.ac.jp
[a] Checker: thonhofer@tugraz.at

2-Azido-2-deoxy building blocks are popular synthons in synthetic carbohydrate chemistry because the azido group shows high stability against various reaction conditions and can be conveniently converted into a free amino group by simple reduction.

The synthesis of 4-*O*-acetyl-2-azido-3,6-di-*O*-benzyl-2-deoxy-α,β-D-glucopyranose (**9**) starts with anomeric deacetylation[1] of acetate **1**,[2] to obtain **2**. The hemiacetal was temporarily protected with the *tert*-butyldimethylsilyl group (**3**),[3] followed by deacetylation,[4] to afford triol **4**. Subsequent 4,6-*O*-benzylidenation afforded syrupy acetal **5**, which crystallized spontaneously after chromatography. (Different melting points—compare experimental and Lit.[5]—may arise from different crystallization conditions. The value of optical rotation was found within the range of data reported from other laboratories.[5-7] It may also be noted that in two references, the optical rotations had been measured from syrupy products[6,7]). Compound **5** was *O*-benzylated to give compound **6**. Subsequently, the benzylidene acetal in **6** was reduced regioselectively[8] to liberate the hydroxyl group at *C*-4 yielding alcohol **7**. Compound **7** is a versatile synthon in carbohydrate chemistry as it can be exploited as a glycosyl acceptor for the synthesis of complex 1,4-linked oligosaccharides. Furthermore, after 4-OH protection to give intermediate **8** followed by removal of the *tert*-butyldimethylsilyl group[3] providing partially protected aldose **9**, the latter may be converted to glycosyl donors. A related synthetic strategy has recently been reported.[9]

EXPERIMENTAL

GENERAL METHODS

Anhydrous chloroform (CHCl₃) and tetrahydrofuran (THF) were purchased from FUJIFILM Wako Pure Chemical Corporation. Flash chromatography was performed on silica gel (Wakogel FC-40, Wako). Melting points were measured with

MP-S3 (Yanaco) and are uncorrected. Optical rotations were determined with a High Sensitive Polarimeter SEPA 300 (HORIBA). NMR spectra were measured with Avance 400 MHz NMR (Bruker). Assignments of ^{1}H signals were determined using COSY technique. Thin-layer chromatography (TLC) was done on silica gel-coated glass plates (TLC Silica gel 60 F_{254}, Merck). Spots were detected under UV light (254 nm), and/or by charring with 5% 12 molybdo(VI) phosphoric acid n-hydrate $(H_3(PMo_{12}O_{40}) \cdot nH_2O)$ in EtOH. Unless otherwise indicated, work-up for each reaction mixture involved dilution with organic solvents; washing the extract consecutively with water, saturated aqueous sodium bicarbonate and brine; drying the combined organic phase with anhydrous sodium sulfate; and concentration under reduced pressure.

3,4,6-Tri-*O*-acetyl-2-azido-2-deoxy-α,β-D-glucopyranose (2)

To a stirred solution of **1**[2] (21.3 g, 57.3 mmol) in DMF (30 mL), hydrazine acetate (6.3 g, 68.7 mmol, 1.2 equiv) was added at room temperature. After stirring for 10 min, the mixture was diluted with EtOAc and processed as described in General Methods. Chromatography (1:2 EtOAc–*n*-hexane) gave known 3,4,6-tri-*O*-acetyl-2-azido-2-deoxy-α,β-D-glucopyranose (**2**)[10] as colorless syrup (15.13 g, 80%); R_f = 0.5 (1:1 EtOAc–*n*-hexane).

tert-Butyldimethylsilyl 3,4,6-tri-*O*-acetyl-2-azido-2-deoxy-β-D-glucopyranoside (3)

Imidazole (3.5 g, 51.53 mmol, 1.2 equiv) and *tert*-butyldimethylchlorosilane (7.8 g, 51.53 mmol, 1.2 equiv) were added at room temperature to a stirred solution of **2** (14.22 g, 42.94 mmol) in anhydrous $CHCl_3$ (40 mL). After stirring for 1 h, the mixture was diluted with EtOAc and processed as described before. Chromatography (1:8 EtOAc–*n*-hexane) gave *tert*-butyldimethylsilyl 3,4,6-tri-*O*-acetyl-2-azido-2-deoxy-β-D-glucopyranoside (**3**)[11] as colorless syrup (14.99 g, 78%). Crystallization from EtOH–*n*-hexane afforded **3** as colorless crystals; mp 64–67°C (Lit.[11] mp 62°C); R_f = 0.3 (1:4 EtOAc–*n*-hexane); ^{1}H NMR (400 MHz, $CDCl_3$): δ 4.99–4.93 (m, 2*H*, H-3, H-4), 4.63 (d, 1*H*, $J_{1,2}$ 7.6 Hz, H-1), 4.19 (dd, 1*H*, $J_{5,6b}$ 6.1 Hz, $J_{6a,6b}$ 12.1 Hz, H-6b), 4.10 (dd, 1*H*, $J_{5,6a}$ 2.5 Hz, $J_{6a,6b}$ 12.1 Hz, H-6a), 3.69–3.64 (m, 1*H*, H-5), 3.45–3.40 (m, 1*H*, H-2), 2.08 (s, 3*H*, Ac), 2.07 (s, 3*H*, Ac), 2.02 (s, 3*H*, Ac), 0.94 (s, 9*H*, *t*-Bu), 0.17 (s, 3*H*, Si*Me*₂), 0.16 (s, 3*H*, Si*Me*₂). For full characterization of **3**, see Lit.[11].

tert-Butyldimethylsilyl 2-azido-2-deoxy-β-D-glucopyranoside (4)

Zemplén deacetylation[4] of compound **3** (13.96 g, 31.33 mmol) in MeOH (80 mL) was carried out conventionally. Chromatography (1:9 MeOH–$CHCl_3$) gave *tert*-butyldimethylsilyl 2-azido-2-deoxy-β-D-glucopyranoside (**4**)[11] as colorless syrup in virtually theoretical yield; R_f = 0.4 (1:9 MeOH–$CHCl_3$); ^{1}H NMR (400 MHz, $CDCl_3$): δ 4.60 (d, 1*H*, $J_{1,2}$ 7.6 Hz, H-1), 3.91–3.80 (m, 2*H*), 3.64–3.57 (m, 2*H*), 3.41–3.31 (m, 3*H*), 3.24 (dd, 1*H*, $J_{1,2}$ 7.6 Hz, $J_{2,3}$ 9.9 Hz, H-2), 2.43–2.40 (m, 1*H*), 0.94

(s, 9H, t-Bu), 0.17 (s, 3H, SiMe_2), 0.16 (s, 3H, SiMe_2). For full characterization of **4**, see Lit.[11].

TERT-BUTYLDIMETHYLSILYL 2-AZIDO-4,6-O-BENZYLIDENE-2-DEOXY-β-D-GLUCOPYRANOSIDE (5)

This compound was prepared as described.[6] The colorless syrup was obtained after chromatography (1:19 EtOAc–n-hexane) crystallized spontaneously; mp 59–61°C (Lit.[5] mp 63–64°C); $[\alpha]_D^{21.5}$–35.2 (c 0.95, CHCl$_3$); Lit.[6] $[\alpha]_D^{25}$ –40 (c 0.40, CHCl$_3$), Lit.[5] $[\alpha]_D^{22}$ –38.6 (c 1, CHCl$_3$), Lit.[7] $[\alpha]_D^{27}$ –35.5 (c 1, CHCl$_3$); ^{1}H NMR (400 MHz, CDCl$_3$): δ 7.49–7.35 (m, 5H, CHPh), 5.54 (s, 1H, CHPh), 4.66 (d, 1H, $J_{1,2}$ 7.6 Hz, H-1), 4.30 (dd, 1H, $J_{5,6b}$ 5.0 Hz, $J_{6a,6b}$ 10.4 Hz, H-6b), 3.79 (dd, 1H, $J_{5,6a}$ 10.0 Hz, $J_{6a,6b}$ 10.4 Hz, H-6a), 3.65 (ddd, 1H, $J_{2,3}$ 9.5 Hz, $J_{3,4}$ 9.2 Hz, $J_{3,OH}$ 2.5 Hz, H-3), 3.58 (dd, 1 H, $J_{3,4}$ 9.2 Hz, $J_{4,5}$ 9.1 Hz, H-4), 3.42 (ddd, 1H, $J_{4,5}$ 9.1 Hz, $J_{5,6a}$ 10.0 Hz, $J_{5,6b}$ 5.0 Hz, H-5), 3.34 (dd, 1H, $J_{1,2}$ 7.6 Hz, $J_{2,3}$ 9.5 Hz, H-2), 2.60 (d, 1H, $J_{3,OH}$ 2.5 Hz, 3-OH), 0.94 (s, 9H, t-Bu), 0.18 (s, 3H, SiMe_2), 0.17 (s, 3H, SiMe_2). Anal. calcd for C$_{19}$H$_{29}$N$_3$O$_5$Si: C, 56.00; H, 7.17; N, 10.31. Found: C, 55.97; H, 7.19; N, 10.38.

TERT-BUTYLDIMETHYLSILYL 2-AZIDO-3-O-BENZYL-4,6-O-BENZYLIDENE-2-DEOXY-β-D-GLUCOPYRANOSIDE (6)

This compound was prepared as described.[6] Crystallization from MeOH afforded **6** as colorless crystals; mp 104–105°C (Lit.[6] mp 100–101°C); ^{1}H NMR (400 MHz, CDCl$_3$): δ 7.49–7.28 (m, 10H, CH$_2Ph$, CHPh), 5.56 (s, 1H, CHPh), 4.90 (d, 1H, J 11.4 Hz, CH_2Ph), 4.79 (d, 1H, J 11.4 Hz, CH_2Ph), 4.58 (d, 1H, $J_{1,2}$ 7.6 Hz, H-1), 4.29 (dd, 1H, $J_{5,6b}$ 5.0 Hz, $J_{6a,6b}$ 10.5 Hz, H-6b), 3.79 (dd, 1H, $J_{5,6a}$ 10.1 Hz, $J_{6a,6b}$ 10.5 Hz, H-6a), 3.71 (dd, 1H, $J_{3,4}$ 9.1 Hz, $J_{4,5}$ 9.5 Hz, H-4), 3.51 (dd, 1H, $J_{2,3}$ 9.6 Hz, $J_{3,4}$ 9.1 Hz, H-3), 3.38 (broad ddd, 1H, $J_{4,5}$ 9.5 Hz, $J_{5,6a}$ 10.1 Hz, $J_{5,6b}$ 5.0 Hz, H-5), 3.37 (dd, 1H, $J_{1,2}$ 7.6 Hz, $J_{2,3}$ 9.6 Hz, H-2), 0.93 (s, 9H, t-Bu), 0.16 (s, 3H, SiMe_2), 0.15 (s, 3H, SiMe_2). For full characterization of **6**, see Ref.[6].

TERT-BUTYLDIMETHYLSILYL 2-AZIDO-3,6-DI-O-BENZYL-2-DEOXY-β-D-GLUCOPYRANOSIDE (7)

Triethylsilane (14.4 mL, 90.0 mmol, 5 equiv) and trifluoroacetic acid (6.9 mL, 90.0 mmol, 5 equiv) were added with stirring at –20°C to a stirred solution of **6** (8.96 g, 18.0 mmol) in CHCl$_3$ (45 mL), and the stirring was continued at room temperature for 30 min. The mixture was diluted with EtOAc and processed as described before. Chromatography (n-hexane → 1:19 → 1:4 EtOAc–n-hexane) gave $tert$-butyldi-methylsilyl 2-azido-3,6-di-O-benzyl-2-deoxy-β-D-glucopyranoside (**7**)[6] as colorless syrup (7.68 g, 85%); R_f = 0.4 (1:4 EtOAc–n-hexane); ^{1}H NMR (400 MHz, CDCl$_3$): δ 7.46–7.28 (m, 10H, CH$_2Ph$), 4.91 (d, 1H, J 11.4 Hz, CH_2Ph), 4.76 (d, 1H, J 11.4 Hz,

CH_2Ph), 4.58 (d, 1*H*, *J* 12.1 Hz, CH_2Ph), 4.55 (d, 1*H*, *J* 12.1 Hz, CH_2Ph), 4.53 (d, 1*H*, $J_{1,2}$ 7.6 Hz, H-1), 3.71 (d, 2*H*, H-6a, H-6b), 3.63 (ddd, 1*H*, $J_{3,4}$ 8.5 Hz, $J_{4,5}$ 9.6 Hz, $J_{4,OH}$ 2.2 Hz, H-4), 3.43–3.38 (ddd, 1*H*, $J_{4,5}$ 9.6 Hz, $J_{5,6a}$ $J_{5,6b}$ 4.7 Hz, H-5), 3.31 (dd, 1*H*, $J_{1,2}$ 7.6 Hz, $J_{2,3}$ 9.9 Hz, H-2), 3.21 (dd, 1*H*, $J_{2,3}$ 9.9 Hz, $J_{3,4}$ 8.5 Hz, H-3), 2.63 (d, 1*H*, $J_{4,OH}$ 2.2 Hz, 4-OH), 0.94 (s, 9*H*, *t*-Bu), 0.16 (s, 6*H*, SiMe_2). For full characterization of the independently synthesized **7**, see Lit.[6]

TERT-BUTYLDIMETHYLSILYL 4-*O*-ACETYL-2-AZIDO-3,6-DI-*O*-BENZYL-2-DEOXY-β-D-GLUCOPYRANOSIDE (8)

N,*N*-Dimethyl-4-aminopyridine (81 mg, 0.66 mmol, 0.1 equiv) and acetic anhydride (10 mL) were added at 0°C to a stirred solution of **7** (3.30 g, 6.61 mmol) in pyridine (5 mL), and the mixture was stirred at room temperature for 30 min, when TLC (1:4 EtOAc–*n*-hexane) showed that the reaction was complete. After concentration and co-evaporation with EtOH, the residue was diluted with EtOAc and processed as described before. Chromatography (1:8 EtOAc–*n*-hexane) gave *tert*-butyldimethylsilyl 4-*O*-acetyl-2-azido-3,6-di-*O*-benzyl-2-deoxy-β-D-glucopyranoside (**8**) as colorless syrup (3.29 g, 92%) which crystallized spontaneously upon standing over several days. Crystallization from cyclohexane–EtOAc gave colorless needles; mp 43–44°C; $[\alpha]_D^{23}$ –27.0 (*c* 8.8, $CHCl_3$); R_f = 0.6 (1:4 EtOAc–*n*-hexane); ^{1}H NMR (400 MHz, $CDCl_3$): δ 7.36–7.25 (m, 10*H*, CH_2Ph), 5.00–4.94 (m, 1*H*, H-4), 4.81 (d, 1*H*, *J* 11.4 Hz, CH_2Ph), 4.61 (d, 1*H*, *J* 11.4 Hz, CH_2Ph), 4.55 (d, 1*H*, $J_{1,2}$ 7.6 Hz, H-1), 4.50 (s, 2*H*, CH_2Ph), 3.55–3.49 (m, 3*H*, H-5, H-6a, H-6b), 3.42–3.33 (m, 2*H*, H-2, H-3), 1.84 (s, 3*H*, Ac), 0.94 (s, 9*H*, *t*-Bu), 0.18 (s, 3*H*, SiMe_2), 0.17 (s, 3*H*, SiMe_2); ^{13}C NMR (100 MHz, $CDCl_3$): δ 169.7 (CH_3CO), 137.8 (*Ph*), 128.4, 128.3, 127.9, 127.8, 127.7, 127.6 (*Ph*), 97.1 (*C*-1), 80.0 (*C*-3), 74.7 (CH_2Ph), 73.5, 73.4 (CH_2Ph, *C*-5), 70.9 (*C*-4), 69.7 (*C*-6), 68.3 (*C*-2), 25.6 (*CMe_3*), 20.8 (CH_3CO), 18.0 (*CMe_3*), –4.3 (SiMe_2), –5.3 (SiMe_2). Anal. calcd for $C_{28}H_{39}N_3O_6Si$: C, 62.08; H, 7.26; N, 7.76. Found: C, 62.23; H, 7.28; N, 7.85.

4-*O*-ACETYL-2-AZIDO-3,6-DI-*O*-BENZYL-2-DEOXY-α,β-D-GLUCOPYRANOSE (9)

To a stirred solution of **8** (887 mg, 1.64 mmol) in THF (5 mL), acetic acid (47 µL, 0.82 mmol, 0.5 equiv) was added at 0°C. Tetra-*n*-butylammonium fluoride solution in THF (1.0 M, 1.96 mL, 1.96 mmol, 1.2 equiv) was added at –40°C to a stirred solution, and the stirring at –40°C was continued for 20 min. The mixture was diluted with EtOAc and processed as described before. Chromatography (1:4 EtOAc–*n*-hexane) gave anomeric mixture **9**[9] as colorless syrup (658 mg, 94%). Despite all efforts, individual anomers could not be separated by chromatography. Crystallization from EtOH afforded **9** as colorless crystals (cubics and needles); mp 77–81°C and 97–103°C, respectively; $[\alpha]_D^{23}$ –1.3 (*c* 10.2, $CHCl_3$); R_f = 0.4 (1:2 EtOAc–*n*-hexane), because preparation of glycosyl donors does not require pure anomers, recrystallization was not attempted. ^{1}H NMR (400 MHz, $CDCl_3$) (1:4 = α:β): δ 7.37–7.27 (m, 50*H*, CH_2Ph),

5.32 (dd, 1H, $J_{1,2}$ $J_{1,OH}$ 2.6 Hz, H-1/α), 5.00 (dd, 1H, $J_{3,4}$ 9.3 Hz, $J_{4,5}$ 10.2 Hz, H-4/α), 4.96–4.92 (m, 4H, H-4/β), 4.82 (d, 1H, J 11.4 Hz, CH_2Ph/α), 4.81 (d, 4H, J 11.4 Hz, CH_2Ph/β), 4.63 (d, 1H, J 11.4 Hz, CH_2Ph/α), 4.61 (d, 4H, J 11.4 Hz, CH_2Ph/β), 4.58 (broad d, 4H, $J_{1,2}$ 5.7 Hz, H-1/β), 4.54–4.47 (m, 10H, CH_2Ph/α,β), 4.29 (broad s, 4H, 1-OH/β), 4.14 (ddd, 1H, $J_{4,5}$ 10.2 Hz, $J_{5,6}$ 3.2 Hz, $J_{5,6}$ 6.0 Hz, H-5/α), 3.99 (dd, 1H, $J_{2,3}$ 10.0 Hz, $J_{3,4}$ 9.3 Hz, H-3/α), 3.87 (broad d, 1H, $J_{1,OH}$ 2.6 Hz, 1-OH/α), 3.57–3.51 (m, 2H, H-5/β, H-6b/β), 3.49–3.42 (m, 5.5H, H-2/α, H-6a/α, H-6b/α, H-6a/β), 3.41–3.37 (m, 2H, H-2/β, H-3/β), 1.87 (s, 3H, Ac/α), 1.84 (s, 12H, Ac/β); ^{13}C NMR (100 MHz, CDCl$_3$): δ 169.73 (CH$_3$CO/α), 169.65 (CH$_3$CO/β), 137.6 (Ph/α,β), 137.31 (Ph/α), 137.27 (Ph/β), 128.5–127.9 (Ph), 96.1 (C-1/β), 91.7 (C-1/α), 80.3 (C-3/β), 77.6 (C-3/α), 74.9 (CH$_2$Ph/α,β), 73.59 (CH$_2$Ph/β), 73.54 (CH$_2$Ph/α), 73.29 (C-5/β), 71.0 (C-4/α), 70.5 (C-4/β), 69.1 (C-6/β), 68.98, 68.95 (C-5/α, C-6/α), 66.9 (C-2/β), 63.4 (C-2/α), 20.75 (CH$_3$CO/α), 20.72 (CH$_3$CO/β). The NMR data agreed with the Lit.[9] Anal. calcd for C$_{22}$H$_{25}$N$_3$O$_6$: C, 61.82; H, 5.90; N, 9.83. Found: C, 61.87; H, 5.92; N, 9.88.

ACKNOWLEDGMENTS

The authors are grateful to Dr. Kimihiko Satoh, Director, Technology Development Division, Koyo Chemical Co., Ltd., Osaka, Japan for providing D-glucosamine hydrochloride. The authors thank to Prof. Dr. Hiroyuki Saimoto and Ms. Miyuki Tanmatsu, Tottori University, Tottori, Japan, for elementary analysis.

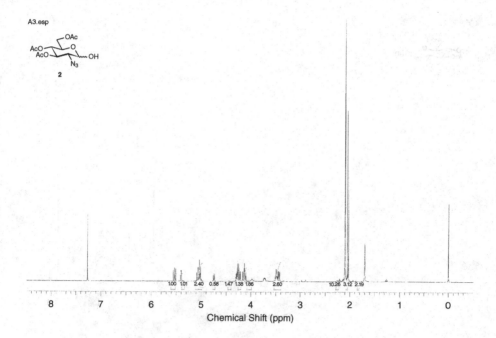

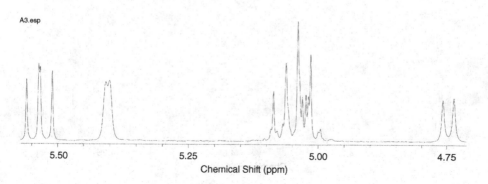

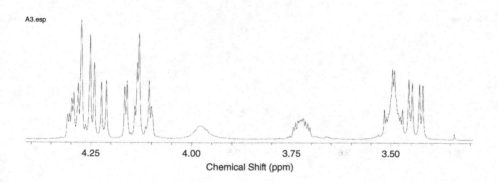

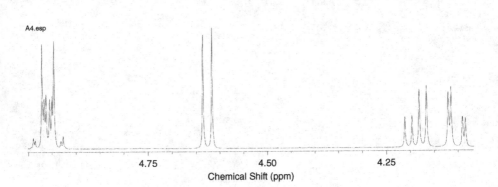

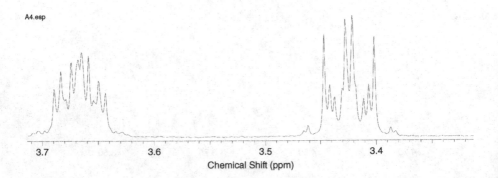

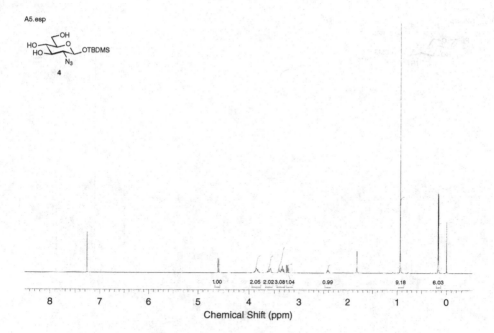

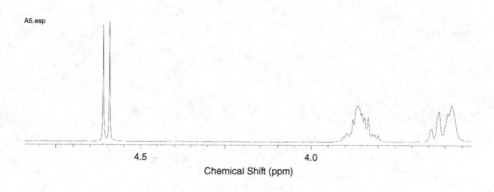

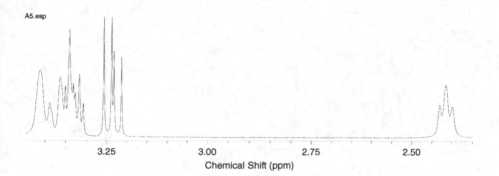

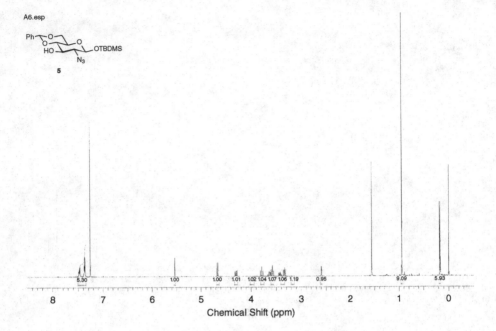

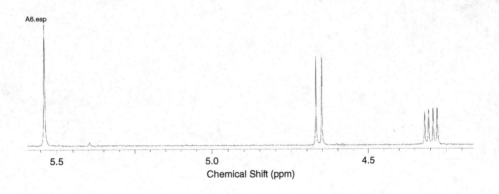

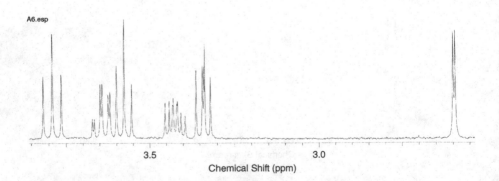

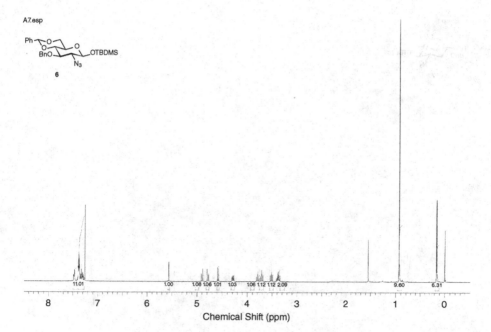

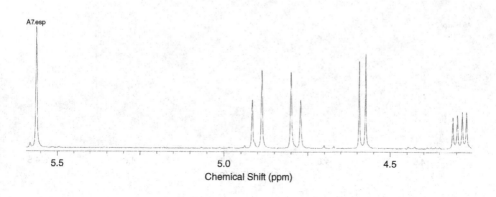

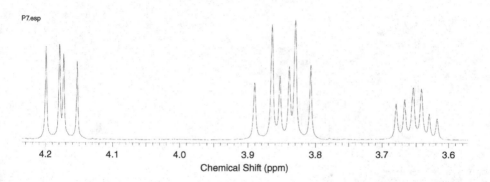

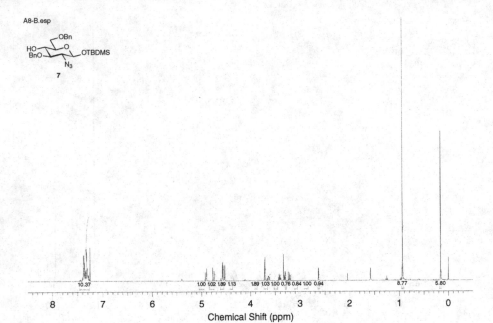

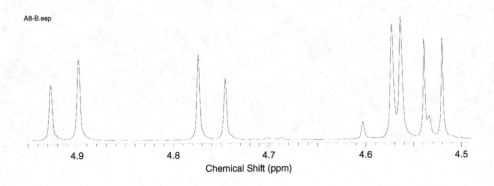

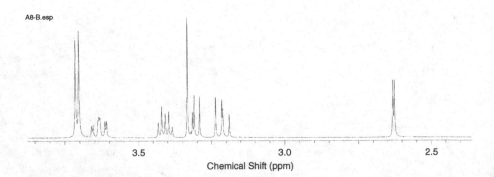

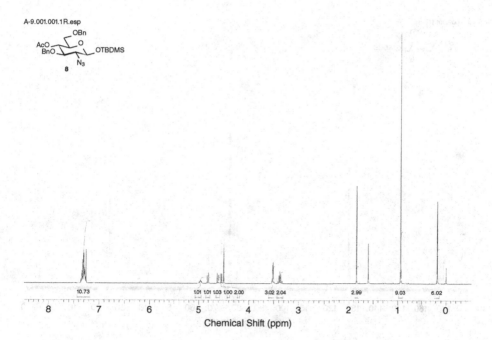

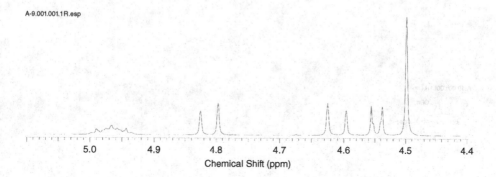

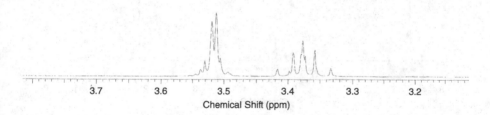

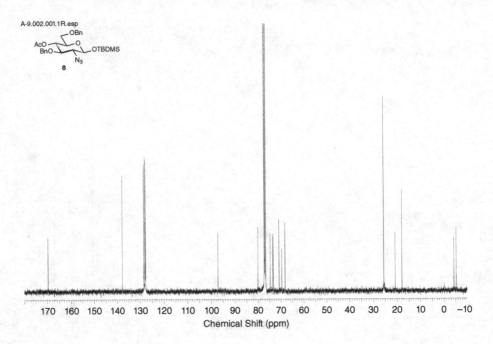

A-9.002.001.1R.esp

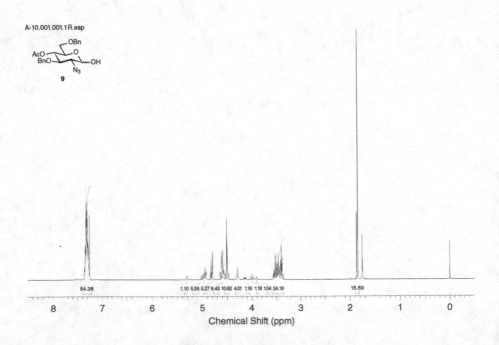

A-10.001.001.1R.esp

A-10.001.001.1R.esp

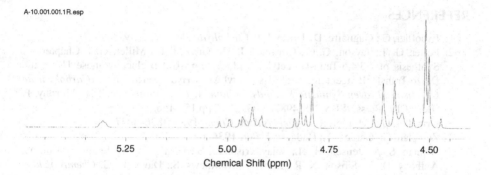

A-10.001.001.1R.esp

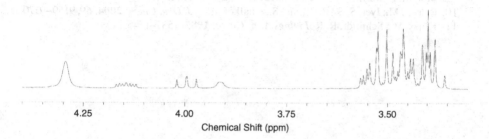

A-10.002.001.1R.esp

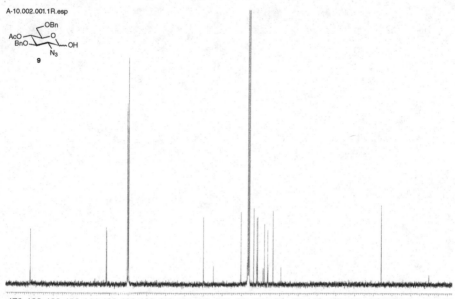

REFERENCES

1. Excoffier, G.; Gagnaire, D.; Utille, J.-P. *Carbohydr. Res.* **1975**, *39*, 368–373.
2. Potter, G. T.; Jayson, G. C.; Gardiner, J. M.; Guazelli L.; Miller, G. J. Chapter 18: Synthesis of 1,3,4,6-Tetra-*O*-acetyl-2-azido-2-deoxy-α,β-D-glucopyranose Using the Diazo-Transfer Reagent Imidazole-1-sulfonyl Azide Hydrogen Sulfate. In *Carbohydrate Chemistry: Proven Synthetic Methods, Volume 4, 1st Edition*, Vogel, D., Murphy, P. (Eds.) CRC Press, ISBN 9781498726917, **2017**, pp 151–156.
3. Grundler, G.; Schmidt, R. R. *Liebigs Ann. Chem.* **1984**, 1826–1847.
4. Zemplén, G.; Gerecs, A.; Hadácsy, I. *Ber.* **1936**, 69, 1827–1829.
5. Allman, S. A.; Jensen, H. H.; Vijayakrishnan, B.; Garnett, J. A.; Leon, E.; Liu, Y.; Anthony, D. C.; Sibson, N. R.; Feizi, T.; Matthews, S.; Davis, B. G. *ChemBioChem* **2009**, 10, 2522–2529.
6. Murakata, C.; Ogawa, T. *Carbohydr. Res.* **1992**, 234, 75–91.
7. Tanaka, H.; Ando, Y.; Wada, M.; Takahashi, T. *Org. Biomol. Chem.* **2005**, 3, 3311–3328.
8. DeNinno, M. P.; Etienne, J. B.; Duplantier, K. C. *Tetrahedron Lett.* **1995**, *36* (5), 669–672.
9. Morelli, L.; Lay, L. *ARKIVOC* **2013**, *2013* (2), 166–184.
10. Rele, S. M.; Iyer, S. S.; Baskaran, S.; Chaikof, E. L. *J. Org. Chem.* **2004**, *69*, 9159–9170.
11. Kinzy, W.; Schmidt, R. R. *Liebigs Ann. Chem.* **1985**, 1537–1545.

31 2,3,4-Tri-O-Benzoyl-6-O-(Tert-Butyldiphenylsilyl)-1-Thio-β-D-Glucopyranose

Jonathan Berry, and Guillaume Despras*
Otto Diels Institute of Organic Chemistry,
Christiana Albertina University of Kiel,
Kiel, Germany

Martin Kurfiřt[a]
Institute of Chemical Process Fundamentals of the CAS,
Praha, Czech Republic

CONTENTS

Glycosyl thiol derivatives are valuable building blocks in carbohydrate chemistry. In contrast to their hemiacetal counterpart, they display excellent stability toward mutarotation, which allows stereocontrol of the C-1 configuration. 1-Thio derivatives may be used in the synthesis of biologically relevant molecules, such as glycosyl sulfenamides/sulfonamides,[1] glycosyl disulfides,[2] glycosyl thionolactones,[3] and also thiosaccharides known to be resistant toward glycosidase-mediated bond cleavage.[4] Glycosyl thiols are also very versatile synthetic intermediates, since they can easily be converted into glycosyl donors such as thioglycosides by substitution of an alkyl halide,[5] or glycosyl thio-acetimidates.[6] Conventionally, glycosyl thiols can be prepared by SN2 reaction between

---

* Corresponding author: gdespras@oc.uni-kiel.de
[a] Checker: kurfirt@icpf.cas.cz

a conveniently protected glycosyl bromide or glycosyl acetate and thiourea[7,8] or potassium thioacetate,[9] followed by thiouronium hydrolysis or S-deacetylation, respectively. Other methods have been reported, such as treatment of glycosyl hemiacetals with the Lawesson's reagent,[10] ring opening of 1,6-anhydro derivatives with bis(trimethylsilyl)sulfide [(TMS)$_2$S],[11] or by bubbling hydrogen sulfide gas in a solution of glycosyl bromide in hydrogen fluoride.[12] However, those methods present several drawbacks: long reaction times, multiple steps, low stereoselectivities, and inconsistent yields. In 2013, Misra *et al.* described a new method to form stereospecifically β-glycosyl thiols from corresponding α-glycosyl bromides upon treatment with disodium sulfide nonahydrate (Na$_2$S·9H$_2$O) and carbon disulfide (CS$_2$) in excellent yields.[5] In addition to a full stereocontrol, this method presents the advantage to be very fast (1–25 min).

In this chapter, we report on a five-step, large-scale synthesis of 2,3,4-tri--*O*-benzoyl-6-*O*-(*tert*-butyldiphenylsilyl)-1-thio-β-D-glucopyranose (**5**) from levoglucosan (**1**) in an overall yield of 56%. Orthogonal silyl protection on the 6 position of the target-building block allows further functionalization, for instance toward the synthesis of (1,6)-thiooligosaccharides[13] or carbohydrate-derived macrocycles.[14]

Synthesis of **5** started with perbenzoylation of levoglucosan, giving **2** in 95% yield. Subsequent cleavage of the 1,6-anhydro bond in **2** with trimethyl(phenylthio) silane (TMSSPh) in the presence of zinc iodide (ZnI$_2$) yielded **3** (96%) as an α/β mixture (1:14). Silylation of the free primary position with *tert*-butyldiphenylchlorosilane (TBDPSCl) in the presence of *N*-methylimidazole (NMI) and iodine (I$_2$) and subsequent bromination of the anomeric carbon with iodine monobromide (IBr) afforded solely α-bromide **4** (84% over two steps). Bromide **4** has shown quite a remarkable stability toward hydrolysis which allowed its purification by column chromatography and its storage for long periods in the fridge. Even after 1 week of storage at room temperature (rt) under ambient atmosphere, no sign of hydrolysis could be observed either by thin layer chromatography (TLC) or NMR analysis. Finally, treatment of bromide **4** with Na$_2$S·9H$_2$O and CS$_2$ in DMF afforded 2,3,4-tri-*O*-benzoyl-6-*O*-(*tert*-butyldiphenylsilyl)-1-thio-β-D-glucopyranose **5** in 73% yield.

EXPERIMENTAL

GENERAL METHODS

Levoglucosan was purchased from Carbosynth and used without further purification. Reactions requiring dry conditions were carried out in flame-dried glassware and under positive pressure of nitrogen. Analytical TLC was performed on silica gel plates (GF 254 Merck). Visualization was achieved by UV light and/or by charring with 10% sulfuric acid in ethanol or with a solution of vanillin (900 mg vanillin and 12-mL H_2SO_4 in 90 mL H_2O and 75 mL EtOH). Solutions in organic solvents were dried ($MgSO_4$) and concentrated *in vacuo*. The solvents used for column chromatography were distilled before use. Dichloromethane (VWR Chemicals, HPLC grade) was dried over aluminum oxide column. Dry 1,2-dichloroethane was purchased from Acros Organics in sealed bottles containing molecular sieves. Optical rotations were measured in solvents indicated using a 10-cm cuvette and a PerkinElmer 241 polarimeter with a sodium D-line (589 nm). Proton (^1H) and carbon (^{13}C) nuclear magnetic resonance spectra (NMR) were recorded with a Bruker DRX-500 spectrometer. Chemical shifts are referenced to internal tetramethylsilane (TMS), or to the signal of the residual nonprotonated NMR solvent. Data are presented as follows: chemical shift, multiplicity (s = singlet, d = doublet, t = triplet, m = multiplet and bs = broad singlet), integration, and coupling constant in Hz. Full signal-nuclei assignment was achieved by 2D NMR techniques (^1H-^1H COSY, ^1H-^{13}C HSQC). ESI mass spectra were recorded on an Esquire-LC instrument from Bruker Daltonics.

1,6-ANHYDRO-2,3,4-TRI-O-BENZOYL-β-D-GLUCOPYRANOSE (2)

Benzoyl chloride BzCl, (4.3 mL, 37.0 mmol) was added with stirring at 0°C to a solution of levoglucosan **1** (1 g, 6.17 mmol) in pyridine (12.3 mL). Upon addition of BzCl, the product (**2**) precipitated (care must be taken to ensure that the stirring remains effective), and the suspension was stirred at rt for 1.5 h. After cooling to 0°C, MeOH (40 mL) was added to destroy excess BzCl and precipitate all the products. The suspension was stirred for 20 min, the solid was filtered-off, washed with MeOH (3 × 15 mL), and dried *in vacuo* to afford **2**[15] (2.77 g, 95%) as a white amorphous solid, which was analytically pure; $[\alpha]_D^{20}$ −42.2 (*c* 1.0 in CHCl$_3$), lit.[15a] $[\alpha]_D^{20}$ −36 (*c* 1 in CHCl$_3$). ^1H NMR (CDCl$_3$, 500 MHz): δ 8.18–8.05 (m, 6*H*, Ar-H$_{ortho}$), 7.64–7.58 (m, 3*H*, Ar-H$_{para}$), 7.52–7.40 (m, 6*H*, Ar-H$_{meta}$), 5.75 (bs, 1*H*, H-1), 5.46–5.43 (m, 1*H*, H-3), 5.10 (bs, 1*H*, H-4), 5.07 (bs, 1*H*, H-2), 4.90 (d, 1*H*, $J_{5,6b}$ 5.7 Hz, H-5), 4.39 (d, 1*H*, $J_{6a,6b}$ 7.8 Hz, H-6a), 3.98 (dd, 1*H*, $J_{6b,6a}$ 7.8 Hz, $J_{6b,5}$ 5.7 Hz, H-6b). ^{13}C NMR (CDCl$_3$, 126 MHz): δ 165.5, 165.3, 164.9 (3 C, Ph*C*O), 133.8, 133.7, 133.7 (3 C, Ar-C$_{para}$), 130.2, 130.2, 130.1 (6 C, Ar-C$_{ortho}$), 129.5, 129.5, 129.2 (3 C, Ar-C$_{ipso}$), 128.8, 128.6, 128.6 (6 C, Ar-C$_{meta}$), 99.6 (C-1), 74.0 (C-5), 70.3 (C-4), 69.9 (C-3), 69.2 (C-2), 65.7 (C-6). ESI HRMS: *m/z* [M + Na$^+$]$^+$ calcd for C$_{27}$H$_{22}$O$_8$ + Na$^+$: 474.1207 found: 474.1209. Anal. calcd for C$_{27}$H$_{22}$O$_8$: C, 68.35; H, 4.67. Found: C, 68.28; H, 4.51.

PHENYL 2,3,4-TRI-O-BENZOYL-1-THIO-α,β-D-GLUCOPYRANOSIDE (3)

To a solution of **2** (2.5 g, 5.27 mmol) in dry 1,2-dichloroethane (26.4 mL), zinc iodide (5.05 g, 15.8 mmol) and (phenylthio)trimethylsilane (2.5 mL, 13.2 mmol) were added at rt. The mixture was stirred at 85°C for 2 h until TLC (8:2 cyclohexane–ethyl acetate) showed disappearance of the starting material (R_f = 0.4), and formation of 6-O-trimethylsilylated intermediates as two very close-moving spots of different intensities, (R_f ~ 0.5, α: minor, β: major). Often, spots corresponding to the 6-OH products may be detected ($R_{f(\alpha)}$ = $R_{f(\beta)}$ ~ 0.2). The suspension was filtered through a pad of Celite®, and the solid washed with CH_2Cl_2 (3 × 20 mL). The filtrate was diluted with MeOH (50 mL) to remove TMS groups, and the mixture was concentrated *in vacuo*. TLC monitoring showed the disappearance of 6-O-TMS products while only the 6-OH products and the excess thiophenol remained. Chromatography (9:1 → 7:3 cyclohexane–ethyl acetate) yielded a 1:14 $\alpha\beta$ mixture of **3** as a white foam (2.95 g, 96%). The NMR data for the major β-**3** product matched those in the literature.[16] α-**3**: ^{1}H NMR (CDCl$_3$, 500 MHz): δ 8.03–7.98 (m, 4H, Ar-H), 7.92–7.88 (m, 2H, Ar-H), 7.56–7.48 (m, 4H, Ar-H), 7.48–7.44 (m, 2H, Ar-H), 7.44–7.36 (m, 4H, Ar-H), 7.35–7.30 (m, 2H, Ar-H), 7.29–7.24 (m, 2H, Ar-H), 6.20–6.14 (m, 2H, H-1, H-3), 5.58–5.52 (m, 2H, H-2, H-4), 4.62–4.57 (m, 1H, H-5), 3.78–3.69 (m, 2H, H-6a, H-6b), 2.55 (dd, 1H, $J_{OH,6a}$ 8.5 Hz, $J_{OH,6b}$ 6.1 Hz, OH). HRMS: m/z [M + Na$^+$]$^+$ calcd for $C_{33}H_{28}O_8S$ + Na$^+$: 607.1397; found: 607.1392. Anal. calcd for $C_{33}H_{28}O_8S$: C, 67.79; H, 4.83; S, 5.48. Found: C, 67.45; H, 4.57; S, 5.27.

2,3,4-TRI-O-BENZOYL-6-O-(*TERT*-BUTYLDIPHENYLSILYL)- α-D-GLUCOPYRANOSYL BROMIDE (4)

To a solution of **3** (2.60 g, 4.45 mmol) and N-methylimidazole (0.390 mL, 4.90 mmol) in pyridine (8.9 mL) were sequentially added at rt *tert*-butyldiphenylchlorosilane (1.27 mL, 4.90 mmoL) and iodine (1.24 g, 4.90 mmol). The mixture was stirred at rt for 20 min, when TLC (9:1 cyclohexane–ethyl acetate) showed disappearance of the starting material (R_f = 0.03), and formation of silylated products (α: R_f = 0.35; β: R_f = 0.3). MeOH (5 mL) was added, and the mixture was stirred for 15 min. The mixture was diluted with EtOAc (100 mL) and washed with aqueous, saturated $Na_2S_2O_3$ (1 × 30 mL) to remove colored material. The colorless organic phase was then washed with 2 M HCl (3 × 30 mL), dried, and concentrated. The residue was dried under high vacuum for 2 h and used for the next step.

The foregoing solid residue was dissolved in dry dichloromethane (44.5 mL), and molecular sieves 3Å (4 g) were added. The mixture was cooled to −20°C and iodine monobromide (1.38 g, 6.68 mmol) was added. The mixture was stirred for 20 min at −20°C until TLC (8:2 cyclohexane–diethyl ether) showed disappearance of the starting material (α: R_f = 0.45; β: R_f = 0.35) and formation of a single α-bromide product (R_f = 0.48). The mixture was filtered through a Celite® pad, and the solid was washed with CH_2Cl_2 (3 × 50 mL). The filtrate was washed with cold 10% aqueous $Na_2S_2O_3$ until colorless. The organic layer was dried and concentrated *in vacuo* at a temperature not exceeding 30°C. Chromatography (1:0 → 8:2 cyclohexane–diethyl ether) to remove any sulfur-containing by-products that could interfere with the next step, afforded bromide **4** as a white foam (2.95 g, 84%), $[\alpha]_D^{20}$ +77.5 (c 1.0 in CHCl$_3$), lit.[17] $[\alpha]_D^{25}$ +64.6 (c 1.0 in CHCl$_3$) for the independently prepared, amorphous material. The NMR data are in agreement with reported literature.[17] HRMS: m/z [M + Na$^+$]$^+$ calcd for C$_{43}$H$_{41}$BrO$_8$Si + Na$^+$: 815.1646; found: 815.1633.

2,3,4-Tri-O-benzoyl-6-O-(tert-butyldiphenylsilyl)-1-thio-β-d-glucopyranose (5)*

To a red solution of sodium sulfide nonahydrate (1.64 g, 6.80 mmol) and carbon disulfide (0.31 mL, 5.10 mmol) in DMF (9 mL) was rapidly added a solution of bromide **4** (2.70 g, 3.40 mmol) in DMF (8.1 mL). The mixture was stirred 1–3 min after addition of the bromide and quenched by diluting with water (20 mL) and 1-M HCl (10 mL). Prolonged reaction time results in the formation of undesired by-products. The mixture was then extracted with EtOAc (3 × 40 mL), and the combined organic layers were dried and concentrated *in vacuo*. TLC (3:1 cyclohexane–diethyl ether) of the crude residue showed disappearance of the starting material (R_f = 0.4) and formation of the desired compound (R_f = 0.3). Chromatography (1:0 → 8:2 cyclohexane–diethyl ether) yielded **5** as a white foam (1.85 g, 73%), $[\alpha]_D^{20}$ +29.6 (c 1.0 in CHCl$_3$). ^{1}H NMR (CDCl$_3$, 500 MHz): δ 7.98–7.93 (m, 2H, Ar-H), 7.89–7.85 (m, 2H, Ar-H), 7.83–7.79 (m, 2H, Ar-H), 7.76–7.71 (m, 2H, Ar-H), 7.63–7.59 (m, 2H, Ar-H), 7.55–7.49 (m, 2H, Ar-H) 7.44–7.20 (m, 13H, Ar-H), 5.80 (t, 1H, $J_{3,2} = J_{3,4}$ 9.5 Hz, H-3), 5.71–5.64 (m, 1H, H-4), 5.44 (t, 1H, $J_{2,3} = J_{2,1}$ 9.6 Hz, H-2), 4.79 (t, 1H, $J_{1,2} = J_{1,SH}$ 9.6 Hz, H-1), 3.91–3.80 (m, 3H, H-5, H-6a, H-6b), 2.33 (d, 1H, $J_{SH,1}$ 9.6 Hz, SH), 1.05 [s, 9H, C(CH$_3$)$_3$]. ^{13}C NMR (CDCl$_3$, 126 MHz): δ 166.0, 165.6, 165.1 (3 C, PhCO), 136.0, 135.8, 133.5, 133.4, 133.2, 133.1, 123.0, 129.9, 129.9, 129.8, 129.8, 129.3, 129.3, 129.0, 128.6, 128.5, 128.4, 127.8, 127.7 (30 C, 30 Ar-C), 79.8 (C-5), 79.0 (C-1), 74.6 (C-2), 74.4 (C-3), 69.1 (C-4), 63.1 (C-6), 27.1, 26.9 [C(CH$_3$)$_3$], 19.4 [C(CH$_3$)$_3$]. ESI HRMS: m/z [M + H$^+$]$^+$ calcd for C$_{43}$H$_{42}$O$_8$SSi + H$^+$: 747.2442; found: 747.2431. Anal. calcd for C$_{43}$H$_{42}$O$_8$SSi: C, 69.14; H, 5.67; S, 4.29. Found: C, 69.43; H, 5.90; S, 4.24.

ACKNOWLEDGMENTS

The authors warmly thank Prof. Dr. T. K. Lindhorst for fundamental support.

Verband der Chemischen Industrie (VCI) is also acknowledged for financial support.

---

* All operations involving carbon disulfide CS$_2$ must be conducted in a well-ventilated hood. This applies also to work up as well to washing the glassware, which should be washed with bleach and thorough rinsing with acetone. Evaporation before purification should also be done in a fume hood.

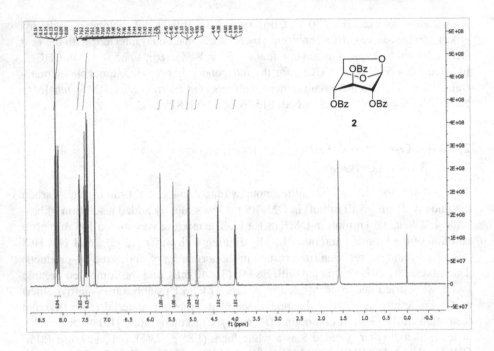

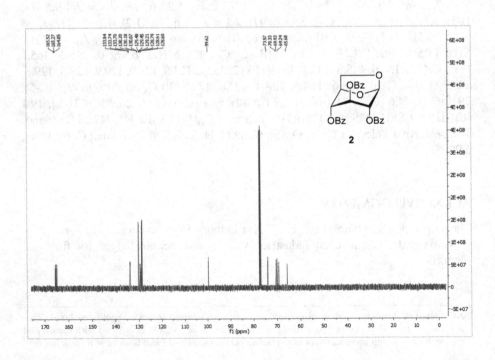

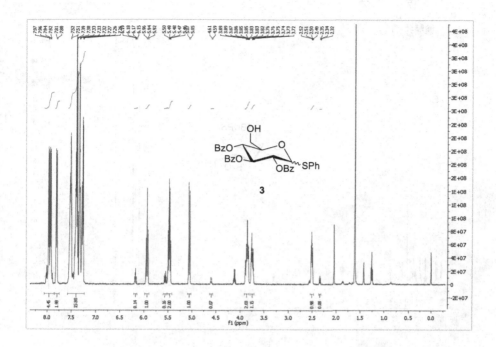

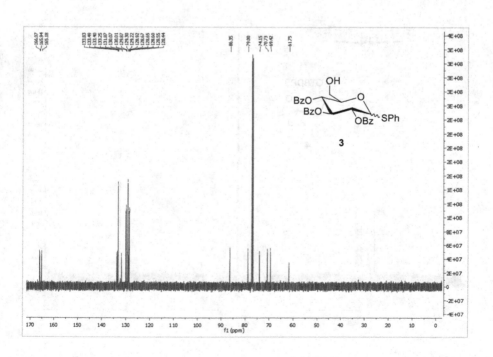

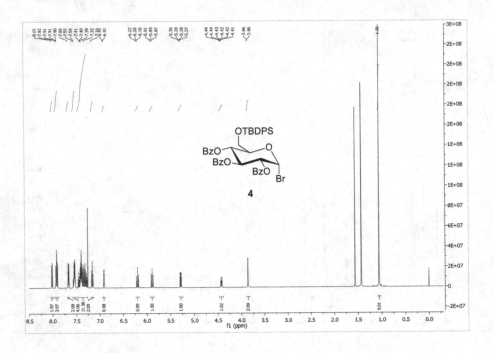

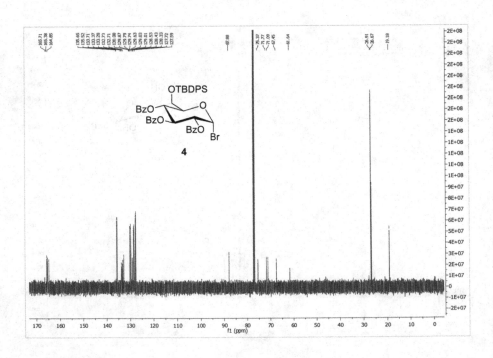

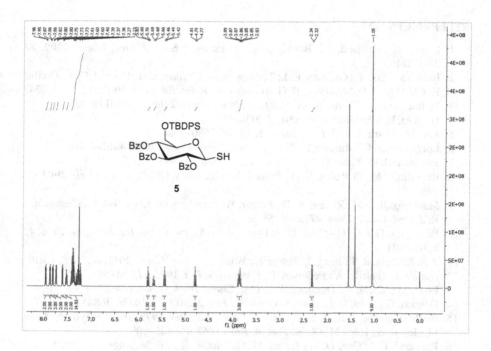

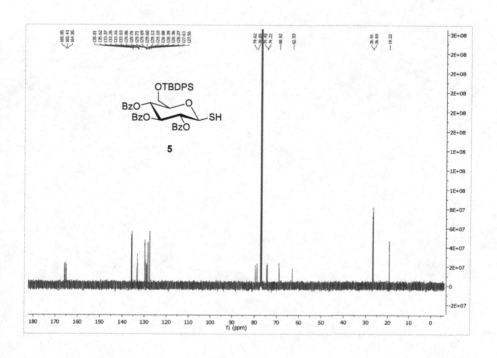

REFERENCES

1. Lopez, M.; Drillaud, N.; Bornaghi, L. F.; Poulsen, S.-A. *J. Org. Chem.* **2009**, *74*, 2811–2816.
2. Bernardes, G. J. L.; Grayson, E. J.; Thompson, S.; Chalker, J. M.; Errey, J. C.; El Qualid, F.; Claridge, T. D. W.; Davis, B. G. *Angew. Chem., Int. Ed. Engl.* **2008**, *47*, 2244–2247.
3. Wilkinson B. L.; Fairbanks A. J. *Tetrahedron Letters.* **2008**, *49*, 4941–4943.
4. Driguez, H. *ChemBioChem* **2001**, *2*, 311–318.
5. Jana M.; Misra A. K. *J. Org. Chem.* **2013**, *78*, 2680–2686.
6. Lucas-Lopez, C.; Murphy, N.; Zhu, X. *Eur. J. Org. Chem.* **2008**, 4401–4404.
7. Johnston, B. D.; Pinto, B. M. *J. Org. Chem.* **2000**, *65*, 4607–4617.
8. Ibatullin, F. M.; Shabalin, K. A.; Jänis, J. V.; Shavva, A. G. *Tetrahedron Lett.* **2003**, *44*, 7961–7964.
9. MacDougall, J. M.; Zhang, X. D.; Polgar, W. E.; Khroyan, T. V.; Toll, L.; Cashman, J. R. *J. Med. Chem.* **2004**, *47*, 5809–5815.
10. Bernardes, G. J. L.; Gamblin, D. P.; Davis, B. G. *Angew. Chem. Int. Ed. Engl.* **2006**, *45*, 4007–4011.
11. Zhu, X.; Dere, R. T.; Jiang, J.; Zhang, L.; Wang, X. *J. Org. Chem.* **2011**, *76*, 10187–10197.
12. Defaye, J.; Gadelle, A.; Pedersen, C. *Carbohydr. Res.* **1991**, *217*, 51–58.
13. Fiore, M.; Marra, A.; Dondoni, A. *J. Org. Chem.* **2009**, *74*, 4422–4425.
14. Despras, G.; Hain, J.; Jäschke, S. O. *Chem. Eur. J.* **2017**, *23*, 10838–10847.
15. (a) Brimacombe, J. S.; Tucker, L. C. N. *Carbohydr. Res.* **1975**, *40*, 387–390. (b) Akagi M.; Tejima S.; Haga M. *Chem. Pharm. Bull.* **1962**, *10*, 905–909.
16. Balmond, E. I.; Coe, D. M.; Galan, M. C.; McGarrigle, E. M. *Angew. Chem. Int. Ed.* **2012**, *51*, 9152–9155.
17. Li, Y.-F.; Yu, B.; Sun, J.-S.; Wang, R.-X. *Tetrahedron Lett.* **2015**, *56*, 3816–3819.

32 Phenyl 2-Azido-4,6-di-O-Benzyl-2,3-Dideoxy-3-Fluoro-1-Thio-α- and β-D-Glucopyranosides

Vojtěch Hamala, and Lucie Červenková Šťastná, Jindřich Karban*
Institute of Chemical Process Fundamentals of the CAS, Praha, Czech Republic

Jonathan Berry [a]
Otto Diels Institute of Organic Chemistry, Christian-Albrechts-University Kiel, Kiel, Germany

CONTENTS

Synthetic, selectively modified carbohydrates are indispensable tools to probe carbohydrate–protein interactions, to obtain their more potent agonists and inhibitors, or to improve stability and bioavailability of glycoconjugates designed for biological and biomedical applications. Deoxyfluoro carbohydrates resulting from OH→F

* Corresponding author: karban@icpf.cas.cz
[a] Checker: under supervision by Guillaume Despras: gdespras@oc.uni-kiel.de

replacement are very useful for this purpose because fluorine has a similar van der Waals radius (1.47 *vs* 1.52) and electronegativity (4.0 *vs* 3.5) as oxygen and, therefore, fluorination leads to minimal steric deviation from parent natural carbohydrates. Access to selectively fluorinated glycosyl donors and acceptors will greatly facilitate synthesis of fluorinated oligosaccharides. Herein we present the synthesis of a C-3 fluorinated, *O*-benzyl-protected phenyl 2-azido-2-deoxy-1-thioglucopyranoside gly-cosyl donor **4**. Compound **4** is a glucosamine donor in which the latent amino group at C-2 is presented by a nonparticipating azide functionality to facilitate synthesis of the 1,2-*cis*-α-glycosidic linkage.

Synthesis of **4** started from the known 1,6-anhydro-2-azido-4-*O*-benzyl-2-deoxy-β-D-glucopyranose **1**.[1,2] Fluorination with DAST using a minor modification of the reported procedure[3] gave 3-deoxyfluoro derivative **2**.[3] The *trans*-diaxial relationship of substituents at C-2, C-3, and C-4 is evident from the large values of geminal coupling constants $^2J_{C-2,F} = 25.3$ Hz and $^2J_{C-4,F} = 26.2$ Hz. Retentive deoxyfluorination of C-3 hydroxyl groups in 1,6-anhydroglucopyranose derivatives has precedents in the literature[4-8] and has been explained by the neighboring group participation of the C-2 and C-4 substituents.[6,8] Cleavage of the 1,6-anhydro ring with *S*-trimethylsilyl thiophenol/ZnI$_2$[9] yielded phenyl thioglycoside **3** as a mixture of anomers **α-3** and **β-3**, which were separated by column chromatography. Subsequent benzylation of each anomer gave the target 3-fluoro thioglycosides **α-4** and **β-4**.

EXPERIMENTAL

GENERAL METHODS

DAST was purchased from Acros Organics. THF and toluene were dried by distillation from sodium, CH$_2$Cl$_2$ and (CH$_2$Cl)$_2$ were dried by distillation from CaH$_2$ and stored over molecular sieves 3 Å. Ethyl acetate and petroleum ether (fraction boiling between 40 and 65°C) were distilled before use. Thin layer chromatography (TLC) was carried out with Merck DC Alufolien with Kieselgel F254, and spots were detected with an anisaldehyde solution in EtOH–AcOH–H$_2$SO$_4$.[10] UV detection at 254 nm was also used when applicable. Column chromatography was performed with silica gel 60 (70–230 mesh, Material Harvest). Solutions in organic solvents were dried (Na$_2$SO$_4$) and concentrated at <45°C. The ^{1}H (400.1 MHz), ^{13}C (100.6 MHz), and ^{19}F (376.4 MHz) NMR spectra were measured with a Bruker Avance 400 spectrometer at 25°C. The ^{1}H and ^{13}C NMR spectra were referenced to the residual protonated solvent peaks (δ_H/δ_C: CDCl$_3$, 7.26/77.16). The ^{19}F spectra were referenced to internal hexafluorobenzene (δ−163.00). Assignments were based on COSY, HSQC, and HMBC experiments. HRMS analyses were done using Bruker o-QIII, using APCI ionization in positive mode.

1,6-Anhydro-2-azido-4-*O*-benzyl-2,3-dideoxy-3-fluoro-β-d-glucopyranose (2)

Diethylaminosulfur trifluoride (DAST, 1.9 mL, 14.4 mmol) was added dropwise, at room temperature (rt) under argon, to a suspension of 1,6-anhydro-2-azido-4-*O*-benzyl-2-deoxy-β-d-glucopyranose (**1**)[2] (1.00 g, 3.61 mmol) in dry toluene (12 mL), leading to complete dissolution of **1**. The mixture was heated with stirring at 85°C (oil bath) for 30 min, when TLC (1:2 ethyl acetate–petroleum ether) indicated absence of the starting compound (R_f = 0.2) and formation of a product (R_f = 0.5). The mixture was cooled (ice water, formation of two phases was observed), methanol (1.2 mL) was added dropwise, and the stirring was continued for about 15 min. Saturated aq. solution of $NaHCO_3$ was added in portions until gas evolution ceased, the mixture was diluted with water (30 mL) and extracted with dichloromethane (4 × 50 mL). If a difficult emulsion forms during extraction, dilution with brine is recommended. The extracts were combined, washed with 1 M HCl and brine, dried, concentrated, and chromatography (1:5 ethyl acetate–petroleum ether, R_f = 0.25) gave **2** as colorless syrup (895 mg, 89%); $[\alpha]_D^{20}$ +22 (*c* 1.3, $CHCl_3$), lit.[3] $[\alpha]_D^{20}$ +22 (*c* 1.5, $CHCl_3$). NMR data matched those in lit.[3] Anal. calcd for $C_{13}H_{14}FN_3O_3$: C, 55.91; H, 5.05; N, 15.05. Found: C, 55.71; H, 4.84; N, 14.87.

Phenyl 2-azido-4-*O*-benzyl-2,3-dideoxy-3-fluoro-1-thio-α-d-glucopyranoside (α-3) and Phenyl 2-azido-4-*O*-benzyl-2,3-dideoxy-3-fluoro-1-thio-β-d-glucopyranoside (β-3)

S-Trimethylsilyl thiophenol (1.9 mL, 10. 0 mmol) and ZnI_2 (1.7 g, 5.3 mmol) were added sequentially, under argon at rt, to a solution of **2** (885 mg, 3.17 mmol) in dry 1,2-dichloroethane (9.5 mL), and the mixture was stirred vigorously, protected from light by aluminum foil, until TLC (1:5 ethyl acetate–petroleum ether R_f = 0.25) indicated the absence of the starting compound (~30 h) and formation of 6-*O*-trimethylsilylated intermediate (R_f = 0.7, spots of anomeric 6-OH products in varying intensity can also be detected near the origin). The mixture was diluted with dichloromethane (30 mL), filtered, the dichloromethane solution was washed with water (30 mL), and the water phase was extracted with dichloromethane (3 × 30 mL). Combined extracts were dried and concentrated. The residue was dissolved in methanol (50 mL), acetic acid (5 drops) was added, and the mixture was stirred until TLC (1:1 petroleum ether–diethyl ether) indicated disappearance of the trimethylsilylated intermediate (~2 h) and the presence of **β-3** (R_f = 0.47) and **α-3** (R_f = 0.34). After concentration, chromatography of the residue (4:1 → 3:2 petroleum ether–diethyl ether; cyclohexane can also be used instead of petroleum ether with a comparable efficiency) afforded first β-anomer **β-3** (296 mg, 24%), mp 77–79°C (heptane–EtOAc), $[\alpha]_D^{20}$ −1 (*c* 1.1, $CHCl_3$). ^{1}H NMR ($CDCl_3$, 400 MHz): δ 7.56–7.53 (m, 2*H*, CH$_{aromatic}$), 7.38–7.31 (m, 8*H*, CH$_{aromatic}$), 4.85 (dd, 1*H*, $J_{H,H}$ 11.3 Hz, $J_{H,F}$ 1.2 Hz, CH*H* Bn), 4.62 (d, 1*H*, $J_{H,H}$ 11.3 Hz, CH*H* Bn), 4.57 (ddd, 1*H*, $J_{H,F}$ 51.6 Hz, $J_{3,2}$ 9.1 Hz, $J_{3,4}$ 8.6 Hz, H-3), 4.44 (dd, 1*H*, $J_{1,2}$ 10.2 Hz, $J_{1,F}$ 0.9 Hz, H-1), 3.90 (ddd, 1*H*, $J_{6a,6b}$ 12.2 Hz, $J_{6a,OH}$ 6.0 Hz, $J_{6a,5}$ 2.6 Hz, H-6a), 3.72 (ddd, 1*H*, $J_{6b,6a}$ 12.2 Hz, $J_{6b,OH}$ 7.7 Hz, $J_{6b,5}$ 4.3 Hz, H-6b), 3.64 (ddd, 1*H*, $J_{4,F}$ 12.6 Hz, $J_{4,5}$ 9.8 Hz, $J_{4,3}$ 8.6 Hz, H-4),

3.45 (ddd, 1H, $J_{2,F}$ 12.9 Hz, $J_{2,1}$ 10.2 Hz, $J_{2,3}$ 9.1 Hz, H-2), 3.36 (dddd, 1H, $J_{5,4}$ 9.8 Hz, $J_{5,6b}$ 4.3 Hz, $J_{5,6a}$ 2.6 Hz, $J_{5,F}$ 1.3 Hz, H-5), 1.78 (dd, 1H, $J_{OH,6b}$ 7.7 Hz, $J_{OH,6a}$ 6.0 Hz, OH). ^{13}C NMR (CDCl$_3$, 101 MHz): δ 137.4 (C$_q$), 133.7 (2 CH$_{aromatic}$), 130.9 (C$_q$), 129.4 (2 CH$_{aromatic}$), 128.9 (CH$_{aromatic}$), 128.7 (2 CH$_{aromatic}$), 128.3 (3 CH$_{aromatic}$), 97.9 (d, $^1J_{C,F}$ 189.0 Hz, C-3), 85.5 (d, $^3J_{C,F}$ 7.3 Hz, C-1), 78.5 (d, $^3J_{C,F}$ 8.4 Hz, C-5), 74.9 (d, $^2J_{C,F}$ 16.7 Hz, C-4), 74.7 (d, $^4J_{C,F}$ 2.9 Hz, CH$_2$ Bn), 63.6 (d, $^2J_{C,F}$ 18.0 Hz, C-2), 61.9 (d, $^4J_{C,F}$ 1.4 Hz, C-6). ^{19}F NMR (CDCl$_3$, 376 MHz): δ −184.62 (dt, $J_{H3,F}$ 51.6 Hz, $J_{H2/H4,F}$ 12.7 Hz). HRMS-APCI: m/z [M − N$_2$ + H]$^+$ calcd for C$_{19}$H$_{21}$FNO$_3$S, 362.1220; found, 362.1219. Anal. calcd for C$_{19}$H$_{20}$FN$_3$O$_3$S: C, 58.60; H, 5.18; N, 10.79. Found: C, 58.23; H, 5.19; N, 10.52.

Next eluted were minor byproducts, followed by the syrupy α-anomer **α-3** (728 mg, 59%). Separation of anomers is recommended because TLC-detectable impurities eluted between them. Data for **α-3**: $[\alpha]_D^{20}$ +237 (c 0.36, CHCl$_3$). ^{1}H NMR (CDCl$_3$, 400 MHz): δ 7.48–7.45 (m, 2H, CH$_{aromatic}$), 7.39–7.30 (m, 8H, CH$_{aromatic}$), 5.56 (ddd, 1H, $J_{1,2}$ 5.7 Hz, $J_{1,F}$ 3.2 Hz, $J_{1,3}$ 0.6 Hz, H-1), 4.91 (ddd, 1H, $J_{3,F}$ 52.5 Hz, $J_{3,2}$ 10.2 Hz, $J_{3,4}$ 8.4 Hz, H-3), 4.91 (dd, 1H, $J_{H,H}$ 11.2 Hz, $J_{H,F}$ 1.2 Hz, CHH Bn), 4.67 (d, 1H, J_{HH} 11.2 Hz, CHH Bn), 4.23 (dddd, 1H, $J_{5,4}$ 10.0 Hz, $J_{5,6a}$ 3.4 Hz, $J_{5,6b}$ 3.2 Hz, $J_{5,F}$ 0.9 Hz, H-5), 4.02 (ddd, 1H, $J_{2,F}$ 11.5 Hz, $J_{2,3}$ 10.0 Hz, $J_{2,1}$ 5.7 Hz, H-2), 3.81–3.73 (m, 3H, H-4, H-6a,b), 1.53 (br s, 1 H, OH). ^{13}C NMR (CDCl$_3$, 101 MHz): δ 137.5 (C$_q$), 132.7 (2 CH$_{aromatic}$), 132.6 (C$_q$), 129.4, 128.7, 128.42 (3 × 2 CH$_{aromatic}$), 128.36, 128.3 (2 × 1 CH$_{aromatic}$), 95.6 (d, $^1J_{C,F}$ 186.3 Hz, C-3), 86.6 (d, $^3J_{C,F}$ 8.1 Hz, C-1), 75.3 (d, $^2J_{C,F}$ 16.5 Hz, C-4), 74.8 (d, $^4J_{C,F}$ 2.9 Hz, CH$_2$Bn), 71.7 (d, $^3J_{C,F}$ 8.0 Hz, C-5), 62.5 (d, $^2J_{C,F}$ 17.5 Hz, C-2), 61.4 (C-6). ^{19}F NMR (CDCl$_3$, 376 MHz): δ −189.58 (dddd, $J_{H3,F}$ 52.8 Hz, $J_{H4,F}$ 14.2 Hz, $J_{H2,F}$ 11.5 Hz, $J_{H1,F}$ 3.2 Hz). HRMS-APCI: m/z [M − N$_2$ + H]$^+$ calcd for C$_{19}$H$_{21}$FNO$_3$S, 362.1220; found, 362.1225. Anal. calcd for C$_{19}$H$_{20}$FN$_3$O$_3$S: C, 58.60; H, 5.18; N, 10.79. Found: C, 58.46; H, 5.17; N, 10.49.

PHENYL 2-AZIDO-4,6-DI-*O*-BENZYL-2,3-DIDEOXY-3-FLUORO-1-THIO-α-D-GLUCOPYRANOSIDE (α-4)

Sodium hydride (60% suspension in oil, 91 mg, 2.27 mmol) was added (under argon at −25°C) to a stirred solution of **α-4** (590 mg, 1.51 mmol) in dry THF (10 mL). The mixture was stirred for 30 min and benzyl bromide (0.3 mL, 2.5 mmol) and a catalytic amount of tetrabutylammonium iodide (TBAI) were added. The mixture was allowed to warm to rt and stirred overnight with exclusion of moisture. TLC (1:10 EtOAc–petroleum ether) indicated the absence of the starting material and formation of one product (R_f = 0.3). The mixture was cooled to 0°C (ice water) and MeOH (0.5 mL) was added dropwise, to quench the reaction. The mixture was diluted with dichloromethane (30 mL) and washed with water (30 mL). The water phase was extracted with dichloromethane (3 × 30 mL), the organic phases were combined, dried, and concentrated. Chromatography (1:10 ethyl acetate–petroleum ether) afforded syrupy **α-4** (575 mg, 79%), $[\alpha]_D^{20}$ +208 (c 0.4, CHCl$_3$). ^{1}H NMR (CDCl$_3$, 400 MHz): δ 7.50–7.46 (m, 2H, CH$_{aromatic}$), 7.36–7.23 (m, 13H, CH$_{aromatic}$), 5.60 (dd, 1H, $J_{1,2}$ 5.7 Hz, $J_{1,F}$ 3.3 Hz, H-1), 4.87 (ddd, 1H, $J_{3,F}$ 52.8 Hz, $J_{3,2}$ 10.1 Hz,

$J_{3,4}$ 8.4 Hz, H-3), 4.87 (dd, 1H, $J_{H,H}$ 11.0 Hz, $J_{H,F}$ 1.2 Hz, CHH O-4Bn), 4.60 (d, 1H, $J_{H,H}$ 12.0 Hz, CHH O-6Bn), 4.53 (d, 1H, $J_{H,H}$ 11.0 Hz, CHH O-4Bn), 4.44 (d, 1H, $J_{H,H}$ 12.0 Hz, CHH O-6Bn), 4.35 (ddd, 1H, $J_{5,4}$ 10.0 Hz, $J_{5,6a}$ 3.8 Hz, $J_{5,6b}$ 2.1 Hz, H-5), 4.06 (ddd, 1H, $J_{2,F}$ 11.5 Hz, $J_{2,3}$ 10.1 Hz, $J_{2,1}$ 5.7 Hz, H-2), 3.86 (ddd, 1H, $J_{4,F}$ 14.4 Hz, $J_{4,5}$ 10.0 Hz, $J_{4,3}$ 8.4 Hz, H-4), 3.80 (dd, 1H, $J_{6a,6b}$ 10.8 Hz, $J_{6a,5}$ 3.8 Hz, H-6a), 3.64 (dt, 1H, $J_{6b,6a}$ 10.8 Hz, $J_{6b,5/F}$ 2.1 Hz, H-6b). ^{13}C NMR (CDCl$_3$, 101 MHz): δ 137.9 (C$_q$ O-6Bn), 137.6 (C$_q$ O-4Bn), 133.0 (C$_q$), 132.4, 129.3, 128.56, 128.55, 128.23 (5 × 2 CH$_{aromatic}$), 128.15 (CH$_{aromatic}$), 128.13 (CH$_{aromatic}$), 128.0 (CH$_{aromatic}$), 95.7 (d, $^1J_{C,F}$ 186.1 Hz, C-3), 86.7 (d, $^3J_{C,F}$ 8.1 Hz, C-1), 75.9 (d, $^2J_{C,F}$ 16.6 Hz, C-4), 74.8 (d, $^4J_{C,F}$ 2.4 Hz, CH$_2$ O-4Bn), 76.6 (CH$_2$ O-6Bn), 71.1 (d, $^3J_{C,F}$ 8.4 Hz, C-5), 68.1 (C-6), 62.4 (d, $^2J_{C,F}$ 17.3 Hz, C-2). ^{19}F NMR (CDCl$_3$, 376 MHz): δ −189.38 (dddd, $J_{H3,F}$ 52.8 Hz, $^3J_{H4,F}$ 14.4 Hz, $J_{H2,F}$ 11.5 Hz, $J_{H1,F}$ 3.3 Hz). HRMS-APCI: m/z [M − N$_2$ + H]$^+$ calcd for C$_{26}$H$_{27}$FNO$_3$S, 452.1695; found, 452.1701. Anal. calcd for C$_{26}$H$_{26}$FN$_3$O$_3$S: C, 65.12; H, 5.46; N, 8.76. Found: C, 64.95; H, 5.53; N, 8.53.

PHENYL 2-AZIDO-4,6-DI-O-BENZYL-2,3-DIDEOXY-3-FLUORO-1-THIO-β-D-GLUCOPYRANOSIDE (β-4)

Benzylation of **β-3** (235 mg, 0.60 mmol), as described before, gave, after chromatography, **β-4** (206 mg, 71%), mp 51–54°C (from heptane–MTBE), [α]$_D^{20}$ +0.5 (c 1.6, CHCl$_3$). ^{1}H NMR (CDCl$_3$, 400 MHz): δ 7.60 (dd, 2 H, $J_{H,H}$ 8.1 Hz, $J_{H,H}$ 1.6 Hz, CH$_{aromatic}$), 7.38–7.24 (m, 13H, CH$_{aromatic}$), 4.80 (dd, 1H, $J_{H,H}$ 11.1 Hz, $J_{H,F}$ 0.8 Hz, CHH O-4Bn), 4.61 (d, 1H, $J_{H,H}$ 12.0 Hz, CHH O-6Bn), 4.56 (d, 1H, $J_{H,H}$ 11.1 Hz, CHH O-4Bn), 4.54 (d, 1H, $J_{H,H}$ 12.0 Hz, CHH O-6Bn), 4.53 (ddd, 1H, $J_{3,F}$ 51.6 Hz, $J_{3,2}$ 9.0 Hz, $J_{3,4}$ 8.8 Hz, H-3), 4.40 (dd, 1H, $J_{1,2}$ 10.2 Hz, $J_{1,F}$ 0.9 Hz, H-1), 3.79 (dt, 1H, $J_{6a,6b}$ 11.0 Hz, $J_{6a,5/F}$ 1.9 Hz, H-6a), 3.73 (dd, 1H, $J_{6a,6b}$ 11.0 Hz, $J_{6b,5}$ 4.3 Hz, H-6b), 3.71 (ddd, 1H, $J_{4,F}$ 12.9 Hz, $J_{4,5}$ 9.8 Hz, $J_{4,3}$ 8.8 Hz, H-4), 3.47 (ddd, 1H, $J_{2,F}$ 12.9 Hz, $J_{2,1}$ 10.2 Hz, $J_{2,3}$ 9.0 Hz, H-2), 3.44 (ddd, 1H, $J_{5,4}$ 9.8 Hz, $J_{5,6b}$ 4.3 Hz, $J_{5,6a}$ 1.9 Hz, H-5). ^{13}C NMR (CDCl$_3$, 101 MHz): δ 138.2 (C$_q$ O-6Bn), 137.6 (C$_q$ O-4Bn), 133.8 (2 CH$_{aromatic}$), 131.1 (C$_q$), 129.2 (2 CH$_{aromatic}$), 128.7 (CH$_{aromatic}$), 128.58, 128.55, 128.22 (3 × 2 CH$_{aromatic}$), 128.15, 127.84 (2 × 1 CH$_{aromatic}$), 127.79 (2 CH$_{aromatic}$), 97.9 (d, $^1J_{C,F}$ 188.9 Hz, C-3), 85.4 (d, $^3J_{C,F}$ 7.2 Hz, C-1), 78.3 (d, $^3J_{C,F}$ 8.7 Hz, C-5), 75.2 (d, $^2J_{C,F}$ 16.7 Hz, C-4), 74.7 (d, $^4J_{C,F}$ 2.6 Hz, CH$_2$ O-4Bn), 73.6 (CH$_2$ O-6Bn), 68.6 (d, $^4J_{C,F}$ 1.6 Hz, C-6), 63.5 (d, $^2J_{C,F}$ 17.8 Hz, C-2). ^{19}F NMR (CDCl$_3$, 376 MHz): δ −184.66 (dt, $J_{H3,F}$ 51.6 Hz, $J_{H2/H4,F}$ 12.9 Hz). HRMS-APCI: m/z [M − N$_2$ + H]$^+$ calcd for C$_{26}$H$_{27}$FNO$_3$S, 452.1695; found, 452.1693. Anal. calcd for C$_{26}$H$_{26}$FN$_3$O$_3$S: C, 65.12; H, 5.46; N, 8.76. Found: C, 64.87; H, 5.47; N, 8.38.

ACKNOWLEDGMENTS

Support of The Czech Science Foundation is gratefully acknowledged (grant no. 17-18203S)

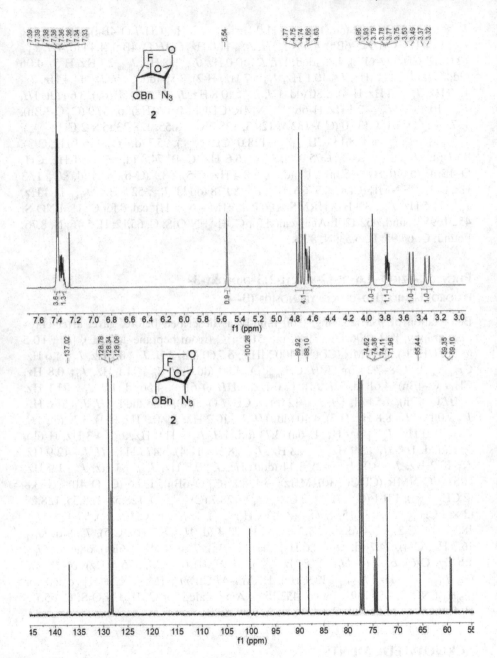

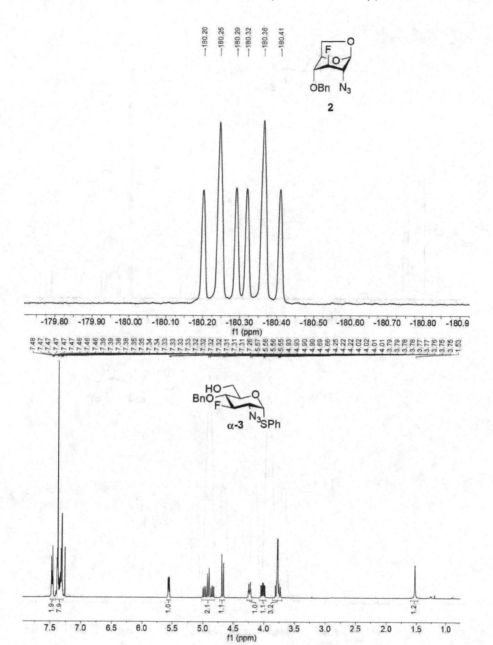

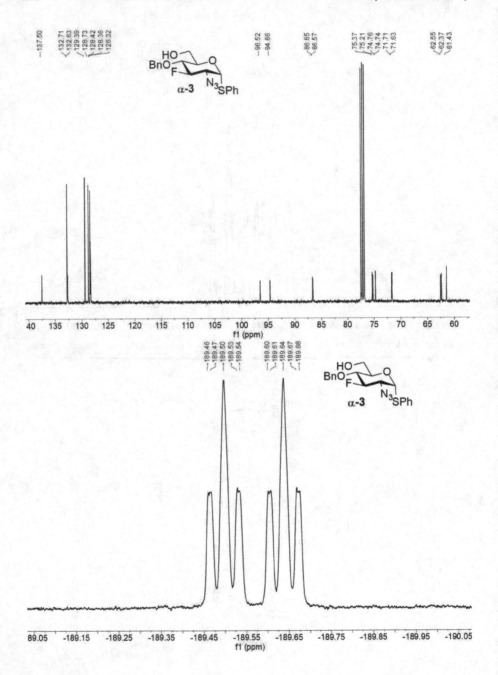

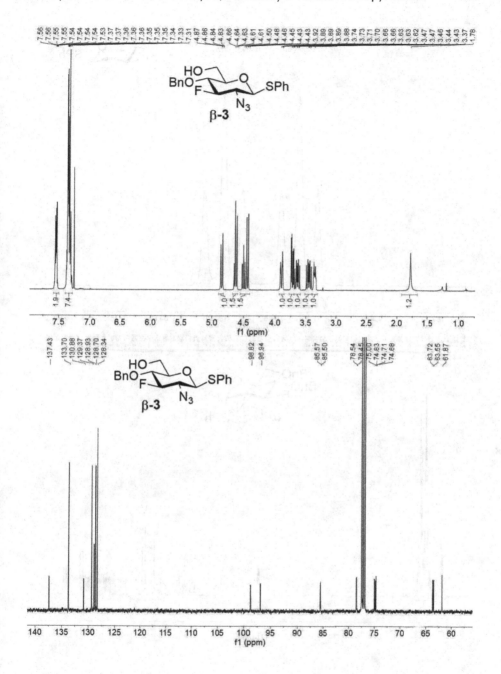

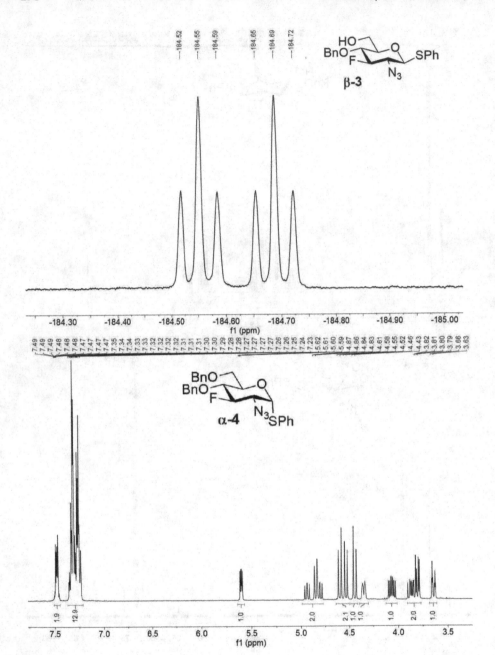

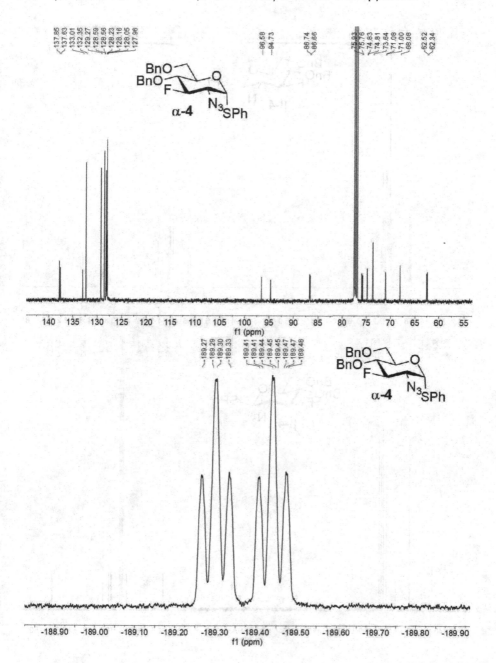

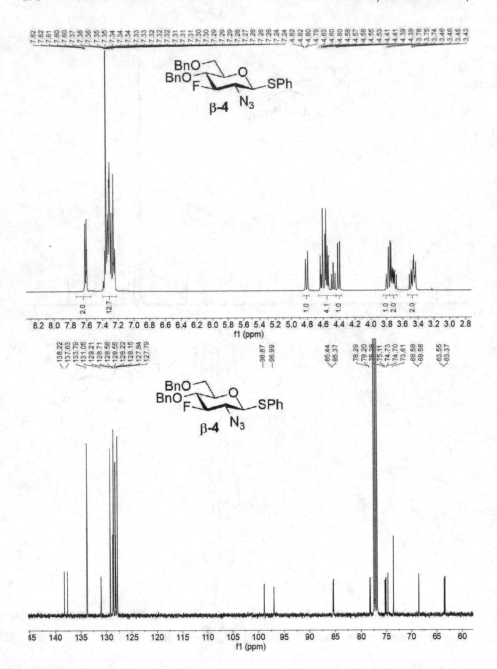

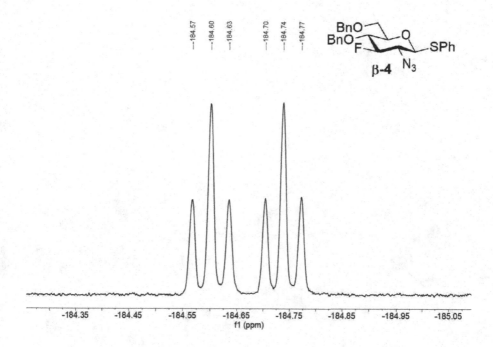

REFERENCES

1. Paulsen, H.; Stenzel, W. *Chem. Ber.* **1978**, *111*, 2334–2347.
2. Karban, J.; Buděšínský, M.; Černý, M.; Trnka, T. *Collect. Czech. Chem. Commun.* **2001**, *66*, 799–819.
3. Faghih, R.; Escribano, F. C.; Castillon, S.; Garcia, J.; Olesker, A.; Thang, T. T. *J. Org. Chem.* **1986**, *51*, 4558–4564.
4. Sarda, P.; Escribano, F. C.; Alves, R. J.; Olesker, A.; Lukacs, G. *J. Carbohydr. Chem.* **1989**, *8*, 115–123.
5. Mtashobya, L.; Quiquempoix, L.; Linclau, B. *J. Fluorine Chem.* **2015**, *171*, 92–96.
6. Horník, Š.; Červenková Šťastná, L.; Cuřínová, P.; Sýkora, J.; Káňová, K.; Hrstka, R.; Císařová, I.; Dračínský, M.; Karban, J. *Beilstein J. Org. Chem.* **2016**, *12*, 750–759.
7. Denavit, V.; Lainé, D.; St-Gelais, J.; Johnson, P. A.; Giguère, D. *Nat. Commun.* **2018**, *9*, 4721.
8. Quiquempoix, L.; Wang, Z.; Graton, J.; Latchem, P. G.; Light, M.; Le Questel, J.-Y.; Linclau, B. *J. Org. Chem.* **2019**, *84*, 5899–5906.
9. Wang, L.-X.; Sakairi, N.; Kuzuhara, H. *J. Chem. Soc., Perkin Trans.* **1990**, *1*, 1677–1682.
10. Pirrung, M. C. *The Synthetic Organic Chemist's Companion*, John Wiley and Sons: **2006**; p 171.

33 An Alternative Synthesis of 3-Azidopropyl 2,4,6-Tri-*O*-Benzyl-β-D-Galactopyranosyl-(1→4)-2,3,6-Tri-*O*-Benzyl-β-D-Glucopyranoside

Mark Reihill, Aisling Ní Cheallaigh*,
and Stefan Oscarson
Centre for Synthesis and Chemical Biology,
University College Dublin,
Belfield, Dublin 4, Ireland

Caecilie M. M. Benckendorff[a]
Lennard-Jones Laboratory,
Keele University,
Keele, United Kingdom

CONTENTS

---

* Corresponding author: aisling.nicheallaigh@ucd.ie; present contact: aisling.nicheallaigh@manchester.ac.uk

[a] Checker: under supervision of Dr. Gavin Miller: c.m.m.benckendorff@keele.ac.uk

Lactose is most commonly known for its presence in milk. The lactose disaccharide is also a prevalent downstream component of human milk oligosaccharides (HMOs).[1,2] HMOs have various biological functions such as modulation of immune response, antimicrobial effects, and probiotic activity.[3] Lactose is also commonly displayed as a component in mammalian and bacterial surface antigens.[4–6] Some of these structures, such as Globo-H, are known to be overexpressed in cancerous cells providing potential candidates for anticancer vaccine development.[7–10] Others are associated with increased susceptibility to certain pathogens, for example, type 1 Lewis b and H-antigens, which are receptors for *Helicobacter pylori* and Norwalk virus.[11,12] Synthetic efforts toward such structures have provided invaluable structure–function tools, in particular for lectin binding interaction studies.[13] Extension at the 3′-position of lactose have been utilized in a number of syntheses toward complex oligosaccharide targets.[14–16] Methyl and benzyl glycosides of lactose are commonly used in synthetic routes.[17–22] However, the azidopropyl aglycone in **4** provides advantages over the corresponding methyl and benzyl derivatives as it can be easily further functionalized at a later stage. Due to this additional flexibility and the prevalence of 3′-O-linked lactose, a reliable, undemanding synthetic route to **4** would be advantageous.

The synthesis of **4** has previously been achieved through tin activation of **1** followed by regioselective allylation of the 3′-position.[23] Subsequent benzylation, followed by deallylation provided **4** in an overall yield of 18%.[14] This route was later improved by Cao and coworkers who used a PMB group for the 3′-protection, delivering **4** in 41% overall yield.[15] The highest yielding route toward **4** is a 5-step synthesis from **1**, with an overall yield of 53%. This approach utilizes a 3′,4′-O-benzylidene acetal which subsequently undergoes reductive ring-opening with $NaCNBH_3$ and $HCl–Et_2O$ to give **4**.[16] Although higher yielding, this route is more cumbersome and laborious. The introduction of a 3′,4′-O-benzylidene acetal leads to a mixture of *endo* and *exo* diastereomers. This, coupled with the lower regioselectivity observed during reductive ring-opening of dioxolane versus dioxanes, results in a mixture of products.[24–27]

Our current approach generates highly regioselectively protected material that is easily purified. Tin activation with Bu_2SnO was performed in methanol followed by regioselective introduction of a naphthylmethyl ether to the 3′-position with 2-(bromomethyl)naphthalene (1.7 equiv) and CsF in DMF.[28,29] Compound **2** was isolated in a 45% yield following flash chromatography on silica gel. Although the yield for this step was relatively low, the advantage of the naphthylmethyl group is the large R_f differences between the unchanged starting material, a di-naphthylmethylated side product, and **2**. This greatly simplifies the isolation of the desired material. Benzylation of the remaining free positions in **2** with NaH and BnBr in DMF proceeded smoothly to furnish **3** in an 83% yield. Whilst performing oxidative cleavage of naphthylmethyl ethers with DDQ in $CH_2Cl_2–H_2O$, we commonly observed the loss of benzyl ethers as side products. When phosphate-buffered saline (PBS) buffer (pH 7.5) was used in place of water in the reaction, greater control of the pH could be achieved, thus limiting the loss of benzyl groups. These conditions produced **4** in a 69% yield which was fully characterized for the first time. This reaction sequence, when performed on 1 g scale, afforded the title compound in a 26% overall yield over 4 steps.

EXPERIMENTAL

GENERAL METHODS

All purchased chemicals were used without further purification unless otherwise stated. Anhydrous solvents were dispensed from a PureSolv™ Solvent Purification System and were dried over activated 4 Å molecular sieves. Concentration of solutions in organic solvents was performed under reduced pressure at 40–60°C. Optical rotations were recorded on automatic polarimeter Rudolph autopol I (concentration in g/100 mL) I or PerkinElmer polarimeter (Model 343) at the sodium D-line (589 nm) using a 1-dm cell. $[\alpha]_D^{26}$ values are given in units of 10^{-1} deg cm^2 g^{-1}. Reactions were monitored by thin-layer chromatography on Merck pre-coated plates (5 × 10 cm, layer thickness 0.25 mm, silica gel 60F254). Spots were detected by charring with 5% H_2SO_4 in methanol or ethanol. Column chromatography was performed by gradient elution from columns of silica gel (0.040–0.063 mm). NMR spectra were recorded (400 and 500 MHz for ^{1}H NMR, 101 and 126 MHz for ^{13}C NMR) using a Varian Inova spectrometer or a Bruker AVIII 400 instrument and standard Bruker NMR software. ^{1}H NMR spectra were referenced to $\delta = 4.87$ ppm for spectra in CD$_3$OD or $\delta = 7.26$ ppm in CDCl$_3$. ^{13}C NMR spectra were referenced to $\delta = 49$ ppm for spectra in CD$_3$OD or to $\delta = 77.00$ ppm for solutions in CDCl$_3$. Full structural assignments were performed with the assistance of 2-D NMR experiments: COSY, DEPT, HSQC, and HMBC experiments and MestReNova processing software. High-resolution mass spectrometry (HRMS) data were recorded on a Waters Micromass LCT LC-TOF instrument using electrospray ionization (ESI) in positive mode. PBS solution (pH 7.5) was prepared by dissolving 1 PBS tablet (purchased from Sigma-Aldrich) in 200-mL deionized water.

3-AZIDOPROPYL β-D-GALACTOPYRANOSYL-(1→4)-β-D-GLUCOPYRANOSIDE (1)

3-Azidopropyl 2,3,4,6-tetra-*O*-acetyl-β-D-galactopyranosyl-(1→4)-2,3,6-tri-*O*-acetyl-β-D-glucopyranoside (1.00 g, 1.39 mmol) was deacetylated following literature methodology.*[30] Crude compound **1** (583 mg, 96%), obtained as a hygroscopic white foam,

---

* Deviation from literature procedure: Cotton wool plug was used in absence of C_{18} reverse phase silica gel and, following filtration, the solution was simply concentrated to give crude **1**; lyophilization was not performed.

was used in the next step without further purification. $R_f = 0.3$ (7:2 EtOAc–MeOH); $[\alpha]_D^{26}$ 1.0 (c 2.0, H_2O); Lit.[30, 31] $[\alpha]_D$ not reported.

^1H NMR (400 MHz, CD_3OD) δ 4.31 (d, $1H$, H-1, J 7.5 Hz), 4.24 (d, $1H$, H-1', J 7.8 Hz), 3.91 (dt, $1H$, $OCH_{2(Linker)}$, J 10.0, 6.0 Hz), 3.85 (dd, $1H$, H6'a, J 12.1, 2.5 Hz), 3.79 (dd, $1H$, H6'b J 12.1, 4.1 Hz), 3.76–3.70 (m, $2H$, H4', H6a), 3.66–3.58 (m, $2H$, H6b, $OCH_{2(Linker)}$), 3.56–3.39 (m, $5H$, H2, H3, H3', H4, H5'), 3.35 (ddd, $1H$, H5, J 9.4, 4.0, 2.6 Hz), 3.31 (m, $2H$, $CH_2N_{3(Linker)}$), 3.19 (dd, $1H$, H2', J 8.9, 8.0 Hz), 1.82 (p, $2H$, $CH_{2(Linker)}$, J 6.6 Hz,) ^{13}C NMR (101 MHz, CD_3OD) δ 105.1 (C1), 104.3 (C1'), 80.5, 77.1, 76.4, 74.8, 72.5 (C2, C3, C3', C4, C5'), 76.4 (C5), 74.7 (C2'), 70.3 (C4'), 67.6 ($OCH_{2(Linker)}$), 62.5 (C6'), 61.8 (C6), 49.3 (CH_2N_3), 30.2 ($CH_{2(Linker)}$). HRMS (ESI) m/z calcd for $C_{15}H_{27}N_3O_{11}$ $[M+Na]^+$, 448.1538; found, 448.1552. The ^1H NMR data listed previously were confirmed by 2D ^1H–^{13}C HSQC experiments and are more complete and corrected, compared to those reported.[30]

3-AZIDOPROPYL 3-*O*-NAPHTHYLMETHYL-β-D-GALACTOPYRANOSYL-(1→4)-β-D-GLUCOPYRANOSIDE (2)

Bu_2SnO (580 mg, 2.33 mmol) was added under N_2 to a suspension of 1 (583 mg, 1.37 mmol) in dry methanol (7 mL)[14] and the mixture was refluxed for 16 h. After concentration, the residue was dried at 18°C and reduced pressure for 7 h and, with exclusion of moisture, was dissolved in anhydrous DMF (7 mL). CsF (270 mg, 1.78 mmol) and 2-(bromomethyl)naphthalene (515 mg, 2.33 mmol) were added, and the mixture was stirred at 60°C for 16 h, when no further consumption of the starting material was observed TLC, 7:2 EtOAc–MeOH; 1: $R_f = 0.2$, 2: $R_f = 0.7$, di-naphthylmethylated side product: $R_f = 0.9$). After cooling to room temperature, a white precipitate formed. The suspension was filtered through Celite® and the Celite® was washed with MeOH (10 mL). The filtrate and the washings were combined and concentrated. Dry-loaded chromatography (8 g silica gel, EtOAc→95:5 EtOAc–MeOH) gave 2 (348 mg, 45%) as white foam, $R_f = 0.7$ (7:2 EtOAc/MeOH); $[\alpha]_D^{26}$ +6.5 (c 1.24, MeOH); ^1H NMR (400 MHz, CD_3OD) δ 7.86 (s, $1H$, Ar-H), 7.79 (d, $3H$, J 7.8 Hz, Ar-H), 7.54 (dd, $1H$, J 8.5, 1.6 Hz, Ar-H), 7.42 - 7.40 (m, $2H$, Ar-H), 4.95–4.73 (m, $2H$, OCH_2Ar), 4.34 (d, $1H$, J 7.8 Hz, H-1'), 4.23 (d, $1H$, J 7.8 Hz, H-1), 4.01 (d, $1H$, J 2.8 Hz, H-4'), 3.90 (dt, $1H$, J 8.2, 5.1 Hz, $OCH_{2(A)Linker}$), 3.87–3.81 (m, $2H$, H-6'), 3.74 (dd, $1H$, J 11.5, 7.5 Hz, H-6a), 3.69 (d, $1H$, J 1.8 Hz, H-5), 3.64 (dd, $1H$, J 11.6, 4.9 Hz, H-6b), 3.61–3.55 (m, $1H$, $OCH_{2(B)Linker}$), 3.55–3.44 (m, $2H$, H-3, H-5'), 3.43–3.33 (m, $3H$, H-4, $CH_2N_{3(Linker)}$), 3.19 (t, $1H$, H-2), 1.81 (p, $2H$, J 6.6 Hz, $-CH_2-_{(Linker)}$); 13C NMR (126 MHz, CD_3OD) δ 135.9 (Ar-C_{quat}), 133.4 (Ar-C_{quat}), 133.1 (Ar-C_{quat}), 127.6 (Ar-CH), 127.5 (Ar-CH), 127.3 (Ar-CH), 126.3 (Ar-CH), 125.8 (Ar-CH), 125.7 (Ar-CH), 125.5 (Ar-CH), 103.7 (C-1'), 102.9 (C-1), 80.8 (C-3'), 79.4 (C-4), 75.5 (C-3), 75.0 (C-5'), 75.0 (C-5), 73.3 (C-2), 71.3 (ArCH_2), 70.5 (C-2'), 66.2 ($OCH_{2(Linker)}$), 65.7 (C-4'), 61.1 (C-6), 60.5 (C-6'), 48.0 ($CH_2N_{3(Linker)}$), 28.8 ($CH_{2(Linker)}$); HRMS (ESI) m/z calcd for $C_{26}H_{35}N_3O_{11}$ $[M+Na]^+$, 588.2169; found 588.2189. Anal. calcd for $C_{26}H_{35}N_3O_{11}$: C, 55.21; H, 6.24; N, 7.43. Found: C, 55.23; H, 6.31; N, 6.87.

3-AZIDOPROPYL 2,4,6-TRI-O-BENZYL-β-D-GALACTOPYRANOSYL-(1→4)-2,3,6-TRI-O-BENZYL-β-D-GLUCOPYRANOSIDE (4)

Compound **2** (566 mg, 1.00 mmol) was dissolved in dry DMF (10 mL); the solution was cooled to 0°C and NaH (60% dispersion in mineral oil, 602 mg, 15.7 mmol) was added. After stirring for 20 min, BnBr (1.80 mL, 15.0 mmol) was added and the mixture was allowed to warm to room temperature, at which it was stirred for 2 h. The mixture was then cooled to 0°C and the reaction was quenched with MeOH. The mixture was concentrated and water (100 mL) was added. The product was extracted with EtOAc (3× 100 mL) and the combined organic phase was washed with brine (100 mL), dried over MgSO$_4$, filtered, and concentrated. Flash chromatography (toluene→4:1 toluene–EtOAc) gave 3-azidopropyl 2,4,6-tri-O-benzyl-3-O-naphthylmethyl-β-D-galactopyranosyl-(1→4)-2,3,6-tri-O-benzyl-β-D-glucopyranoside (**3**) as white waxy solid (917 mg, 83%). $R_f = 0.5$ (9:1 toluene–EtOAc,; $[\alpha]_D^{20}$ +12 (c 1.0, CHCl$_3$); ^{1}H NMR (500 MHz, CDCl$_3$) δ 7.82 (m, 1H, Ar-H), 7.80–7.74 (m, 2H, Ar-H), 7.68 (m, 1H, Ar-H), 7.50–7.40 (m, 3H, Ar-H), 7.37–7.08 (m, 30H, Ar-H), 5.05–4.98 (m, 2H, 2× OCH$_2$Ar), 4.89–4.67 (m, 7H, 7× OCH$_2$Ar), 4.59 (d, 1H, J 11.5 Hz, OCH$_2$Ar), 4.52 (d, 1H, J 12.1 Hz, OCH$_2$Ar), 4.44 (d, 1H, J 7.7 Hz, H-1'), 4.41–4.30 (m, 3H, H-1, OCH$_2$Ar), 4.25 (d, 1H, J 11.8 Hz, OCH$_2$Ar), 4.02–3.89 (m, 3H, H-4, H-4', OCH$_{2(A)Linker}$), 3.84–3.75 (m, 2H, H-2', H-6a), 3.71 (dd, 1H, J 10.8, 1.8 Hz, H-6b), 3.65–3.51 (m, 3H, H-3', H-6'a, OCH$_{2(B)Linker}$), 3.45 (dd, 1H, J 9.7, 2.9 Hz, H-3'), 3.42–3.30 (m, 6H, H-2, H-5, H-5', H-6'b, CH$_2$N$_{3(Linker)}$), 1.96–1.81 (m, 2H, CH$_{2(Linker)}$); ^{13}C NMR (126 MHz, CDCl$_3$) δ 139.3 (Ar-C$_{quat}$), 139.2 (Ar-C$_{quat}$), 138.9 (Ar-C$_{quat}$), 138.8 (Ar-C$_{quat}$), 138.5 (Ar-C$_{quat}$), 138.2 (Ar-C$_{quat}$), 136.1 (Ar-C$_{quat}$), 133.4 (Ar-C$_{quat}$), 133.1 (Ar-C$_{quat}$), 128.5, 128.5, 128.4, 128.4, 128.3, 128.2, 128.2, 128.1, 128.0, 128.0, 128.0, 128.0, 127.9, 127.8, 127.8, 127.7, 127.7, 127.6, 127.6, 127.5, 127.2, 126.2, 126.2, 126.0, 125.7 (Ar-CH), 103.7 (C-1), 102.9 (C-1'), 83.2 (C-3), 82.6 (C-3'), 81.9 (C-2), 80.1 (C-2'), 76.8 (C-4), 75.5 (Ar-CH$_2$), 75.4 (Ar-CH$_2$), 75.3 (Ar-CH$_2$), 75.2 (C-5), 74.9 (Ar-CH$_2$), 73.7 (C-4'), 73.6 (Ar-CH$_2$), 73.24 (Ar-CH$_2$), 73.15 (C-5'), 72.7 (Ar-CH$_2$), 68.4 (C-6), 68.2 (C-6'), 66.6 (OCH$_{2(Linker)}$), 48.5 (CH$_2$N$_{3(Linker)}$), 29.4 (CH$_{2(Linker)}$); HRMS (ESI) m/z calcd for C$_{68}$H$_{71}$N$_3$O$_{11}$ [M + Na]$^+$, 1128.4986; found 1128.5029. Anal. calcd for C$_{68}$H$_{71}$N$_3$O$_{11}$: C, 73.82; H, 6.47; N, 3.80. Found: C, 73.37; H, 6.58; N, 3.51.

Compound **3** (1.11 g, 1.00 mmol) was dissolved in CH$_2$Cl$_2$ (90 mL) and PBS buffer (pH 7.5, 9 mL) was added. The mixture was cooled to 0°C followed by addition of DDQ (442 mg, 1.94 mmol). The reaction vessel was covered with aluminum foil and the mixture was stirred at 0°C for 2 h and diluted with CH$_2$Cl$_2$. Ten percent aq. Na$_2$S$_2$O$_3$ (100 mL) was added and the stirring was continued for 30 min. The organic layer was separated, and the aqueous phase was extracted with CH$_2$Cl$_2$ (2× 100 mL). The combined organic layer was washed with brine (100 mL), dried over MgSO$_4$, filtered, concentrated, and chromatography (toluene → 4:1 toluene–EtOAc) gave title compound **4** as colorless syrup (668 mg, 69%). $[\alpha]_D^{20}$ +2.1 (c 1.0, CHCl$_3$); literature: $[\alpha]_D^{25}$ +3.4 (c 1.0, CHCl$_3$);[14] ^{1}H NMR (500 MHz, CDCl$_3$) δ 7.42–7.21 (m, 27H, Ar-H), 7.19–7.14 (m, 3H, Ar-H), 5.02 (d, 1H, J 10.7 Hz, OCH$_2$Ar), 4.87–4.70 (m, 5H, 5× OCH$_2$Ar), 4.67 (d, 1H, J 11.4 Hz, OCH$_2$Ar), 4.61 (d, 1H, J 11.6 Hz, OCH$_2$Ar), 4.57

(d, 1*H*, *J* 12.1 Hz, OC*H*$_2$Ar), 4.47–4.32 (m, 4*H*, H-1, H-1′, 2× OC*H*$_2$Ar), 4.27 (d, 1*H*, *J* 11.8 Hz, OC*H*$_2$Ar), 4.04–3.91 (m, 2*H*, H-4, OC*H*$_{2(A)Linker}$), 3.83 (d, 1*H*, *J* 2.5 Hz, H-4′), 3.79 (dd, 1*H*, *J* 10.9, 4.2 Hz, H-6a), 3.73 (dd, 1*H*, *J* 10.9, 1.8 Hz, H-6b), 3.67–3.53 (m, 3*H*, H-3, H-6′a, OC*H*$_{2(B)Linker}$), 3.52–3.47 (m, 2*H*, H-2′, H-3′), 3.44–3.36 (m, 6*H*, H-2, H-5, H-5′, H-6′b, C*H*$_2$N$_{3(Linker)}$), 2.17 (d, 1*H*, *J* 5.1 Hz, O*H*), 1.94–1.82 (m, 2*H*, C*H*$_{2(Linker)}$); ^{13}C NMR (126 MHz, CDCl$_3$) δ 139.2 (Ar-C_{quat}), 138.8 (Ar-C_{quat}), 138.7 (Ar-C_{quat}), 138.6 (Ar-C_{quat}), 138.3 (Ar-C_{quat}), 138.2 (Ar-C_{quat}), 128.6, 128.5, 128.5, 128.5, 128.4, 128.2, 128.1, 128.0, 127.9, 127.9, 127.8, 127.8, 127.7, 127.7, 127.7, 127.3 (Ar-*C*H), 103.7 (C-1), 102.8 (C-1′), 83.0 (C-3), 81.9 (C-2), 80.8 (C-2′), 76.8 (C-4), 76.1 (C-4′), 75.5 (Ar-*C*H$_2$), 75.3 (C-5), 75.2 (Ar-*C*H$_2$), 75.2 (Ar-*C*H$_2$), 75.1 (Ar-*C*H$_2$), 74.2 (C-3′), 73.5 (Ar-*C*H$_2$), 73.4 (C-5′), 73.3 (Ar-*C*H$_2$), 68.4 (C-6), 68.1 (C-6′), 66.6 (OC*H*$_{2(Linker)}$), 48.5 (*C*H$_2$N$_{3(Linker)}$), 29.4 (*C*H$_{2(Linker)}$). NMR data are consistent with those reported;[14] HRMS (ESI) *m/z* calcd for C$_{57}$H$_{63}$N$_3$O$_{11}$ [*M* + Na]$^+$ 988.4360; found 988.4402. Anal. calcd for C$_{57}$H$_{63}$N$_3$O$_{11}$: C, 70.86; H, 6.57; N, 4.35. Found: C, 70.69; H, 6.44; N, 4.07.

ACKNOWLEDGMENTS

We thank Science Foundation Ireland for project grant funding (SFI IvP grant 13/IA/1959 and SFI BEACON grant 16/RC/3889) and Dr. G. J. Miller, Keele University for hosting ANC during a visiting researcher placement.

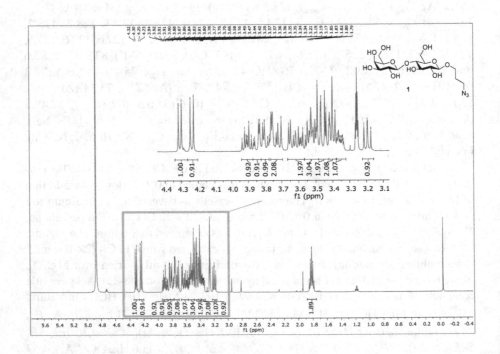

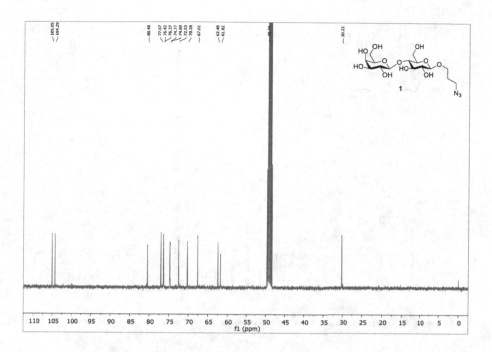

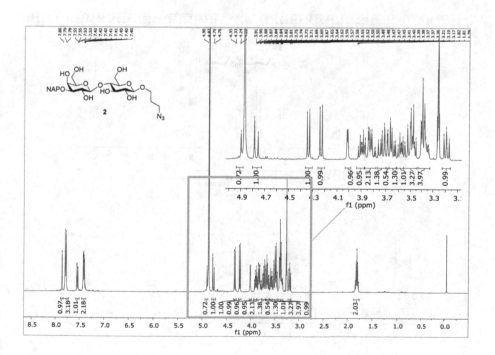

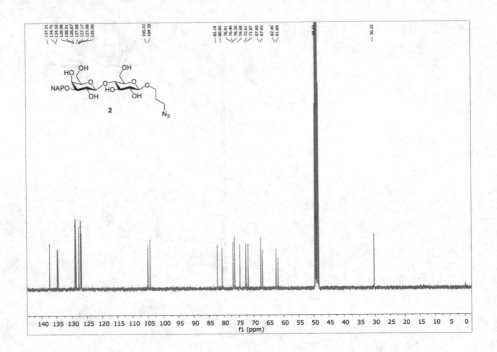

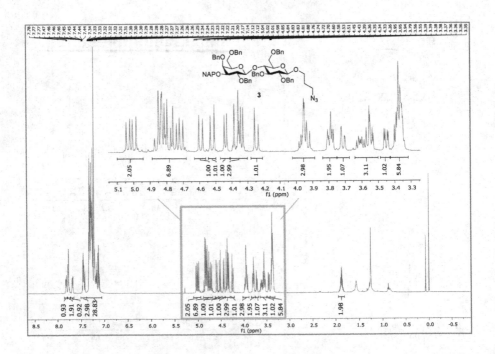

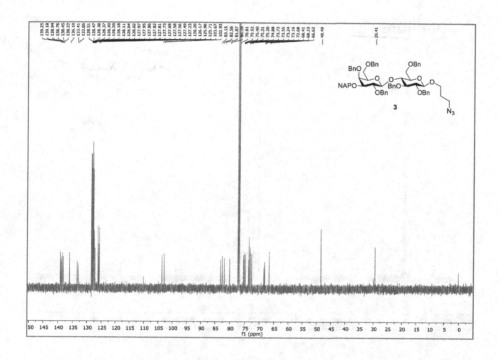

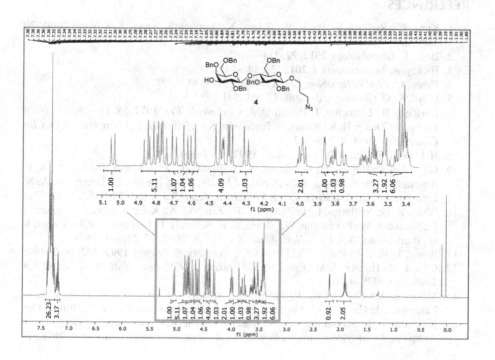

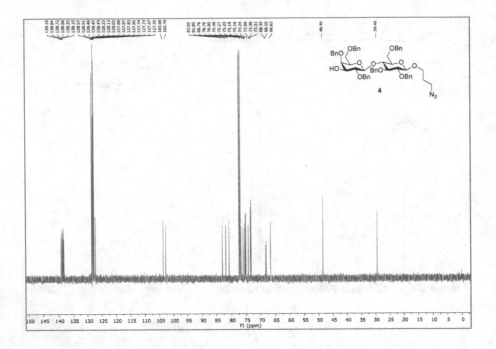

REFERENCES

1. Kunz, C.; Rudloff, S.; Baier, W.; Klein, N.; Strobel, S. *Annu. Rev. Nutr.* **2000**, 20, 699–722.
2. Bode, L. *Glycobiology* **2012**, 22, 1147–1162.
3. Hickey, R. M. *Int. Dairy J.* **2012**, 22, 141–146.
4. Green, C. *FEMS Microbiol. Lett.* **1989**, 47, 321–330.
5. Lloyd, K. O. *Glycoconj. J.* **2000**, 17, 531–541.
6. Lindberg, B.; Lönngren, J.; Powell D. A. *Carbohydr. Res.* **1977**, 58, 177–186.
7. Zhang, S.; Zhang, H. S.; Reuter, V. E.; Slovin, S. F.; Scher, H. I.; Livingston, P. O. *Clin. Cancer Res.* **1998**, 4, 295–302.
8. Hakomori, S.; Zhang, Y. *Chem. Biol.* **1997**, 4, 97–104.
9. Gilewske, T.; Ragupathi, G.; Bhuta, S.; Williams, L. J.; Musselli, C.; Zhang, X.-F.; Bencsath, K. P.; Panageas, K. S.; Chin, J.; Hudis, C. A.; Norton, L.; Houghton, A. N.; Livingston, P. O.; Danishefsky, S. J. *Proc. Natl. Acad. Sci. U.S.A.* **2001**, 98, 3270–3275.
10. Wang, Z. G.; Williams, L. J.; Zhang, X. F.; Zatorski, A.; Kudrya-shov, V.; Ragupathi, G.; Spassova, M.; Bornmann, W.; Slovin, S. F.; Scher, H. I.; Livingston, P. O.; Lloyd, K. O.; Danishefsky, S. J. *Proc. Natl. Acad. Sci. U.S.A.* **2000**, 97, 2719–2724.
11. Boren, T.; Falk, P.; Roth, K.A.; Larson, G.; Normark, S. *Science* **1993**, 262, 1892–1895.
12. Choi, J. M.; Hutson, A. M.; Estes, M. K.; Prasad, B. V. V. *Proc. Natl. Acad. Sci. U.S.A.* **2008**, 105, 9175–9180.
13. Moonens, K.; Gideonsson, P.; Subedi, S.; Mendez, M.; Bugaytsova, J.; Romaõ, E.; Lahmann, M.; Castaldo, G.; Coppens, F.; Lo, A.; Nordén, J.; Shevtsova, A.; Brännström, K.; Solnick, J.; Hammarström, L.; Vandenbussche, G.; Oscarson, S.; Berg, D. E.; Arnqvist, A.; Muyldermans, S.; Borén, T.; Remaut, H. *Cell Host Microbe* **2016**, 19, 55–66.

14. Miermont, A.; Zeng, Y.; Jing, Y.; Ye, X-S.; Huang, X. *J. Org. Chem.* **2007**, *23*, 8958–8961.

15. Yao, W.; Yan, J.; Chen, X.; Wang, F.; Cao, H. *Carbohydr. Res.* **2014**, *401*, 5–10.

16. Lahmann, A.; Lahmann, M.; Oscarson, S. *Beilstein J. Org. Chem.* **2010**, *6*, 704–708.

17. Liakatos, A.; Kiefel, M. J.; von Itzstein, M. *Org. Lett.* **2003**, *5*, 4365–4368.

18. Dahmén, J.; Gnosspelius, G.; Larsson, A-C.; Lave, T.; Noori, G.; Palsson, K.; Frejd, T.; Magnusson, G. *Carbohydr. Res.* **1985**, *138*, 17–28.

19. Fernandez-Mayoralas, M.; Martin-Lomas, M.; Villanueva, D. *Carbohydr. Res.* **1985**, *140*, 81–91.

20. Alonso-Lopez, M.; Bernabe, M.; Fernandez-Mayoralas, A.; Jimenez-Barbero, J.; Martin-Lomas, M.; Penades, S. *Carbohydr. Res.* **1986**, *150*, 103–109.

21. Santos-Benito, F. F.; Nieto-Sampedro, M.; Fernández-Mayoralas, A.; Martín-Lomas, M. *Carbohydr. Res.* **1992**, *230*, 185–190.

22. Koeman, F. A. W.; Meissner, J. W. G.; van Ritter, H. R. P.; Kamerling, J. P.; Vliegenthart, J. F. G. *J. Carbohyd. Chem.* **1994**, *13*, 1–25.

23. Wagner, D.; Verheyden, J. P. H.; Moffatt, J. G. *J. Org. Chem.* **1974**, *39*, 24-30.

24. Ní Cheallaigh, A.; Oscarson, S. *Can. J. Chem.* **2016**, *11*, 883–893.

25. Lipták, A.; Imre, J.; Harangi, J.; Násási, P.; Neszmélyi, A. *Tetrahedron* **1982**, *38*, 3721–3727.

26. Sherman, A. A.; Mironov, Y. V.; Yudina, O. N.; Nifantiev, N. E. *Carbohydr. Res.* **2003**, *338*, 697–703.

27. Tanaka, N.; Ogawa, I.; Yoshigase, S.; Nokami, J. *Carbohydr. Res.* **2008**, *343*, 2675–2679.

28. Nagashima, N.; Ohno, M. *Chem. Lett.* **1987**, 141–144.

29. Danishefsky, S. J.; Hungate, R. *J. Am. Chem. Soc.* **1986**, *108*, 2486–2487.

30. Hsieh, H-W.; Davis, R. A.; Hoch, J. A.; Gervay-Hague, J. *Chem. Eur. J.* **2014**, *20*, 6444–6545.

31. Calatrava-Pérez, E.; Bright, S. A.; Achermann, S.; Moylan, C.; Senge, M. O.; Veale, E. B.; Williams, D. C.; Gunnlaugsson, T.; Scanlan, E. M. *Chem. Commun.* **2016**, *52*, 13086–13089.

Index

A

Acetals
- benzaldehyde dimethyl, **II**, 191
- 1,2-*O*-benzylidene, **IV**, 141
- 1,3-*O*-benzylidene, **II**, 232
- 4,6-*O*-benzylidene, **I**, 199, 205; **II**, 9, 10, 162, 176, 189, 191; **III**, 168, 169
- chlorobenzylidene, **II**, 72
- diphenylmethylene, **II**, 10
- dithio-, **II**, 270
- kinetic formation, **II**, 231
- *p*-methoxybenzylidene, **II**, 11
- pseudo-C2-symmetric *bis*-, **II**, 231
- stannylene, **II**, 59, 61

2-Acetamido-4,6-*O*-benzylidene-2-deoxy-D-glucopyranose
- preparation of, **I**, 200
- spot visualization, **I**, 200

4-*O*-(2-Acetamido-6-*O*-benzyl-2-deoxy-3,4-*O*-isopropylidene-β-D-talopyranosyl)-2,3:5,6-di-*O*-isopropylidene-*aldehydo*-D-glucose dimethyl
- preparation of, **III**, 74

4-*O*-(2-Acetamido-3,6-di-*O*-benzyl-2,4-dideoxy-α-L-*erythro*-hex-4-enopyranosyl)-2,3:5,6-di-*O*-isopropylidene-*aldehydo*-D-glucose dimethyl acetal
- preparation of, **III**, 77

(2-Acetamido-3,4,6-tri-*O*-acetyl-2-deoxy-β-D-glucopyranosyl)benzene
- preparation of, **III**, 221

4-(2-Acetamido-3,4,6-tri-*O*-acetyl-2-deoxy-β-D-glucopyranosyl)bromobenzene
- preparation of, **III**, 222

2-Acetamido-3,4,6-tri-*O*-acetyl-2-deoxy-1-*O*-*p*-nitrophenoxycarbonyl-α-D-glucopyranose
- preparation of, **III**, 216

Acetolysis
- deoxysugars of, **I**, 3

Acetone elimination, **III**, 73

***N*-Acetyl-D-galactosamine**, **III**, 174

***N*-Acetyl-D-glucosamine**
- benzylidenation of, **I**, 200
- conversion into *N*-acetylgalactosamine, **III**, 175

***N*-Acetylneuraminic acid**
- acetylated glycal, preparation of, **I**, 245
- chloride, NMR data, **I**, 246, 249

methyl α-glycoside, **II**, 197
substituted benzyl glycosides, **I**, 251, 255

Acylation
- anomeric, **IV**, 125
- selective, **II**, 61
- by tributylstannyl ether method, **II**, 60

Alcohols
- oxidation of, **II**, 21

Aldonolactones
- synthesis from aldose hemiacetals, **III**, 33

Alkenylamines, **II**, 67
- precursors, **II**, 68
- Vasella-reductive amination, **II**, 48

Alkylation of carbohydrate derivatives
- anomeric, **IV**, 127
- enhanced rate of Purdie methylation, **I**, 27
- selective, **II**, 61

α-D-Allofuranose
- *C*-nitromethyl derivative, **V**, 142

Allyl 2-acetamido-2-deoxy-4,6-di-*O*-pivaloyl-β-D-galactopyranoside, **III**, 175

Allyl 2-acetamido-2-deoxy-3,6-di-*O*-pivaloyl-β-D-glucopyranoside, **III**, 174

Allyl glycoside
- from anomeric orthoester, **II**, 150, **V**, 101
- via Fischer glycosidation, **V**, 55

4-Amino-4-deoxy-L-arabinose (Ara4N), **III**, 193

2-Aminoethyl diphenylborinate, **II**, 62

***p*-{*N*-[4-(4-Aminophenyl α-D-glucopyranosyl)-2,3-dioxocyclobut-1-enyl]amino} phenyl α-D-glucopyranoside**, **III**, 108

***p*-{*N*-[4-(4-Aminophenyl β-D-glucopyranosyl)-2,3-dioxocyclobut-1-enyl]amino} phenyl β-D-glucopyranoside**, **III**, 109

***N*-(3-Aminopropyl)-2,3,4,6-tetra-*O*-benzyl-D-gluconamide**, **III**, 29

***N*-(3-Aminopropyl)-2,3,4-tri-*O*-benzyl-D-xylonamide**, **III**, 28

Analytical samples
- preparation of, **II**, xviii

1,6-Anhydrosugars, **I**, 56, 59, 271; **II**, 98; **V**, 176, 259, 269

Anomeric
- allylation, **V**, 55
- azide, **II**, 257
- S-deacetylation, **III**, 89
- dihalides, **II**, 245

Anthracenylmethylene, **II**, 10

Printed in the United States
by Baker & Taylor Publisher Services

Printed in the United States
by Baker & Taylor Publisher Services